T0392440

Storage of LPG in Large Rock Caverns

This book uses actual construction projects as examples to elaborate the various technologies regarding groundwater control and air-tightness guarantees for the construction of large storage rock caverns in complicated geological conditions. It introduces the latest technologies to give hands-on experience of what to do in unexpected geological conditions, and provides insight into the design, construction and operation of underground facilities. The book is hence useful in feasibility studies, developments and other research into these underground facilities.

Drawing on the experience gained from building Japan's largest LPG storage caverns, *Storage of LPG in Large Rock Caverns* is a must-read for engineers, academics and students in the fields of rock mechanics, geotechnical engineering and related disciplines.

ISRM Book Series

The ISRM Book Series comprises significant scientific publications in rock mechanics and related disciplines. The aim of the Series is to promote the scientific output of ISRM. The Series is published on a regular basis, according to the emergence of noteworthy scientific developments. Each volume in the Series is prepared independently and focuses on a topical theme. Contributions are rigorously peer-reviewed by experts in the field, ensuring strong scientific standards, and are expected to be leading in rock mechanics and related disciplines. The Series may include edited volumes, textbooks and reference works.

Discontinuous Deformation Analysis in Rock Mechanics Practice
Yossef H. Hatzor, Guowei Ma, and Gen-hua Shi

Back Analysis in Rock Engineering
Shunsuke Sakurai

Rock Reinforcement and Rock Support
Ömer Aydan

Continuum and Computational Mechanics for Geomechanical Engineers
Ömer Aydan

Storage of LPG in Large Rock Caverns
Kenji Aoki

For more information about this series, please visit: www.routledge.com/ISRM-Book-Series/book-series/ISRM

Storage of LPG in Large Rock Caverns

Kenji Aoki

CRC Press is an imprint of the
Taylor & Francis Group, an **informa** business
A BALKEMA BOOK

Cover image: © JOGMEC (Japan Organization for Metals and Energy Security)

First published 2023
by CRC Press/Balkema
Schipholweg 107C, 2316 XC Leiden, The Netherlands
e-mail: enquiries@taylorandfrancis.com
www.routledge.com – www.taylorandfrancis.com

CRC Press/Balkema is an imprint of the Taylor & Francis Group, an informa business

Library of Congress Cataloging-in-Publication Data
Names: Aoki, Kenji, author.
Title: Storage of LPG in large rock caverns / Kenji Aoki.
Description: Boca Raton : CRC Press, 2023. | Includes bibliographical references and index.
Identifiers: LCCN 2022028326 (print) | LCCN 2022028327 (ebook) | ISBN 9780367420772 (hardback) | ISBN 9781032383132 (paperback) | ISBN 9780367822163 (ebook)
Subjects: LCSH: Liquefied petroleum gas--Underground storage--Japan. | Underground construction--Japan. | Caves--Japan.
Classification: LCC TP761.P4 A5 2023 (print) | LCC TP761.P4 (ebook) | DDC 665.7/73--dc23/eng/20221207
LC record available at https://lccn.loc.gov/2022028326
LC ebook record available at https://lccn.loc.gov/2022028327

ISBN: 9780367420772 (hbk)
ISBN: 9781032383132 (pbk)
ISBN: 9780367822163 (ebk)

DOI: 10.1201/9780367822163

Typeset in Times New Roman
by Deanta Global Publishing Services, Chennai, India

Contents

About the Principal Author

Kenji Aoki, PhD
Professor Emeritus, Kyoto University

Kenji Aoki, the principal author, was born in 1946. Graduating from the Department of Engineering of Kyoto University with a bachelor's degree in 1969 and a master's in 1971, he joined Kajima Institute of Construction Technology, where, as well as research and development, he participated in design, construction and measurement management in a number of civil engineering structures including rock caverns and long tunnels.

Through over 20 years' experience in the construction of large-scale caverns, he developed many testing and measurement techniques and numerical analysis methods of the mechanical properties of rocks during excavation. He researched the mechanisms of groundwater flow in and around deep caverns and presented the results as papers in the *JSCE* and the *ISRM* and through lectures. He has been active in these committees. In the early 1980s, he introduced the technology of storing crude oil in an unlined rock cavern in Japan utilising groundwater, which had been developed in the Scandinavian countries.

Over more than ten years from 1981, Aoki contributed to construction of Japan's first pilot plant of rock cavern storage of crude oil and to clarify the role of groundwater in cavern storage with cracks (this project was led by the former Japan National Oil Corporation). Based on the site measurements, he contributed to establishing technical standards for the design and construction of these sorts of facilities. These standards were used in constructing three large-scale rock cavern crude oil storages in Japan in the early 1990s with a total capacity of five million kilolitres. Their ability for earthquake endurance is well recognised and they are still in operation.

Aoki was awarded a PhD from Kyoto University in 1989 and he became professor of Kyoto University in 1999.

In 2004. Aoki was a co-organiser of the third Asian Rock Mechanics Symposium (ISRM) in Kyoto.

The LPG rock storage project of this book owes to Aoki's long-established knowledge in surveying, testing and analysing groundwater flow in rocks, resulting from his research and development in controlling groundwater flow in rocks of various properties. He gained this extensive practical experience in the Japan Organization for Metals and Energy Security (JOGMEC) as chief technical officer.

Publication Committee

Committee Chairman

Kenji Aoki	Ph.D.		
	Professor Emeritus,	Kyoto University	Principal Author

Committee

Toshio Maejima	Dr Eng	Formerly JOGMEC	Storage Construction Manager
Tatsuya Iwahara	JOGMEC	Director, JOGMEC	
Yuriko Okazaki	Dr Eng	Formerly JOGMEC	Responsible for air-tightness tests
Shuichi Okubo	P.E.Jp	TEPSCO* (Formerly JOGMEC)	Editing
Hiroki Kurose	Dr Eng	TEPSCO* (Formerly JOGMEC)	Editing

*Tokyo Electric Power Services Co., Ltd.

Acknowledgements

The data used in this book are mainly from the database of construction of rock cavern storages of National LPG Stockpiling Project at Kurashiki and Namikata. The project was carried out by JOGMEC commissioned by the Agency for Natural Resources and Energy. The Publication Committee owes the JOGMEC staff concerned for their support from construction to publication.

The authors thank Professor Hidekazu Yoshida of Nagoya University and Professor Shizuo Yoshida, formerly of the University of Tokyo, for discussions about the geology of both sites.

Special thanks are due to Professor Toshifumi Matsuoka, formerly of Kyoto University, for suggestions and guidance on this publication, and Mr Koya Suto of Terra Australis Geophysica Pty Ltd for discussion on and translation of the text.

On publication of this volume, the invaluable cooperation of Alistair Bright and Marjanne Bruin of Taylor & Francis is gratefully acknowledged.

Chapter 1

Introduction

1.1 Background

Under Japan's energy policy, the Japan Organization for Metals and Energy Security (JOGMEC) operates the national stockpiling project of crude oil and liquefied petroleum gas (LPG). In addition to three existing above-ground LPG tank facilities, in 2002, construction began on two of the world's largest facilities, Kurashiki (capacity 400,000 t) and Namikata (450,000 t). These rock cavern storage systems were completed in 2012 allowing the long-term stockpiling operation to continue.

For the first time in Japan, these two high pressure gas storages adopted the hydrodynamic containment system, in which the groundwater pressure holds the gas in unsealed rock tunnels. They are of the world's largest class.

This book describes new knowledge and experiences in rock engineering and rock mechanics learnt from surveying, planning, designing and constructing storage caverns to testing the air-tightness of the storage caverns, taking the two facilities as examples.

To prevent leakage of high pressure gas from storages of this type, the vital issues are to maintain a stable groundwater level and ensure a specified water pressure in the rocks surrounding the storage cavern, considering the presence of fractures in the rocks throughout the construction and operation phases.

The only means to confirm the air-tightness of a storage cavern of this scale is to test the air-tightness of the actual storage on completion of construction. If a gas leak is found at that stage, the choice of the methods to repair it is extremely limited. Therefore, each stage from survey to design to construction needs to devise ways to address the above issue. In practice, an effective construction system must incorporate monitoring based on hydrogeology, the behaviour of the groundwater and the mechanical behaviour of the rocks, predictive analysis, assessment and decision-making based on the result of measurements.

Especially, the important countermeasure for fractures in rocks aims to minimise the risk of leaking. It utilises a combination of injection of additional water and improvement of water-tightness by pre-grouting the fractures which control groundwater and installation of necessary water curtains. This book describes the details construction methodology of the hydrodynamic containment system along practical examples.

On completion of construction of the storage cavern, compressed air was injected into the entire cavern at the required pressure and examined for empirical variation of the pressure in the storage cavern. A quality air-tightness was ascertained by this manner.

During the actual air-tightness test, several important factors affect the internal pressure such as temperature variation and dissolution of the injected air into the groundwater.

DOI: 10.1201/9780367822163-1

Therefore, a detailed consideration of measurement plan, placement of sensors and analysis and evaluation criteria is necessary prior to the test. To contribute to similar projects in the future, the actual air-tightness tests, for example, are described in detail.

1.2 Underground Storage of Oil and Gas

The constituent of LPG is hydrocarbon gases such as propane and butane liquefied by cooling and compressing.

There are two ways to store LPG in a rock formation: under low temperature-normal pressure (boiling point of propane: –42°C; butane: –0.5°C) and normal temperature-high pressure (vapour pressure at 20°C; propane: 7.3 atm; butane: 1.1 atm). The low temperature-normal pressure storage method utilises a rock cavern excavated from a relatively shallow part of the rock under the water table, forming a frozen layer in the surrounding rock. This method causes several issues including difficulty in forming a frozen layer where the groundwater flows fast or in large volume; opening fractures by freezing water; and increasing the boil-off gas (BOF). On the other hand, the normal temperature-high pressure storage method stores LPG under normal temperature and high pressure in a storage cavern not sealed with concrete or steel but by hydrodynamic containment of the groundwater pressure around the cavern. It requires water pressure higher than the internal pressure of the storage cavern, and the storage tends to be deep: commonly over 100 m for propane and over 40 m for butane. The groundwater level must be kept stable from construction to operation. This is achieved by constructing water curtains of tunnels and boreholes.

As LPG rock cavern storage is deep underground, it has several advantages: there is little exposure to natural disasters such as earthquakes and typhoons; little chance of fire or explosion as it is separated from the atmosphere; and it has a large capacity without requiring a large installation on the surface. The last issue is particularly important to Japan which is prone to earthquake disasters. The tsunami that hit the Kuji Underground Oil Stockpile Base in Iwate Prefecture following the 2011 Great East Japan Earthquake caused damage to the above-ground facilities while the underground storage was not affected. This event proved the safety of underground storage is superior to above-ground tanks (Figure 1.1).

Figure 1.1 Aftermath of tsunami at Kuji Underground Oil Stockpile Base: devastation of above-ground facility (left) and after restoration (right).

The concept of storing oil in an underground cavern without a steel or concrete lining originated in the 1930s. In 1930, Herman Jansson obtained a patent for an idea to store oil in an unlined vertical shaft under the water table. However, this idea was only realised in 1951 when an oil storage test plant in an unlined cavern was built in Saltsjobaden, a suburb of Stockholm. This was a prototype of the present hydrodynamic containment system. This facility stored 30 m^3 of gasoline for five years and confirmed there was no leak.

The first hydrodynamic containment LPG storage cavern of practical use was built in Goteborg, Sweden, in 1968. This 2000 m^3 rock cavern storage was built in hard gneiss rock 90 m deep, and confirmed the efficacy of the hydrodynamic containment of high pressure gas. In 1971, the first hydrodynamic containment cavern storage with water curtains on top was constructed in Marseille, France. This storage was dug in hard limestone and consisted of three tunnels with a cross section area of 230 m^2 and a total capacity of 12,000 m^3.

In the United States, a large number of caverns have been excavated in thick impermeable sedimentary rock formation of sandstones and shales since the 1950s. North America characteristically utilises the caverns of a "room and pillar" system, rather than the hydrodynamic containment system which relies on the existence of groundwater.

The LPG rock storages using the hydrodynamic containment system originated in Sweden. Since 1980, many storages have been built in Europe including Sweden, Norway, Finland, France, the United Kingdom and Portugal. More recently, storages were built in Asian countries such as Korea, China and India. These facilities have rock caverns with a capacity of 100,000 to 300,000 m^3. The rocks are mainly gneiss, limestone and chalk, which are dense and impermeable. Some facilities were built in hard and relatively permeable rock such as granite.

In Japan, the applicability of rock storage for crude oil was first considered in 1976. A pilot plant was built in 1980 and experiments continued for the next ten years to address various technical issues including the behaviour of groundwater. These experiments confirmed the applicability of rock cavern storage of the hydrodynamic containment system to the geological conditions of Japan. Following the result, three national crude oil stockpile bases for underground storage were constructed with a total capacity of 5 million kL between 1987 and 1992. They continue to operate stably: Kuji in Iwate Prefecture, Kikuma in Ehime Prefecture and Kushikino in Kagoshima Prefecture. The case studies in this book, the facilities of the Kurashiki (820,000 m^3) and Namikata (910,000 m^3) bases, are Japan's first projects in high pressure LPG rock storage.

1.3 Projects for Case Study

1.3.1 Japan's National Oil Gas Stockpile Facilities

LPG stored in several facilities is a distributed energy resource. It is used as a source of energy in an emergency response during a disaster and it is a clean energy source that generates relatively small amounts of CO_2. Japan imports 74% of its LPG, 87% of which is from the Middle East. This weak supply structure prompted Japan to implement national stockpiling projects from the viewpoint of energy security.

Japan's policy of stockpiling LPG states the national stockpile should be 1,500,000 t, equivalent to 40 times the average daily import and 50 times that in the private sector. Accordingly, the JOGMEC constructed five facilities, as shown in Figure 1.2. Of those, the three aboveground facilities, Nanao, Ishikawa Prefecture (250,000 t), Fukushima, Nagasaki Prefecture (200,000 t) and Kamisu, Ibaragi Prefecture (200,000 t), were completed in 2005 and are now

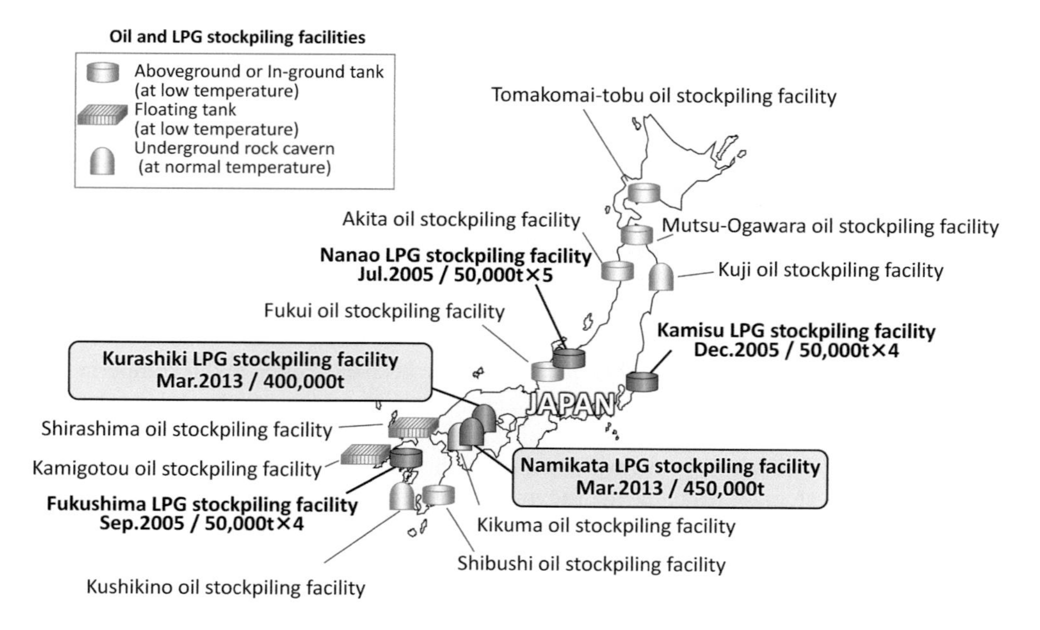

Figure 1.2 National oil and gas stockpiling facilities in Japan.

in operation. Two of the world's largest facilities were constructed in Kurashiki in Okayama Prefecture (400,000 t) and Namikata in Ehime Prefecture (450,000 t). They were the first to adopt the hydrodynamic containment system in Japan.

Underground storage commonly uses the hydrodynamic containment system overseas (Section 1.2). In Japan, three facilities, Kuji, Kikuma and Kushikino, use this system for oil storage at normal temperature-normal pressure. However, the facilities at Kurashiki and Namikata are the first underground rock cavern storage for high pressure LPG in Japan.

1.3.1.1 The Kurashiki Facility

The Kurashiki facility was constructed adjacent to an oil refinery on a large (7 km NS, 6 km EW, 5 m above sea level) area of reclaimed land on the shore of Mizushima Bay in the Seto Inland Sea (Figures 1.3 and 1.4). The purpose of the national LPG stockpiling facility is to prepare for emergency situations, and it is not expected to be used frequently. Therefore it is not economical to build a new reception/despatch facility. For this reason, an existing LPG import facility of the private sector in the vicinity is utilised.

The geology under the reclaimed land of the Kurashiki facility site is comprises top layer of approximately 70 m of Neocene sediments overlying approximately 50 m of Late Cretaceous weathered granite on fresh granite. The storage is planned in the fresh granite and the elevation of its ceiling is 160 m below sea level. The cross section of the storage cavern is egg-shaped measuring 18 m wide and 24 m high. The storage is made up of four caverns with lengths from 488 to 640 m totalling 2.2 km (Figure 1.5). Its cavern volume is 820,000 m^3 and the designed storage capacity is 400,000 t, which is the largest in the world. Table 1.1 shows its design parameters.

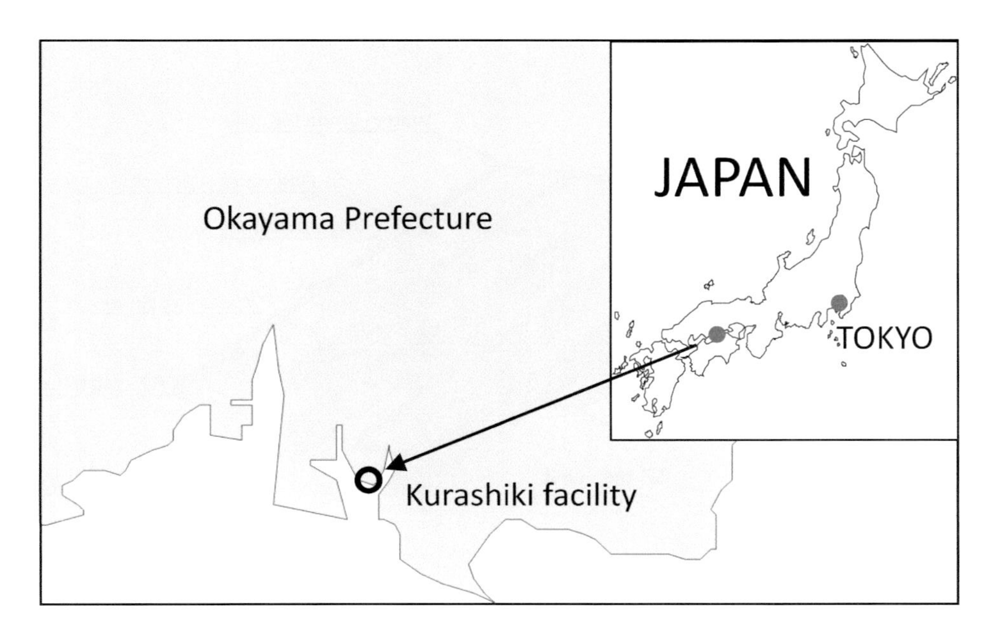

Figure 1.3 Location of the Kurashiki national LPG stockpiling facility.

Figure 1.4 Layout of storage caverns at the Kurashiki facility.

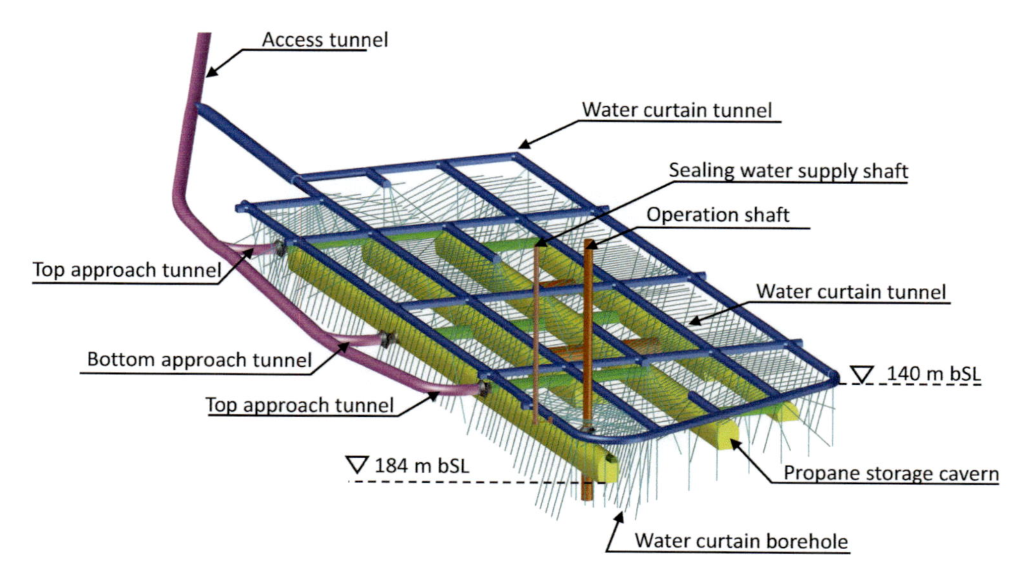

Figure 1.5 Perspective view of the underground facility at the Kurashiki facility.

Table 1.1 Design parameters of the Kurashiki facility

Storage capacity	Propane 400,000 t (cavern volume: 820,000 m^3)
Design storage pressure	950 kPaG
Storage dimension	24 × 18 × 488–640 m (H×W×L) × four caverns
Cavern arrangement	Water curtain tunnel: 140 m bSL
	Top of cavern: 160 m bSL; bottom of cavern: 184 m bSL
Rock temperature	22°C

Water curtain tunnels were drilled 20 m above the storage caverns at 140 m bSL. Water curtain boreholes were drilled to form water curtains, the pressure of which can be controlled by water injection. Faults and fractures were present in the rocks. It is necessary to prevent loss of groundwater pressure in the vicinity at the time of excavation of the caverns. In order to maintain a secure water pressure around the storage caverns through from construction to operation phases, the water curtain boreholes were drilled vertically on the sides of the caverns in addition to the horizontal ones on the top.

1.3.1.2 The Namikata Facility

The Namikata facility is 150 m below ground level of a cape which is 0.8 km wide (north-south) and 3 km long (east-west), located at the tip of the Takanawa Peninsula in the Namikata district of Imabari City, Ehime Prefecture. It was built adjacent to an LPG facility belonging to the private sector (Figures 1.6 and 1.7).

The storage at the Namikata facility consists of three caverns of the world's largest class:

- One combined butane-propane storage: length 430 m, volume 280,000 m^3, capacity 150,000 t.

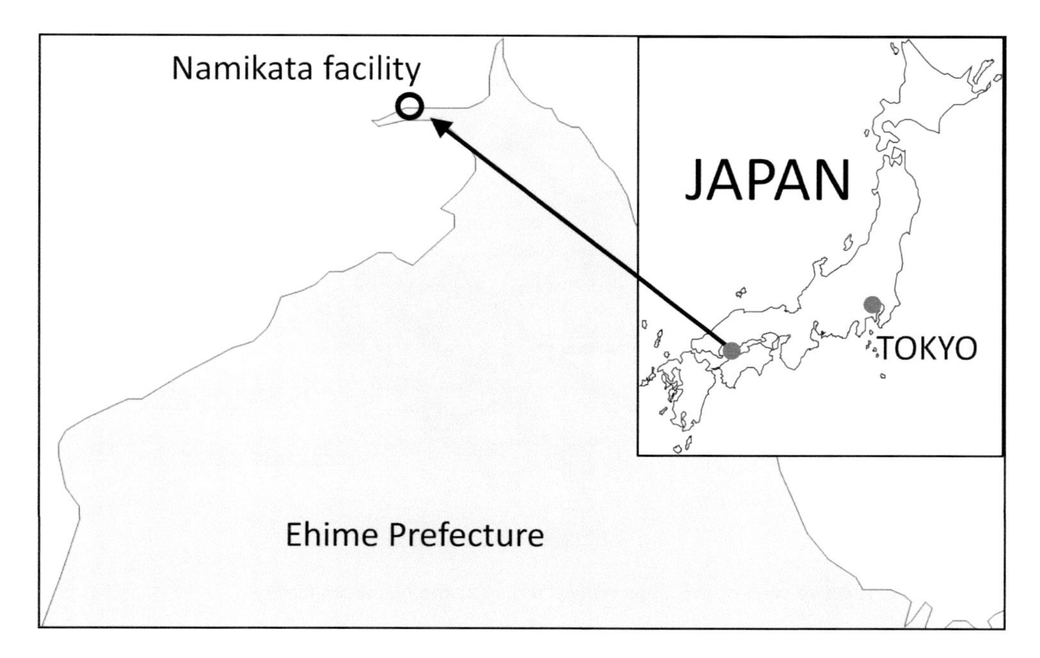

Figure 1.6 Location of the Namikata national LPG stockpiling facility.

Figure 1.7 Layout of storage caverns at the Namikata facility.

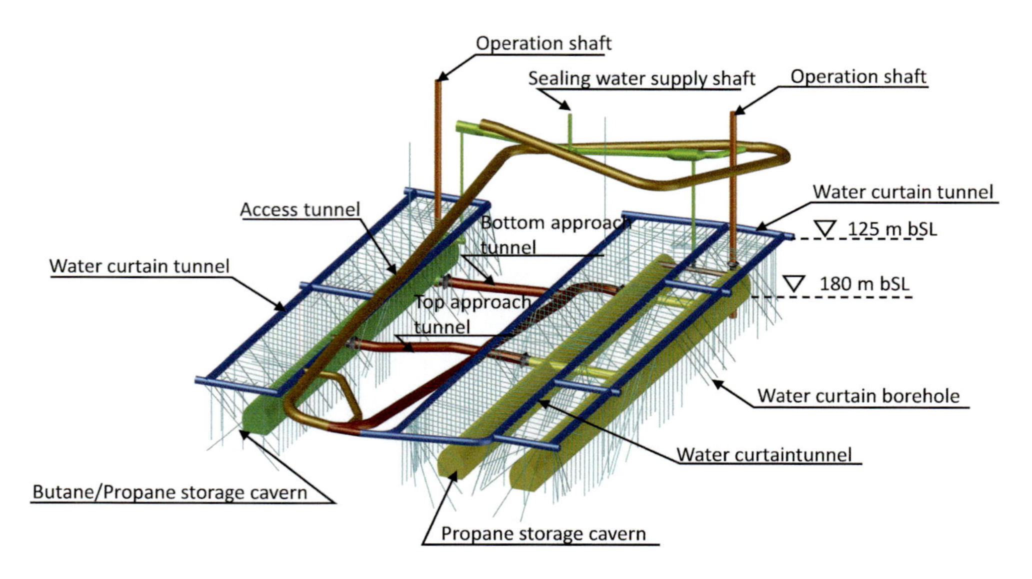

Figure 1.8 Perspective view of the underground facility at the Namikata facility.

Table 1.2 Design parameters of the Namikata facility

Storage capacity	Butane-propane storage cavern: 150,000 t (cavern volume: 280,000 m³)
	Propane storage cavern: 300,000 t (cavern volume: 630,000 m³)
Storage pressure	Propane: 970 kPaG; butane: 240 kPaG
Storage dimension	Butane-propane storage cavern: 30 × 26 × 430 m (H×W×L)
	Propane storage cavern: 30 × 26 × 485 m (H×W×L) × two caverns
Cavern arrangement	Water curtain tunnel: 125 m bSL
	Top of cavern: 150 m bSL; bottom of cavern: 180 m bSL
Rock temperature	21°C

- Two propane storages: length 485 m each, combined volume 630,000 m³ and combined capacity 300,000 t.

The cross section of the caverns is egg-shaped measuring 26 m wide and 30 m high (Figure 1.8). Table 1.2 shows its design parameters. As the caverns were designed with different storage pressures – 240 kPaG for combined butane-propane and 970 kPaG for propane – they were located 200 m apart to prevent cross-contamination of the stored gas.

1.3.1.3 Technical Issues on Construction

Three important technical issues addressed at the two sites are (1) establishing techniques for evaluating and controlling the behaviour of groundwater; (2) establishing a technology for the systematic excavation and construction of rock caverns for LPG; and (3) developing air-tightness test technology for large-scale rock caverns.

(1) Establishing techniques for evaluating and controlling the behaviour of groundwater.

The hydrodynamic containment system of rock storage is a way to confine the LPG which is in equilibrium between the liquid and gas phases under normal temperature and high pressure in the storage cavern by groundwater steadily flowing toward a storage set in deep underground. In Japan, the development of faults and fractures is prominent in bedrocks. Therefore, the biggest issues are reliable control and maintenance of the groundwater flow in such bedrocks. Two techniques for controlling the behaviour of groundwater were established: (a) reducing seep by grouting into highly permeable zones around the storage caverns and (b) maintaining groundwater pressure by injecting water into water curtains placed above and to the sides of the cavern.

To control the complex water pressure field in the bedrocks and evaluate the air-tightness of the storage cavern, it is crucial to understand and accurately predict the water pressure field around the storage cavern. A 3D hydrogeological model was constructed based on a detailed hydrogeological survey and high-precision prediction of the groundwater level. This model progressively incorporated the results of geology and groundwater data accumulated by drilling and improved by geostatistical treatment. This model led to an understanding of the behaviour of groundwater in the entire storage area.

(2) Establishing a method for systematic excavation and construction of rock caverns for LPG.

A construction system was established for this rock cavern storage project. This system compares the measurement data and the prediction by 3D heterogeneous hydrogeological model at each drilling stage, judges the effectiveness of the grout improvement and water curtains and devises appropriate response measures. In this system, the water pressure around the cavern is maintained by the water curtains during excavation, and the effect of the groundwater pressure must be taken into consideration. This highly sophisticated hydrogeological model analyses the distribution of pressure of the surrounding water which is ignored in ordinary underground excavation. The system to evaluate the stability of the cavern incorporates the effect of the mechanical improvement of the grouting. Thus, excavation considering the effect of water pressure is a technology specific to LPG storage caverns unlike the excavation of traditional caverns used for tunnels and underground power plants.

(3) Development of air-tightness test technology for large-scale rock caverns

On completion of excavation, the air-tightness of the storage caverns must be confirmed by an air-tightness test. The test assesses the variation in air pressure by pressurised injection of air into the storage cavern. A large-scale storage cavern is subject to the effects of variation of temperature in the gas phase, a reduction in the void space volume by seepage and the dissolution of air into the seepage. To address these effects, the evaluation criteria need to be based on strictly defined high-precision measurement and evaluation procedures. To define the procedure, a preliminary experiment using a small cavern at the same location was carried out and its applicability to a large scale was confirmed using high-precision instruments including a thermometer. This test was applied to the Kurashiki and Namikata facilities and confirmed excellent air-tightness in both. The result proved the efficacy of the air-tightness evaluation technology and the construction system controlling the groundwater.

1.4 Composition of This Book

This book is composed of the following seven chapters.

Chapter 1: Introduction.
Chapter 2: Basic Plan of LPG Rock Storage and Geology. This chapter

This chapter describes the geology of the Kurashiki and Namikata areas. In particular, the complex hydrogeological structure of the Kurashiki area, which is dominated by discontinuities, is detailed. It also outlines the structures and arrangements of facilities to ensure the stability of the groundwater level and the pore water pressure during excavation under hydrogeological conditions. The basic specification of the rock cavern storage is described including the shape of the cross section, the support pattern and the arrangement of water curtains. The important groundwater control system is introduced and an outline of the air-tightness test is provided.

Chapter 3: 3-Dimensional Hydrogeological Model.

An accurate understanding and prediction of the behaviour of groundwater are crucial to evaluate the functionality of the groundwater around the storage cavern of a hydrodynamic system. This chapter introduces the case history of the construction and use of a high-precision 3D hydrogeological model incorporating the results of detailed surveys, tests and monitoring during excavation. Comparing the geostatistical prediction of the behaviour of groundwater using a 3D hydrogeological model with the actual results of monitoring, the chapter explains the interpreted complex behaviour of groundwater and provides an assessment of the applicability of the groundwater control measures.

Chapter 4: Groundwater Controlling System.

This chapter details the practical application of water curtains and water curtain grouting at two sites with different hydrogeological properties, Kurashiki and Namikata. Surveys and tests in water curtain boreholes took place for the initial basic pattern, and prediction by 3D analysis of groundwater behaviour was incorporated to evaluate the water-tightness of the facilities. Following the surveys, additional water curtain boreholes were placed if found necessary. This procedure is explained in the chapter. Also included are the results and the effects of grouting in accordance with the characteristics of the cracks at each site and the properties of the grouting materials injected.

Chapter 5: Mechanical Stability of Storage Cavern.

To discuss the mechanical stability of a cavern requires some consideration of the effect of the groundwater pressure and grouting, as LPG rock storage caverns are excavated maintaining the groundwater pressure. This chapter describes details of necessary additional supports and post-grouting according to an analysis of the data on the displacement of the rocks.

Chapter 6: Air-tightness Test.

On completion of the excavation of the storage caverns, air-tightness tests were carried out to confirm the functionality of the storage. The test judges air-tightness by observing the preservation of the pressure of compressed air injected into the cavern. The pressure is greatly affected by a variation of the temperature in the gas phase, a variation of the capacity of the gas due to seepage and the dissolution of the gas into the water. These effects must be measured with high precision. For this purpose, the air-tightness test was tried in a small-scale cavern to examine the applicability of the high-precision instruments and the testing method. To conclude the chapter, the actual air-tightness tests at the Kurashiki and Namikata facilities are explained.

Chapter 7: Cavern Storage and Surrounding Groundwater during Operation.

During the six years of operation, the variation of the groundwater level and the pore water pressure was scrupulously analysed. The results did not find a clear anomaly and the cavern is regarded stable. The water-tightness was confirmed by comparing the behaviour of the groundwater with the values predicted by the 3D heterogeneous model. The long-term variation of the groundwater level was predicted. This chapter also outlines the operational equipment and safety management system. The system includes measurements of acoustic emissions (AE) and seismic acceleration. These measurements showed very limited variation and stable operation is assured.

References and Further Readings

Aoki K. (2007) Rock engineering in underground space utilization in Japan-experience of underground energy storage projects. *3rd Iranian Rock Mechanics Conference*, Tehran, 1031–1055.

Aoki K., Hibiya K. (1994) Groundwater control during the construction of large underground crude oil storage caverns. Integral approach to applied rock mechanics. *ISRM International Symposium*, Vol.1, 567–577.

Aoki K., Hibiya K., Yoshida T. (1989) Storage of refrigerated liquefied gases in rock caverns-Characteristics of rock under very low temperatures. *Proceedings of the International Conference on Storage of Gases in Rock Caverns*, Trondheim, 221–227.

Aoki K., Hibiya K., Yoshida T. (1990) Storage of refrigerated liquefied gases in rock caverns: Characteristics of rock under very low temperatures. *Tunneling and Underground Space Technology*, Vol.5, No.4, 319–325.

Aoki K., Ismail A., Uno H., Chang C.S., Maejima T., Nakamura Y. (2010) Hydraulic behaviors characterization for the design of unlined underground LPG storage cavern in fractured rock mass. *Rock Mechanics in Civil and Environmental Engineering, EUROCK2010*, Switzerland, 713–716.

Aoki K., Mito Y., Maejima T. (2006) Japanese underground storage facilities. *Proceedings of the International Workshop of the Underground Storage Facilities in Conjunction with 4th ARMS, ISRM International Symposium 2006, 4th ARMS*, Singapore, 37–51.

Aoki K., Miyashita K., Hanamura T., Tajima T. (1986) The first test plant of underground crude oil storage in unlined cavern. *Large Rock Caverns*, Vol.1, 3–14.

Bergman M. (1977) Storage in excavated rock caverns. *Proceedings of the First International Symposium*, Stockholm, p. 832.

Lindblom U.E. (1989) The development of hydrocarbon storage in hard rock caverns. *Proceedings of the International Conference on Storage of Gases in Rock Caverns*, Trondheim, 15–30.

Nilsen B., Olsen J. (1989) Storage of gases in rock caverns. *Proceedings of the International Conference on Storage of Gases in Rock Caverns*, Trondheim, p. 398.

Chapter 2

Basic Plan of LPG Storage Cavern and Geology

Liquefied petroleum gas (LPG) rock storage is a system for storing gas in a storage cavern while maintaining the groundwater pressure higher than the internal pressure of the cavern. The cavern is built deep underground where the piezometric head is expected to be higher than the internal pressure of the cavern with room for a safety margin. Where the hydrogeological structure of the deep bedrock is complex and the groundwater supply from the ground surface is insufficient, the water pressure around the storage cavern cannot be maintained. Water curtains are installed to stabilise the groundwater level.

When a deep storage cavern is excavated, it is exposed to atmospheric pressure and the groundwater seeps, lowering the water pressure around the cavern. The water level drops, particularly in areas where permeable faults and joints occur, resulting in unsaturated zones. If such unsaturated zones form around the storage cavern, it is difficult to maintain the water pressure higher than the storage cavern. Once an unsaturated zone is formed, it is extremely difficult to re-saturate it using groundwater. Therefore, it is necessary to maintain the water saturation of the rocks around the storage cavern at all times from construction to operation. To ensure saturation during excavation, it is necessary to inject pressurised water through the water curtain boreholes, which are drilled from the water curtain tunnels at the top of the cavern prior to excavation, and to grout in order to reduce the flow rate of seepage influx through the permeable faults across the storage cavern.

The Kurashiki site of this case study has a complex heterogeneous hydrogeological structure with (1) permeable faults and microfractures; (2) less permeable zones; and (3) zones where fault clays cut off the water flow. The Namikata site is less permeable than the Kurashiki site, but has irregular continuous cracks that provide a passage for water. To deal with these complex heterogeneous hydrogeological structures, additional water curtains are constructed, monitoring the data along with the progress of excavation. The groundwater in highly permeable zones and faults around the cavern, is controlled by pre-grouting to improve water-tightness prior to excavation, and by post-grouting if the data on completion suggest that it is necessary.

In large-scale rock storage, an understanding of the complex heterogeneous hydrogeological structure is necessary over a large area; however, details of the structure are difficult to obtain before excavation. Such details are provided by analysis of the geological and measurement data during excavation. Hence, only a basic design specification for water curtains and grouting can be planned before excavation, which is continuously improved as details of the hydrogeological properties found during excavation and an analysis of the measurement data and prediction are provided. The measured data are compared with the prediction based on a 3-dimensional (3D) hydrogeological model and the difference is

DOI: 10.1201/9780367822163-2

analysed to critically review the basic plan and provide feedback during excavation to control the groundwater.

A basic pattern was devised for the water curtains prior to excavation. While monitoring the variation of the groundwater level and other test data during arch and bench excavations, if the sealing functionality was found inadequate, the pattern was modified with additional water curtain boreholes.

The shape of cross section of the cavern was determined to minimise looseness associated with excavation according to the surrounding heterogeneous geology including crack zones and faults with clay. The stability of the cavern was analysed from in-situ testing of faults and shatter zones in the access tunnels and water curtain tunnels. The support pattern of the storage cavern was designed prior to excavation for a model assuming the position and properties of faults. During excavation, the stability of the cavern was re-evaluated with geological and measurement data and additional supports were installed as necessary to ensure stability.

On completion of excavation, the air-tightness of the storage cavern was checked by an air-tightness test. Air-tightness was judged by the variation of pressure in response to injection of compressed air into the cavern. For a large-scale storage cavern, this test requires a testing methodology and high-precision measuring instruments because many factors affect the pressure of the injected air, including a temperature variation in the gas phase, a change in the available space due to seepage and the dissolution of air into the water. A preliminary experiment of the air-tightness test was carried out in a small tunnel to confirm the applicability of the high-precision instruments and the testing methodology. These were adopted in the actual tests used in the Kurashiki and Namikata facilities, and confirmed the excellent air-tightness of these facilities.

Figure 2.1 shows the flow of the groundwater control system for LPG rock storage, the subject of this book.

2.1 Structure of Rock Cavern Storage

An LPG rock storage facility consists of underground storage and a management block on the ground. A shaft connects the aboveground management block and underground storage and several pipes for various functions are installed. A cylindrical bottom water discharge tank is installed immediately below the shaft.

The piping in the shaft includes plumbing for the receipt and dispatch of LPG, a draining pipe for seepage, pipes for measuring the fluid level, boundary and gas pressure and vent-purge pipes. The ground facilities include the receipt and dispatch facility for injected LPG in the gas phase to and from the storage cavern; a dehydrator to reduce the water that the LPG absorbs from seepage during storage; a drainage controller to manage the quality of the water from the bottom drainage tank; and a water utility to supply sealing water (Figures 2.2 and 2.3).

2.2 Design and Construction System of LPG Storage Caverns

2.2.1 Geology of the Construction Sites

2.2.1.1 The Kurashiki Facility

The Kurashiki facility is situated in a large area of reclaimed land in Mizushima Bay, west of the Kojima Peninsula and south of Okayama Prefecture. Several hills, formerly islands, lower than

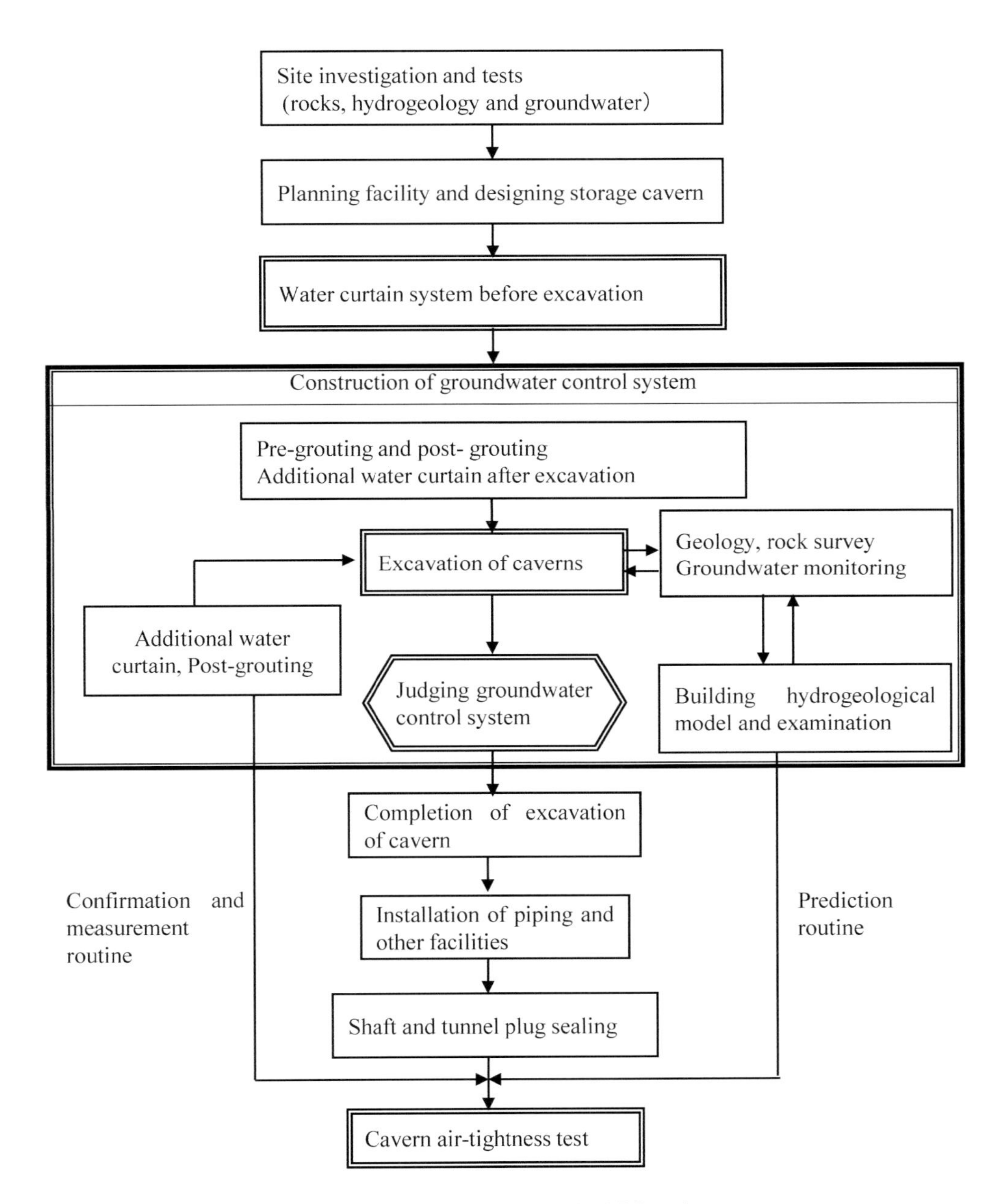

Figure 2.1 Flow chart of the groundwater control system for LPG rock storage.

50 m are scattered over the reclaimed land. East of the reclaimed land across the Yobimatsu waterway is the 200–250 m high mountain range of Kojima Peninsula with Washuzan (113 m) at its southern end (Figure 2.4). The sea bottom topography of Mizushima Port, which surrounds the Korashiki site, is at a relatively constant depth of 10–20 m below sea level to Mizushima Sea 3 km to the south-southeast. The seafloor is covered with thick sediment.

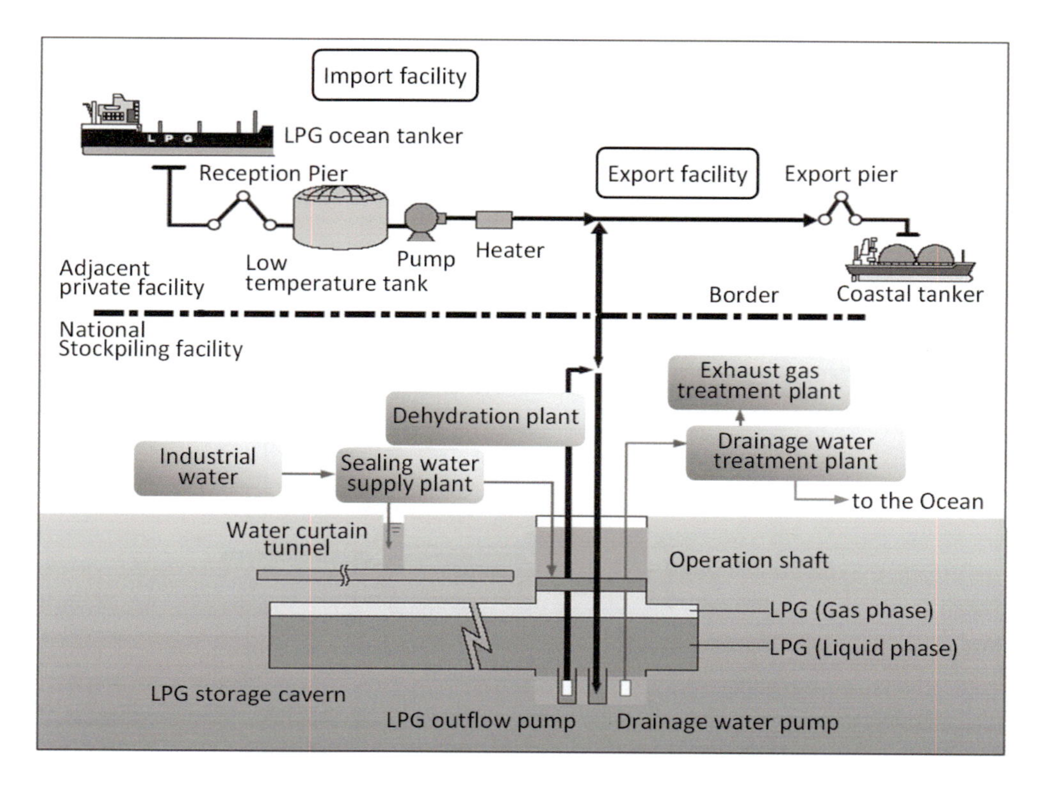

Figure 2.2 Flow of the LPG reception and dispatch, and flow of the water supply and drainage.

Figure 2.3 Photos of the Kurashiki facility: inside the storage cavern (left) and piping in the shaft (right).

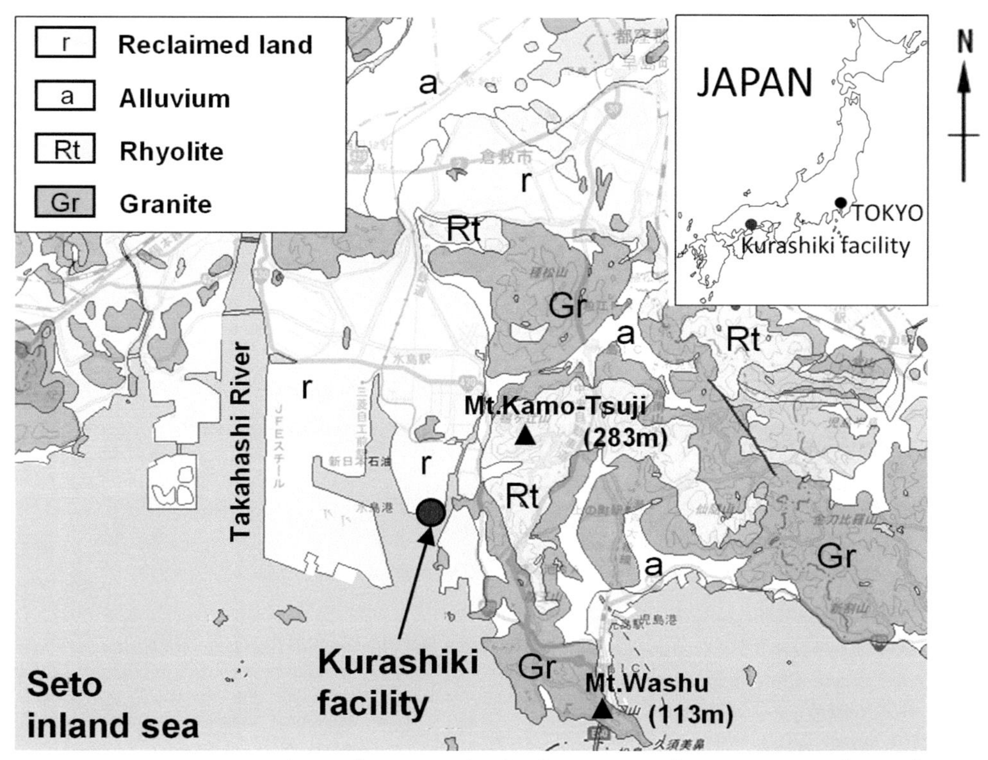

Figure 2.4 Surface geology map around the Kurashiki site.

The reclaimed land of the Kurashiki site is underlain by sediments from the Pliocene to Holocene age to a depth of 70 m below sea level. The sediment is made from silt, pebbles and interbedded sandstone-mudstone. These sediments overlay Cretaceous granite, weathered to 120 m below sea level and fresh granite below (Figure 2.5).

The storage cavern was constructed in the fresh granite 160 m below ground level to ensure water-tightness and mechanical stability. Most of the fresh granite is hard with a uniaxial compressive strength of around 115 MPa and an average permeability coefficient of about 0.35 Lu. (Note: In the permeability coefficient, 1 Lu is equivalent to 1.0×10^{-5} cm/sec.)

The dominant orientations of the fractures in the fresh granite are NNW-SSE and ENE-WSW. The geological structure is classified into five zones (I–V) by its fracture orientation and density (Figure 2.5). Within the planned storage cavern sites, five faults were recognised crossing orthogonally and diagonally: F2 and F3 faults of the NNW-SSE system near parallel to the caverns with a 50° to 70° dip; and F4–F6 faults of the ENE-WSW system intersecting the caverns with a steep dip. These faults were deformed before the granite fully solidified by the uprise of the granite after forming mylonite with a NNW strike. It was subject to hydrothermal alteration and the faults contain clay veins.

Very fine microfractures are found in the southern side of the storage caverns. These features are considered to be formed from friction between blocks surrounded by mylonite and cleavages at the uprise of the granitic body.

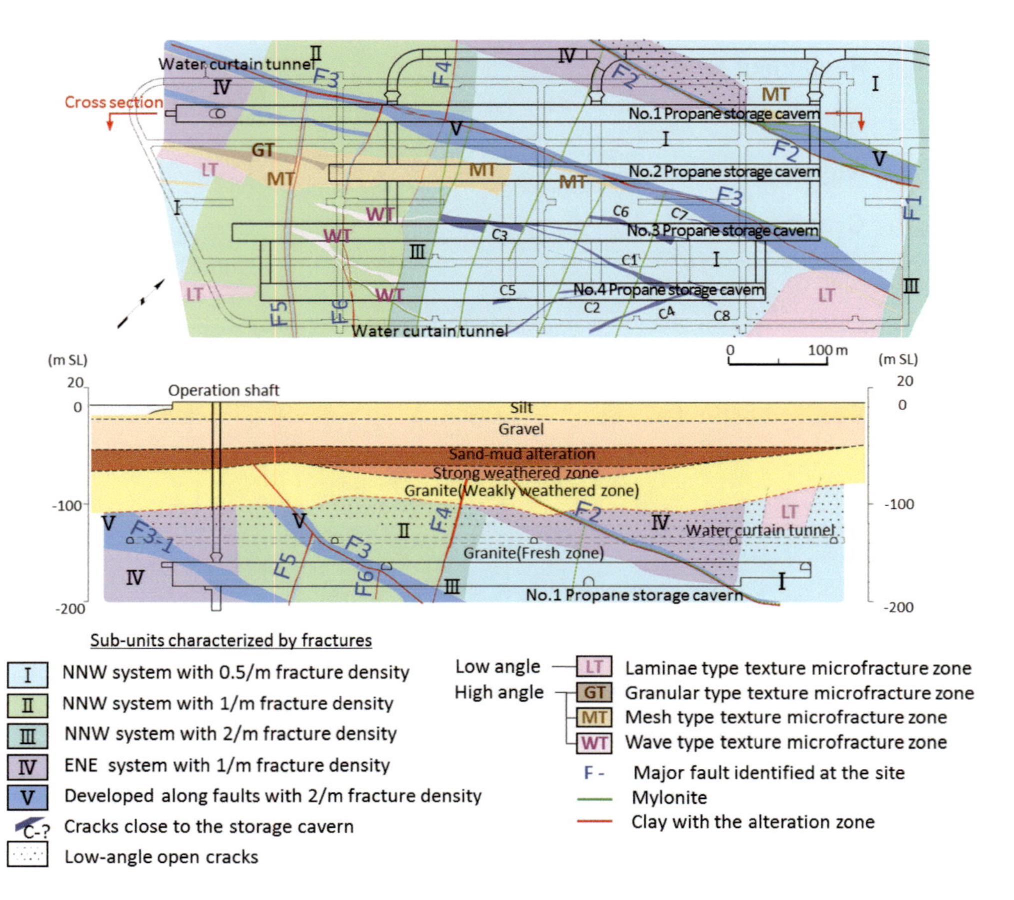

Figure 2.5 Regional geological map (top) and cross section (bottom).

The density of the fractures in the fresh granite is about 0.5 m and the average permeability is 0.3 Lu (maximum 4 Lu). In the F3 fault system, the fracture density is about 2 m and the average and maximum permeability are about 0.9 and 13 Lu, respectively. There are small fault rocks (mylonite and others) and areas with very fine microfractures with an average and maximum permeability of about 0.6 and 3 Lu, respectively.

As the fresh granite over the storage cavern is relatively thin at about 40 m, cracks develop in the rock around the cavern. Some parts are affected by weathering and hydrothermal alteration resulting in weal permeable faults and fractures. In these areas, the rocks are degraded from Class L to M.

2.2.1.2 The Namikata Facility

The Namikata site is situated in Namikata-machi, Imabari-shi in Ehime Prefecture. It is on a western-oriented cape at the tip of Takanawa Peninsula (Figure 2.6). The storage cavern is overlain by 150–250 m of rocks. The thinnest part is under the reclaimed land.

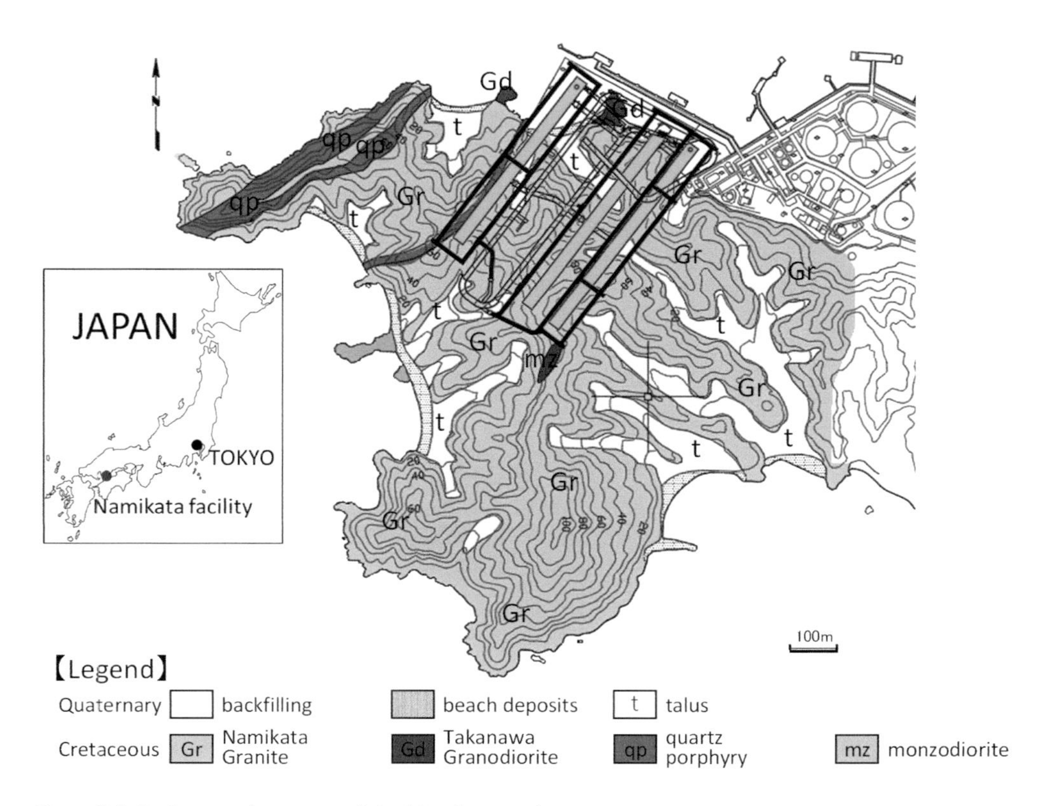

Figure 2.6 Surface geology map of the Namikata region.

The rocks around the Namikata storage site are largely Cretaceous granite. Most of the land area is covered with Namikata Granite and Takanawa Granodiorite and is abundant in biotite but poor in quartz near the northern coast. In the north-western part of the land area, quartz porphyry veins intrude into the Namikata Granite. To the south of the storage cavern is a hydrothermal shatter zone which has been weakened by hydrothermal alteration. The depth of weathering is 40–60 m from the ground surface. The storage cavern is in fresh granite with a minimum and maximum depth of 140 and 250 m, respectively, from ground level.

The uniaxial compressive strength of the fresh rock is, on average, about 118 MPa, and its average permeability coefficient is about 0.1 Lu (1×10^{-6} cm/s).

The granites surrounding the storage cavern are divided into seven geological zones by a shatter belt, an alteration belt and the density and dominant orientation of cleavages. The dominant orientation of the cracks near the lithological boundary between the Namikata Granite and the Takanawa Granodiorite (Zone V) is N20W and N60W across the cavern. The density of the cracks is high at 5–30 cm, but the cracks are closed. In addition to these orientations, Zones II and IV have open cracks in a N70E orientation close to parallel to the cavern. These cracks are generally highly permeable. Zone VI has similar crack orientations to Zone IV with medium density. The orientation of the cracks of Zone I is parallel to the cavern with a high dip angle. The density of Zone I is low and permeability is also low. The average Lugeon value is generally around 0.1 Lu for all zones, but some local permeable areas are present: over 1 Lu in 10% of Zone IV and 6% of Zone V. From the analysis of the data of a permeability test of the water curtain boreholes and the geological structure, some highly permeable areas are present in Zones II and IV. They are considered to be influenced by the cracks and low-angle cleavages that change the N60W system. A saline-fresh boundary is found near the boundary between the Takanawa Granodiorite and the Namikata Granite (Figure 2.7).

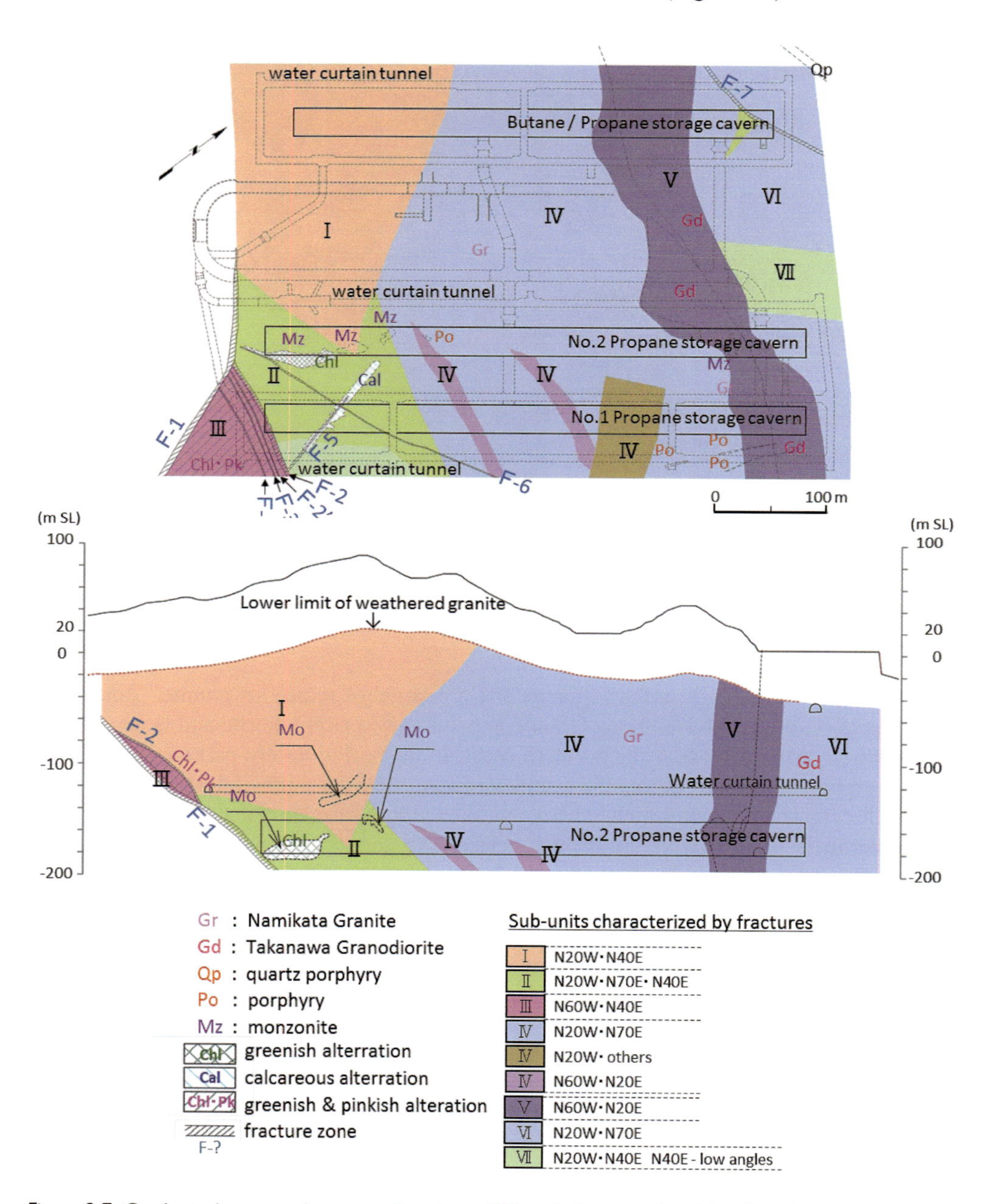

Figure 2.7 Geological map at bottom of arch at 157 m below sea level (top) and cross section of propane storage No. 1 (bottom).

2.2.2 Method of Groundwater Control

Both the Kurashiki and Namikata facilities adopt water curtains and water cut-off grout to control the flow of groundwater. The hydrogeological structure at the Kurashiki and Namikata sites is characterised by its heterogeneity: there are both highly permeable and impermeable faults, and permeability varies from one geological zone to another. The highly permeable areas around the caverns were improved to make them impermeable by forming a grouted rock zone. Water curtain boreholes – horizontal on the caverns and vertical along the side-walls of the caverns – provide a stable supply of groundwater to the water curtains to ensure that the water curtains maintain groundwater pressure to the permeable and impermeable faults around the caverns (Figure 2.8). During excavation, the effect of the water cut-off grout and water curtains was confirmed by monitoring the water pressure in the hydrogeological heterogeneity around the caverns using instruments arranged around the caverns, and the groundwater was controlled based actual measurement.

The basic procedure for controlling groundwater is: (1) pressurised water injection before excavation through the water curtain boreholes systematically arranged above and to the sides of the cavern; (2) measurement and management of the groundwater by arranging good conditions for 3-dimensional groundwater monitoring; and (3) improvement of the rock property in the highly permeable areas by pre-grouting before excavation. This procedure reduces the flow rate of seepage and drop in the groundwater pressure, and maintains the groundwater pressure around the cavern to prevent the occurrence of unsaturated zones. Once the groundwater is lost in the cracks and voids, it is very difficult and time-consuming to re-saturate them naturally or artificially. Therefore, water sealing before excavation is an essential step in constructing a storage cavern.

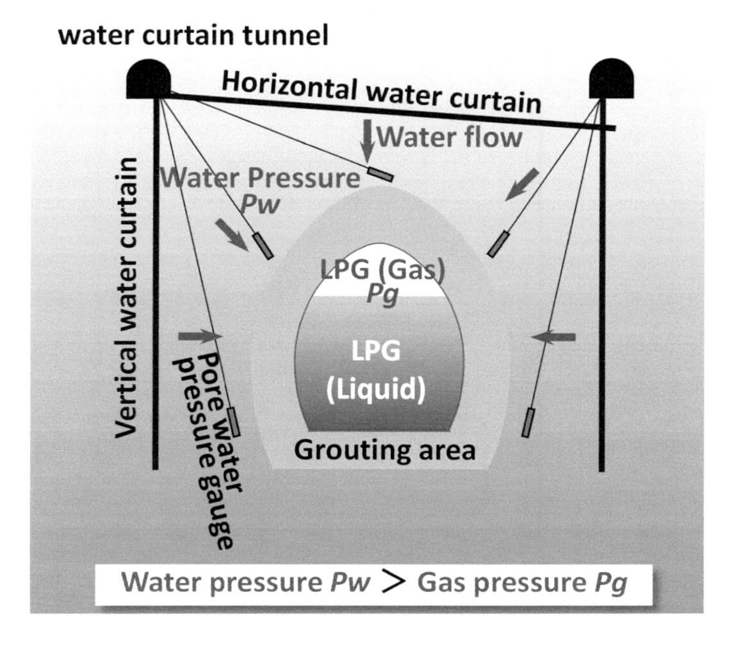

Figure 2.8 Conceptual diagram of rock cavern storage by water curtain.

The complex heterogeneous hydrogeological structure was revealed at the time of cavern excavation. The variation of the groundwater level from actual measurements during excavation showed values different from those predicted using a simple homogeneous model. A 3D heterogeneous model was built and refined, incorporating the data from a hydrogeological analysis and the groundwater measurement, also considering the effect of grouting and the water curtains. This model was used to predict the variation of the groundwater leveland was compared with the actual measurement. The analysis allowed an assessment of the complex behaviour of the groundwater with considerable accuracy.

The process of building the heterogeneous model from excavation to grouting is shown in Figure 2.9. The 3D model was refined by incorporating both the geological and grouting data at each step of the arch and bench excavation to improve the accuracy of prediction.

Before starting excavation of the arch, the initial heterogeneous model (1) was built with knowledge of the hydrogeological structure gained from excavation of the water curtain tunnels using the design parameters of grouting and water curtains. This model was used to predict the controlled behaviour of the groundwater (drop in groundwater pressure and seepage flow rate) during excavation of the arch.

The actual measurement of the groundwater level during excavation of the arch was compared with the analysis values of the initial heterogeneous model (1), a homogeneous model and a heterogeneous model (2) which incorporated the results of grouting data at arch excavation. This comparison allowed the heterogeneity of the groundwater behaviour to be

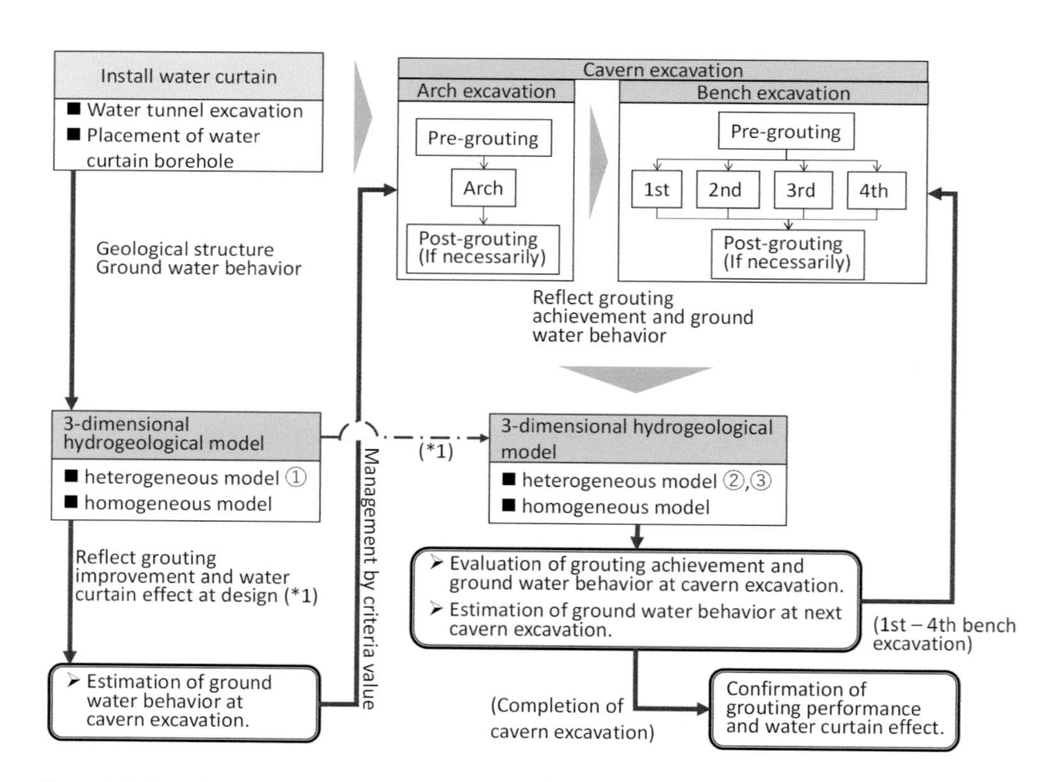

Figure 2.9 Procedure of excavation and grouting.

estimated, and the effectiveness of the grouting and water curtains was confirmed by this analysis.

Before starting excavation of the bench, the initial heterogeneous model (2) was built using the design parameters of grouting of the bench and the water curtains. This model was used to predict the controlled behaviour of the groundwater (drop in groundwater pressure and seepage) during excavation of the bench. The actual measurement of the groundwater behaviour during excavation of the bench was compared with the analysis values of the initial heterogeneous model (2), the homogeneous model and the heterogeneity of the groundwater behaviour during bench excavation. This comparison allowed to estimate the heterogeneity of the groundwater behaviour, and the effectiveness of the grouting and the water curtains was confirmed using this analysis.

On completion of excavation, the groundwater level was estimated from the analysis values of the heterogeneous model (3) which incorporated the results of grouting during bench excavation, and thus the effectiveness of grouting and the water curtains was confirmed.

2.2.3 *Shape of the Storage Caverns*

When a large cavern is excavated, the normal stress around the cavern is released and is concentrated in a tangential direction. This concentration of stress changes the kinetic regime of the rocks around the existing cracks at the cavern opening and creates new cracks. When the concentration of the stress reaches the failure strength of the rock, the stress drops to the stress level corresponding to the failure strength and the excess stress is redistributed in the rocks behind it. The area surrounding the caverns with the altered kinetic regime is called a "loosened zone".

In order to ensure the kinetic stability of large rock caverns, it is very important to curb loosened zones by reducing the concentration of the stress around the cavern wall and avoiding the redistribution of the stress. Checking the kinetic stability using a numerical simulation and assessing loosened zones are particularly important in designing grout for groundwater control in LPG rock storage. To curb loosened zones, the construction site is examined for the rock's kinetic and ground pressure properties, and the shape of the cavern is selected to smoothly distribute the stress of excavation so that the concentration of stress around corners is minimised. Reducing loosened zones is also necessary in designing support and planning operations.

The shape of the cavern was determined considering the required storage capacity and an assessment of the kinetic stability based on numerical analysis (excavation analysis). The deformation of the cavern during excavation, considering the redistribution of the stress using the ground pressure, is checked by an in-situ test and inversion on the model built for a site-specific fault and crack zone.

The rock formation around storage tanks is complex and heterogeneous with areas of faults and crack development. The rock properties are tested in-situ at heterogeneous areas. These sites are selected at the planning stage by geological investigation of the access tunnels and water curtain tunnels, borehole surveys and laboratory tests of cores. The rock properties from in-situ test are compared with the physical property values inverted from the deformation during excavation of the tunnels. This test confirms reasonable values for the physical properties of the rock cavern storage design to perform an excavation analysis of the heterogeneous model.

The deformation of the rock is monitored throughout the excavation of the cavern. The faults and deformation of the crush zone are observed and compared with the prediction by excavation analysis of the cavern. The scale, continuity and characteristics of the faults are studied in detail to ensure the stability of the cavern. Where the measurement differs from prediction, the rock properties are re-examined by an additional survey, and the support pattern is modified according to subsequent excavation analysis using the reviewed physical properties, and the excavation proceeds. This is discussed in detail in Chapter 5.

2.2.3.1 Rock Property

In the evaluation of the kinetic characteristics of rocks for a cavern design, "behaviour at excavation" means "behaviour of the rocks at the time of releasing stress by excavation". As the in-situ test is a test on samples of tens of centimetres long, the result at areas with intensive faults and cracks may be affected by local disturbances. In order to establish the macroscopic properties of such rocks, zoning of the kinetic characteristics of the rocks is carried out by detailed geological investigation and in-situ tests, and rock classes are assigned to each zone. The classes of rock properties are assigned by inversion of the deformation during excavation of the access tunnels and water curtain tunnels, calculation of the macroscopic apparent initial stress of rocks and properties and comparison of these with the results of in-situ tests.

2.2.3.1.1 ROCK PROPERTIES OF THE KURASHIKI SITE

The in-situ rock shear test, deformation test and initial stress measurement of the Kurashiki site were carried out at the locations shown in the geological map of Figure 2.10. The rocks expected around the caverns at the time of excavation of the access tunnels and water curtain tunnels were largely Class H, but there were also several Class L faults with clay and Class M rocks with low-angle cracks around the faults. Hence, the in-situ rock shear test and deformation test were carried out in the area of Class M rocks with cracks around the faults, and an initial stress measurement was performed on Class H.

The kinetic properties of the rocks were estimated from the in-situ rock shear test, in-hole loading test, plate loading test, core strength test in a laboratory and the result of an inversion analysis of the deformation during excavation of the water curtain tunnels. The design parameters are listed in Table 2.1.

The strength of Class M rocks used was established from the results of the rock shear test on Class M rocks with low-angle cracks, while the strength of Class L was established by a strength test of cores from the alteration zone and an estimation of the correlation between a P-wave velocity and the shear strength of granites.

Young's modulus for the design of Class M rocks was determined from the results of the plate loading test of Class M rocks with low-angle cracks and the inversion of the deformation during excavation of the water curtain tunnels. Young's modulus for the design of Class L rocks was determined from the results of the inversion of deformation during excavation of the F1 fault in the water curtain tunnel and the in-hole horizontal in a loading test at faults. As the F1 fault area of the water curtain tunnel underwent improvement by grouting against its high permeability, the deformation during excavation reflected the results of the improvement. Accordingly, Young's modulus for design was set as the constant after grouting.

The validity of these rock property parameters was checked using an excavation analysis of the deformation in the cross section that obliquely crosses the F2 fault and contains faults and Class M rocks. The excavation analysis was carried out using the rock properties set

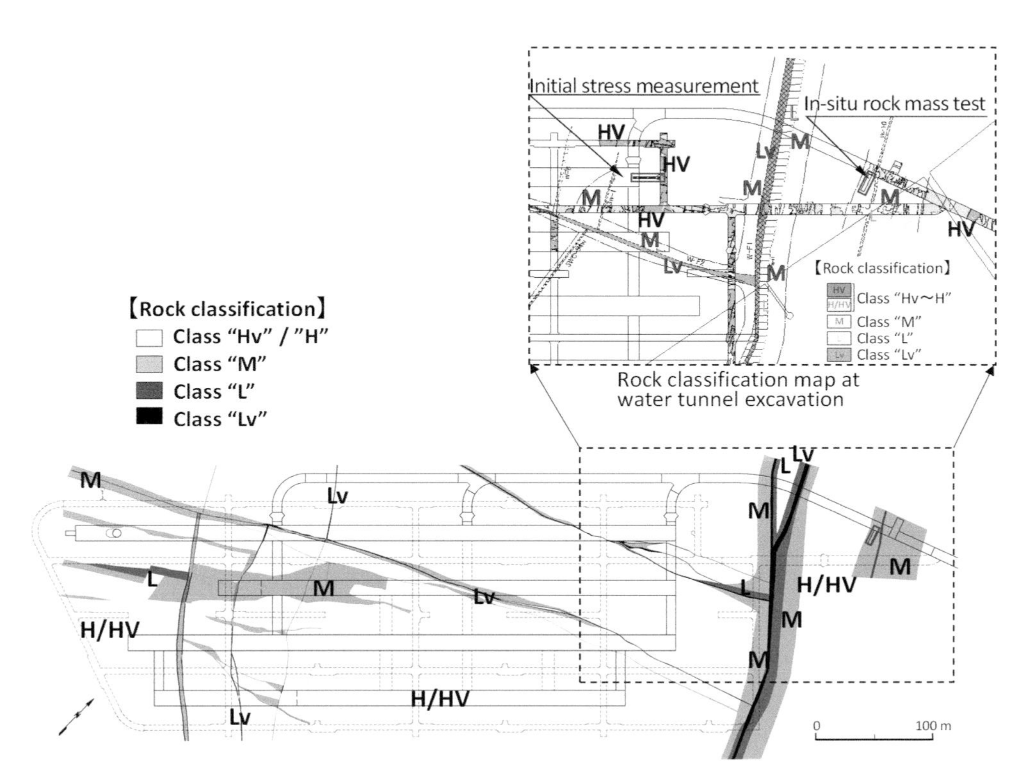

Figure 2.10 Geology map of the Kurashiki site with test and measurement locations.

Table 2.1 Design parameters of rock properties

Rock property	*Young's modulus* **E** *(GPa)*	*Peak strength*		*Residual strength*	
		Cohesion C_p *(MPa)*	*Internal friction angle* φ_p *(deg)*	*Cohesion* C_p *(MPa)*	*Internal friction angle* φ_p *(deg)*
Class H	8.0	3.0	49	0.75	49
Class M	4.5	1.5	46	0.45	49
Class L	1.5	0.7	40	0.45	40

for Classes M and L rocks on the heterogeneous model for the faults and Class M rocks (Figure 2.11). The agreement between the measured behaviour and the analysed values confirmed the validity of the parameters.

The initial stress of rocks is very important in an evaluation of the stability of caverns. However, the geological structure of this site is complex and the stress has been released by various crustal movements and uplifting. Therefore, it is important to set up design

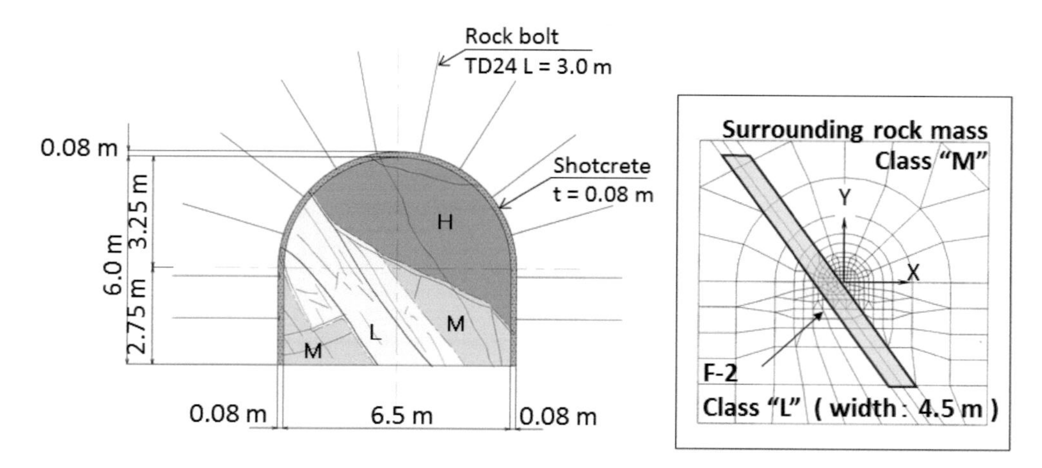

Figure 2.11 Excavation model for confirmation of rock property parameters.

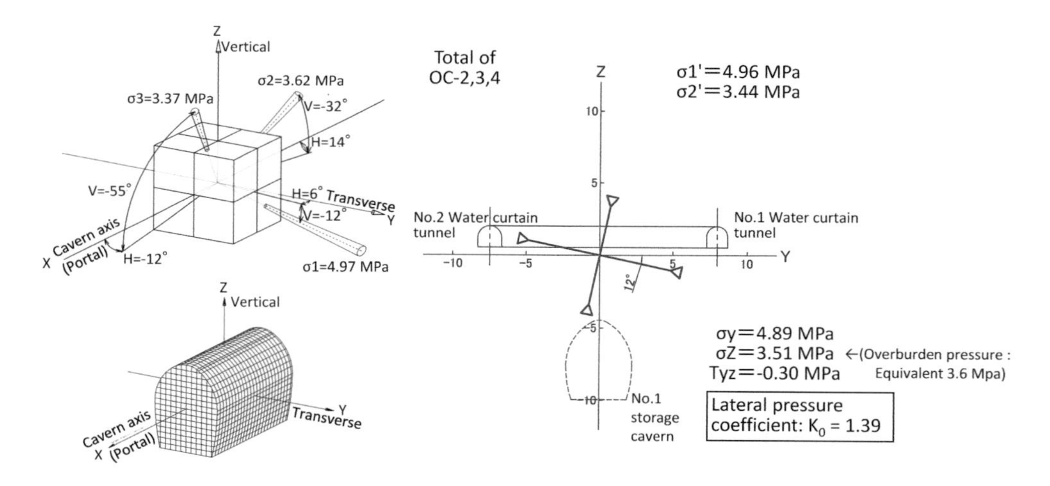

Figure 2.12 Result of initial stress measurement by the compact conical-ended borehole overcoring technique.

parameters by comparing the magnitude and orientation of the macroscopic apparent initial stress of rocks estimated from the inversion of the deformation during excavation of the water curtain tunnels with the result of an in-situ test.

The initial stress of rocks was measured using the compact conical-ended borehole overcoring technique, and it was found that the principal stress of the cross section of the caverns was a maximum of 5.0 MPa and a minimum of 3.4 MPa and the vertical earth stress was 3.5 MPa. Accordingly, the design parameter for vertical stress was set to 3.5 MPa and a lateral pressure coefficient $K_0 = 1.39$. The vertical stress was equivalent to the vertical overburden pressure at 3.55 MPa (Figure 2.12).

2.2.3.1.2 ROCK PROPERTIES OF THE NAMIKATA SITE

Figure 2.13 shows a geology map of the Namikata site during excavation of the water seal tunnels. The rocks are mainly of Classes Hv and H and some are Class M rocks distributed to the south of the propane storage cavern. There are continuous cracks with and without clay.

In-situ tests including the rock shear test and the plate load test were carried out on Classes H and M rocks. The result showed that Class H rocks had hard kinetic properties with an internal friction angle of 62°, cohesion of 4.8 MPa and Young's modulus of 30 GPa, while Class M rocks had an internal friction angle of 50°, cohesion of 4.0 MPa and Young's modulus of 8 GPa, equivalent to Class H rocks in the Kurashiki site.

The initial stress was measured using the hydraulic fracturing method in reconnaissance boreholes and the compact conical-ended borehole overcoring technique in the boreholes of the access tunnels. The compact conical-ended borehole overcoring technique gave a lateral pressure coefficient of $K_0 = 1.6$. Accordingly, vertical stress of 5.4 MPa, which is equivalent to the vertical overburden pressure, and horizontal stress of 8.6 MPa were used as the design parameters (Table 2.2).

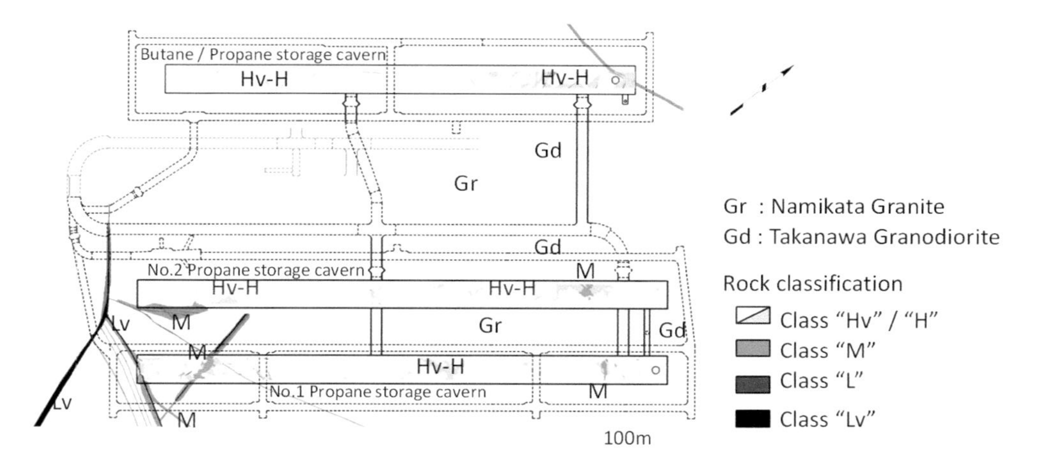

Figure 2.13 Rock class map in plan view of the Namikata site at 125 m bSL.

Table 2.2 Physical properties and initial stress of rocks

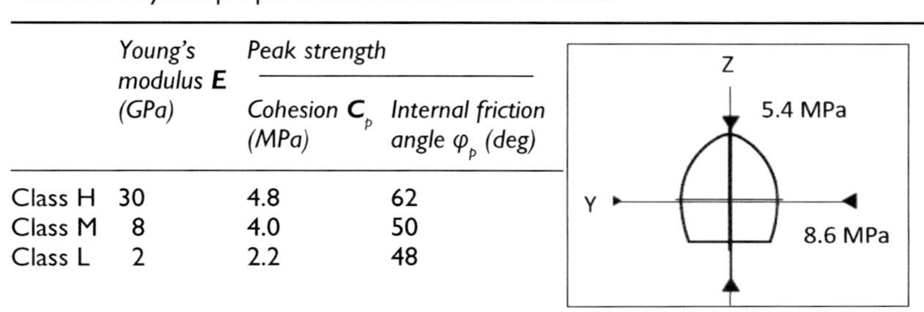

	Young's modulus **E** *(GPa)*	*Peak strength*	
		Cohesion C_p *(MPa)*	*Internal friction angle* φ_p *(deg)*
Class H	30	4.8	62
Class M	8	4.0	50
Class L	2	2.2	48

2.2.3.2 Shape of Cavern and Basic Pattern of Support

When a large-scale cavern is excavated, normal stress is released near the cavern and the stress concentrates in a tangential direction. The deep walls also respond to the change by redistributing the stress. It is necessary to select a cavern shape that is most appropriate to the geological structure, rock property and the characteristics of the rock pressure of the site to ensure the safety of the cavern. The shape of the cavern is designed to curb the occurrence of loosened areas to ensure a smooth redistribution of the stress caused by excavation and to reduce the concentration of stress.

The cross sections of the caverns at the Kurashiki and Namikata facilities are shown in Figure 2.14. Both are egg-shaped, and are 18 m wide and 24 m high for Kurashiki and 26 m wide and 30 m high for Namikata.

2.2.3.2.1 CAVERN AND SUPPORT DESIGN FOR THE KURASHIKI FACILITY

Considering that the Kurashiki site has many faults and cracks along the faults and that its rock class is low, a cross section with a large arch curvature was selected. The effect of the faults and the lower rock class on the stability of the cavern was evaluated using an excavation analysis. The excavation analysis was carried out on a "fault model" with faults crossing the caverns at acute angles and a "Class M rock model" which analyses the effect of Class M rocks with cracks. From its results, the distribution of local safety factors is shown in Figure 2.15. For the fault model, the loosened area in the fault zones next to the caverns was restricted to 2.5 m, while the loosened area in the Class M rocks was restricted to 2 m at the lower part of the side wall.

The support design was based on this result. Figure 2.16 shows the support pattern. The stress loaded onto the support was calculated by excavation analysis with a model of concrete spray and rock bolts. The results confirmed that the stress was under the specified permissible stress and the supports were healthy.

2.2.3.2.2 CAVERN AND SUPPORT DESIGN FOR THE NAMIKATA FACILITY

The geology of the Namikata site mainly consists of Class H rocks with some Class M rocks to the south of the propane storage cavern. The excavation analysis was carried out on Classes H and M rocks. From its results, the distribution of local safety factors is shown in

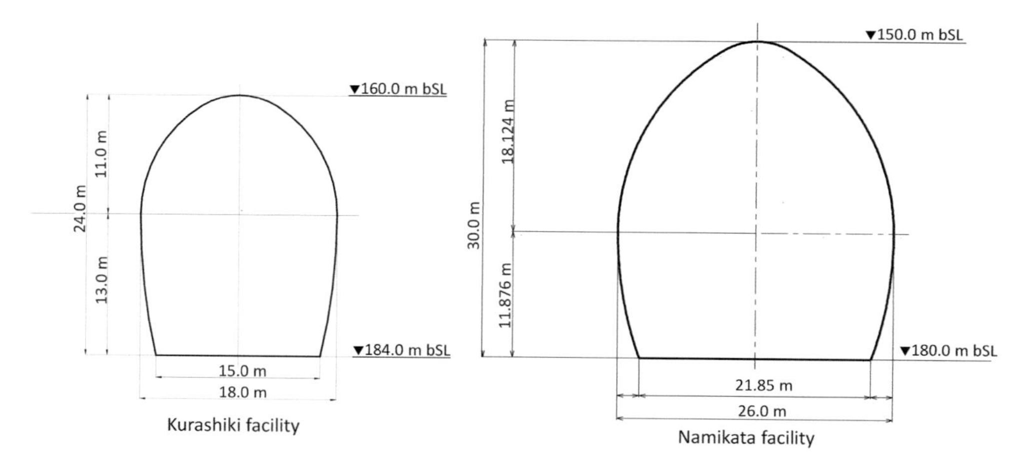

Figure 2.14 Cross sections of storage caverns of the Kurashiki and Namikata facilities.

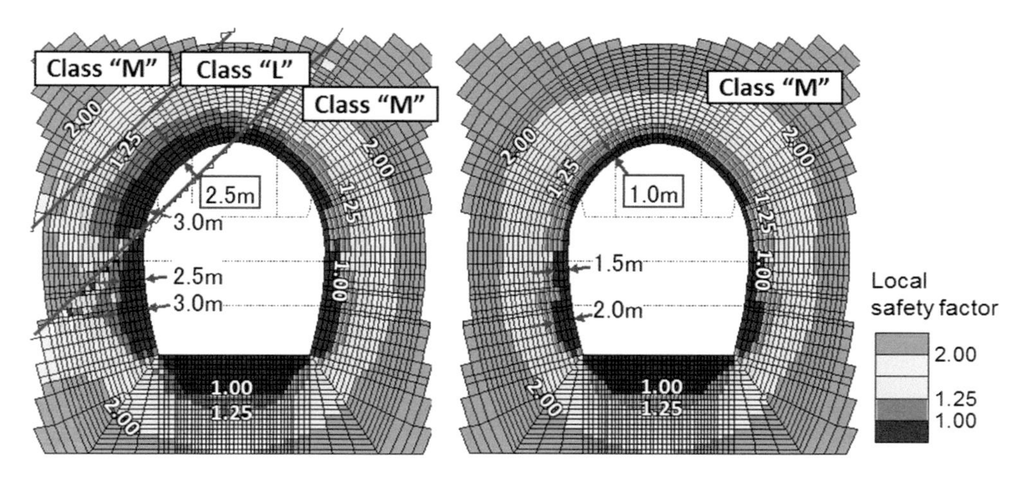

Figure 2.15 Spatial distribution of local safety factors. Fault model (left). Class M rock model (right).

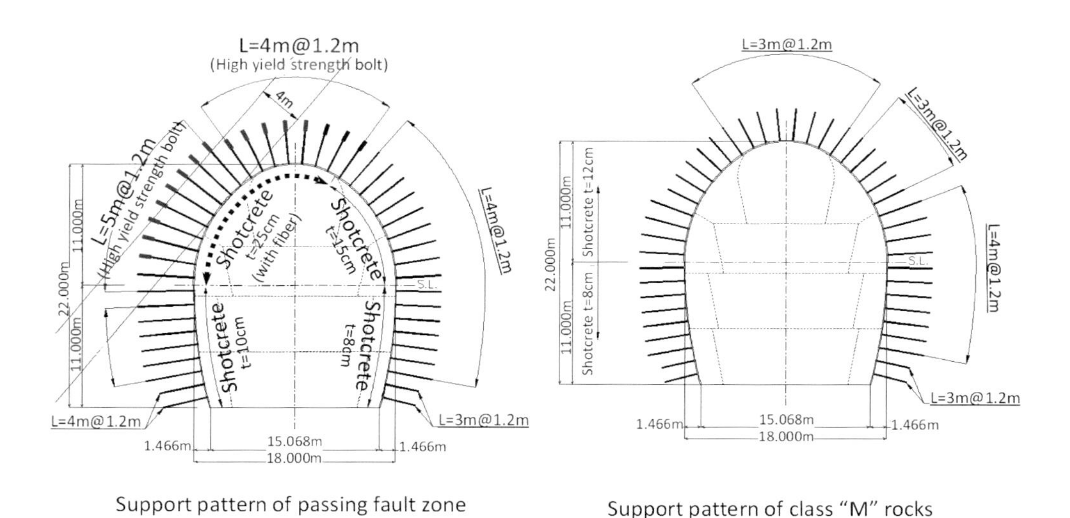

Figure 2.16 Support pattern.

Figure 2.17. The loosened area in Class H rocks was limited to 2–3 m, while the loosened area in Class M rocks was restricted to 3 m, which confirmed the safety of the cavern.

The support design was based on this result. Figure 2.18 shows the support pattern. The stress loaded onto the support was calculated by excavation analysis with a model of concrete spray and rock bolts. The results confirmed that the stress was under the specified permissible stress and the supports were healthy.

2.2.4 Arrangement of Water Curtains

In order to ensure air-tightness, maintaining a stable groundwater level for the hydrogeological characteristics and the groundwater recharge condition of the construction site

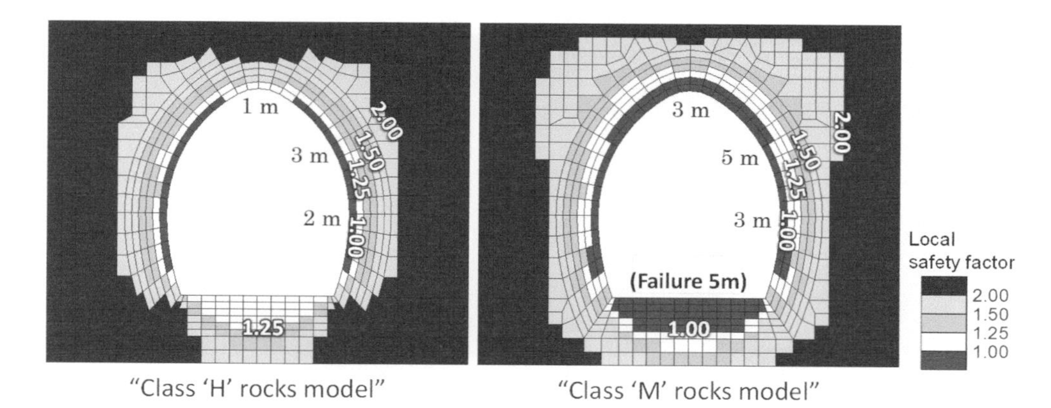

Figure 2.17 Spatial distribution of local safety factors.

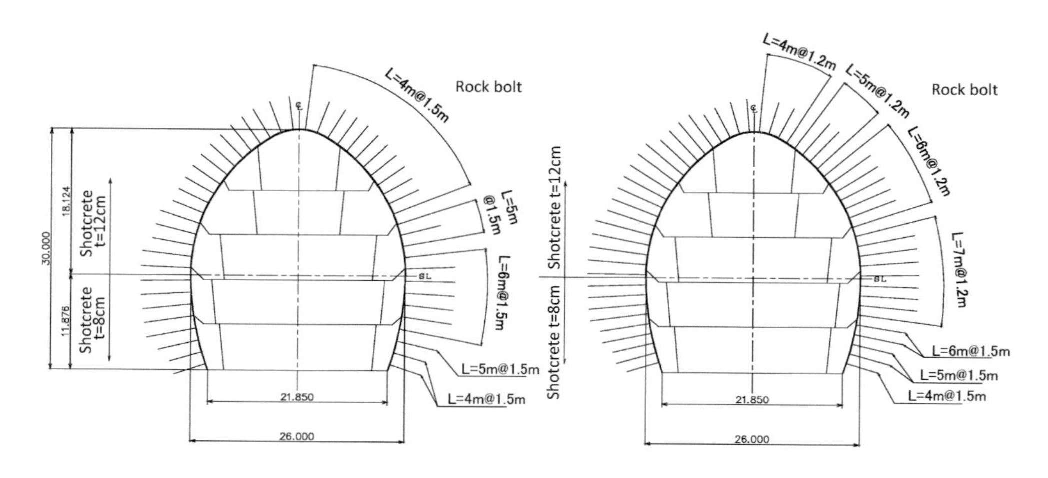

Figure 2.18 Support pattern.

is necessary. If a stable groundwater level is necessary for the condition of the site, water curtains are designed to form artificial hydrological boundaries around the storage cavern by injecting groundwater through the water curtain boreholes at the top of the cavern. In the current sites, it was decided to install such water curtains to maintain a stable groundwater level.

The water curtains are facilities to inject groundwater into the water curtain boreholes drilled from the water curtain tunnels above the cavern. At the planning stage of storage caverns, the arrangement of the water curtain boreholes is designed assuming the uniform permeability of the rocks surrounding the caverns.

Excavation of the caverns revealed the complex heterogeneous nature of rocks with faults, shatter zones and the orientation and density of cracks around the storage cavern. The effect of water curtains was estimated from the data on groundwater behaviour and tests on the

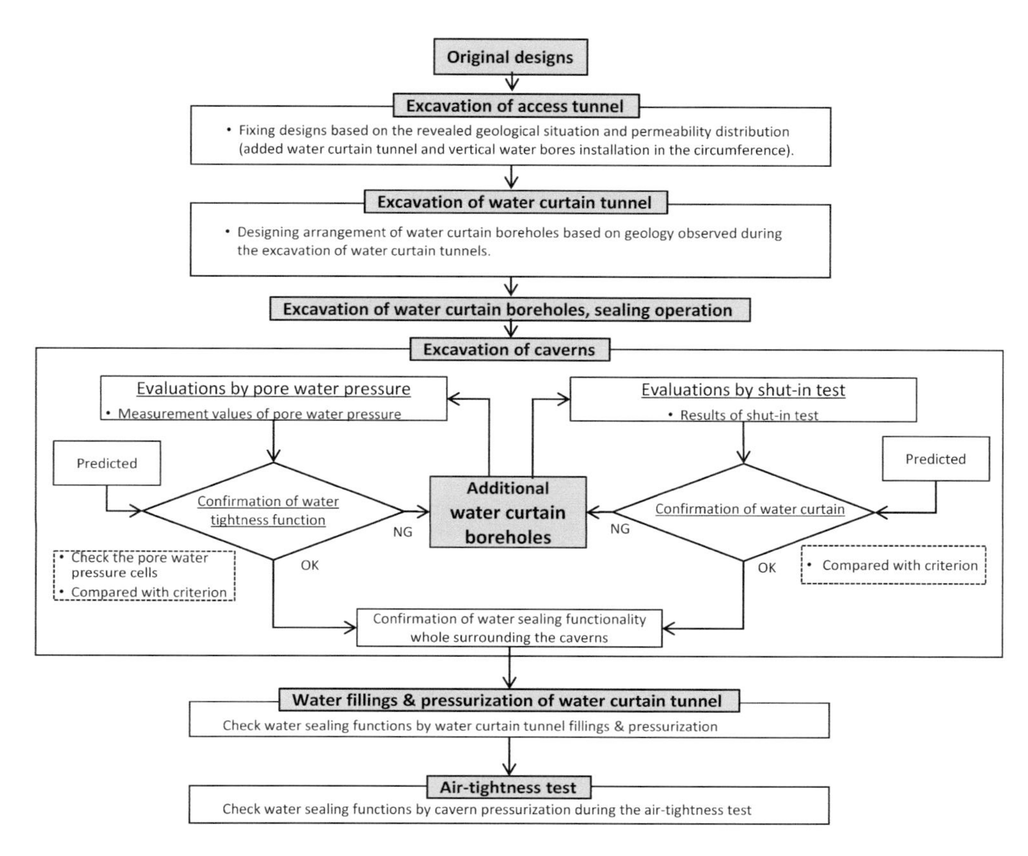

Figure 2.19 Design and construction system of water curtain boreholes.

rocks. The systematic design and construction of additional water curtain boreholes were devised to ensure the final air-tightness of the storage cavern (Figure 2.19).

2.2.4.1 Water Curtains of the Kurashiki Facility

The basic arrangement of the water curtain was a combination of horizontal and vertical water curtain boreholes. The horizontal boreholes were spaced every 10 m and their orientation was 60° against the wall and floor of the water curtain tunnel so that they crossed the dominant orientation of the cracks around the cavern as close to perpendicular as possible. The vertical boreholes were arranged at 10 m intervals between the access tunnel and the adjacent cavern, in a highly permeable area near the outer rim of the cavern and the areas affected by faults, and at 50 m intervals between other caverns for water injection to curb unsaturated zones.

The hydrogeological feature of the Kurashiki site is characterised by F2 and F3 faults, the shatter zone of the F3 system and microfractures.

The F2 fault has impermeable clay veins. Low-dip cracks develop in the hanging wall of the F2 fault and they have low permeability, while the foot wall shows a tendency for higher hydraulic conductivity.

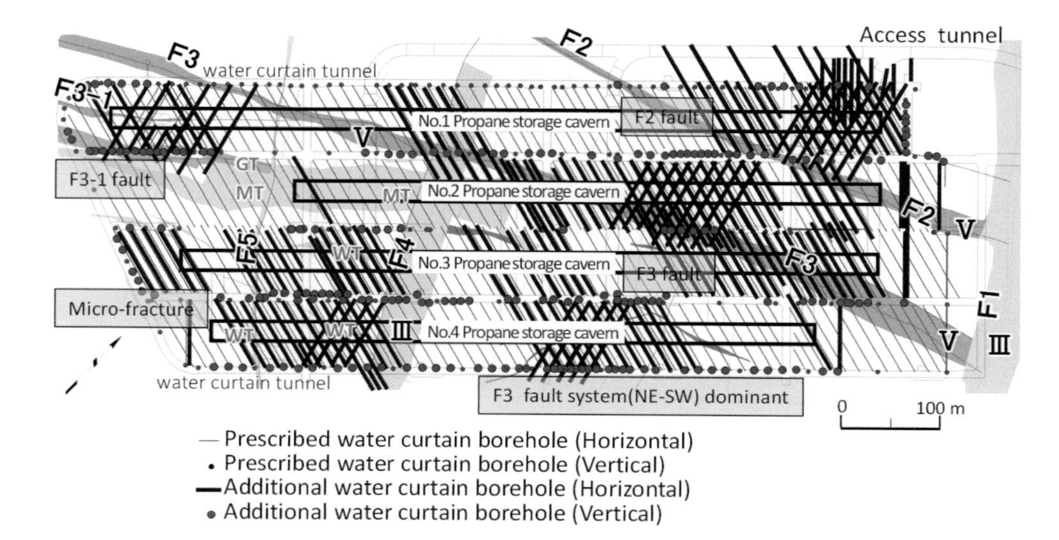

Figure 2.20 Arrangement of water curtain boreholes – plan view.

The F3 fault has a weak sealing capacity as the clay veins are thin. Around the F3 fault, there are permeable cracks in the same orientation as the fault, forming a permeable zone. In the zone of the F3 crack system, the permeability is high along the cracks themselves, but the actual rocks around them are impermeable.

The microfracture belts are a few centimetres to a few metres thick. Their strikes are both along and across the orientation of the storage caverns, and the dips are both low and high. The cracks are fine and highly permeable. Near the F3-1 fault, highly permeable microfractures are present to the east of No. 1 storage cavern. These microfractures are connected to the low-angle cracks with horizontally anisotropic permeability near the bottom of the F3-1 fault, resulting in a complex hydrogeological structure.

For this hydrogeological structure, some water curtain boreholes were added to satisfy the water-tightness, according to the results of hydrogeological surveys and shut-in tests. In some areas around the faults and in the microseating zones, the interval between the water curtain boreholes was reduced to 5 m and oblique water curtain boreholes were added (Figure 2.20).

2.2.4.2 *Water Curtains of the Namikata Facility*

The water curtains of the Namikata facility were designed under the same principle as for the Kurashiki facility. However, as the Namikata facility includes propane and combined propane-butane storage, the distance between the vertical water curtain boreholes was reduced to 10 m to prevent cross-contamination. The orientation of the water curtain boreholes was 90° against the wall and floor of the water curtain tunnels so that they crossed the dominant orientation of the cracks around the cavern as close to perpendicular as possible.

More water curtain boreholes were added to satisfy the water-tightness requirement of the complex hydrogeological structure incorporating information from the geological structure, the groundwater behaviour and the results of water-tightness tests during excavation of the caverns. From knowledge of the geological structure, some water curtain boreholes were

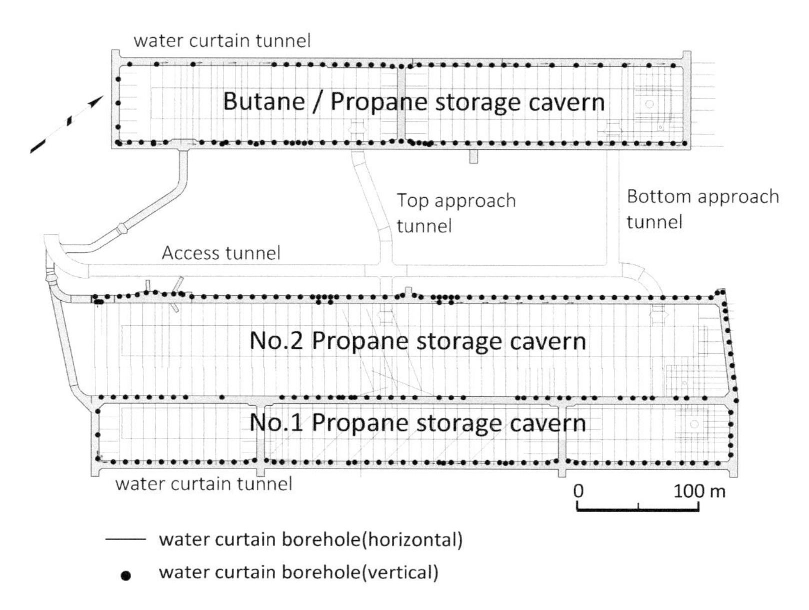

Figure 2.21 Arrangement of water curtain boreholes – plan view.

arranged to obliquely cross the existing water curtain boreholes. Figure 2.21 shows the final borehole arrangement after these modifications.

2.2.5 Confirmation of Air-tightness of LPG Rock Cavern Storage

The LPG storage caverns were constructed deep underground where water-tightness is ensured. The construction involved drilling access tunnels and shafts for piping, and then the storage caverns were excavated. On completion of excavation, concrete plugs were installed at the intersections of the access tunnels, shafts and the storage cavern to isolate them before the air-tightness test. Finally, the access tunnels and shafts were filled with water (Figure 2.22).

The air-tightness test took place on completion and closure of the LPG storage cavern. During the air-tightness test, the cavern was pressurised with air to the test pressure and the test confirmed that the internal pressure of the cavern was maintained to a constant (Figure 2.23).

Compressed air was injected into the cavern to a pressure higher than the designed pressure, and air-tightness was confirmed over 72 h. Details of the air-tightness test and the results are described in Chapter 6.

The result of the air-tightness test was judged by the variation of pressure ΔP (Figure 2.24). An example of the temporal variation of the pore water pressure at the ceiling of the cavern during pressurising is shown in Figure 2.25. The pore water pressure around the cavern increased with pressurised air injection. The difference in the hydraulic head between the internal pressure and the pore water pressure at the ceiling of the cavern was maintained at an adequate level throughout the air-tightness test. Thus, the air-tightness of the storage caverns and the surrounding area was confirmed.

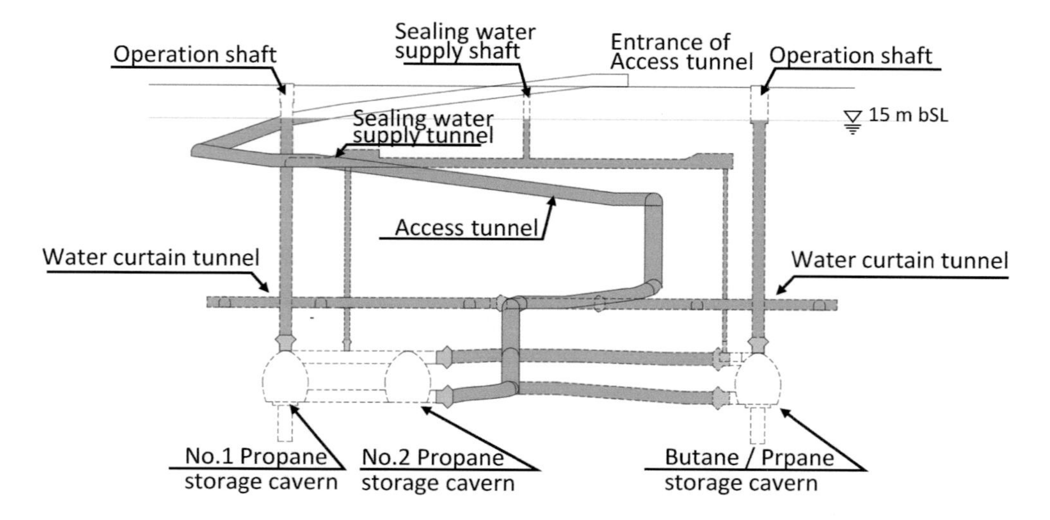

Figure 2.22 Status of the Namikata facility at air-tightness test.

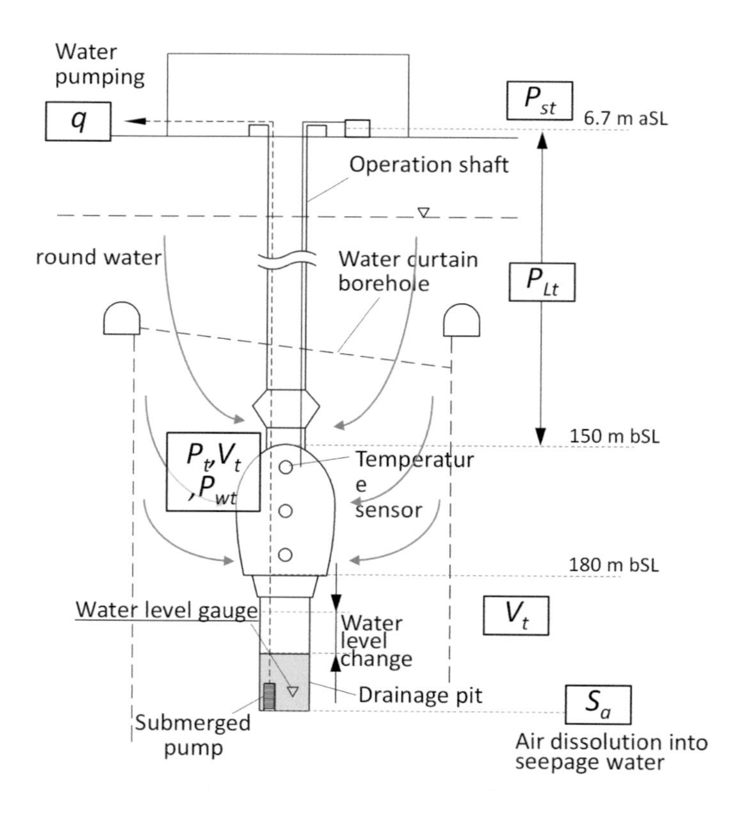

Figure 2.23 Conceptual diagram of the air-tightness test.

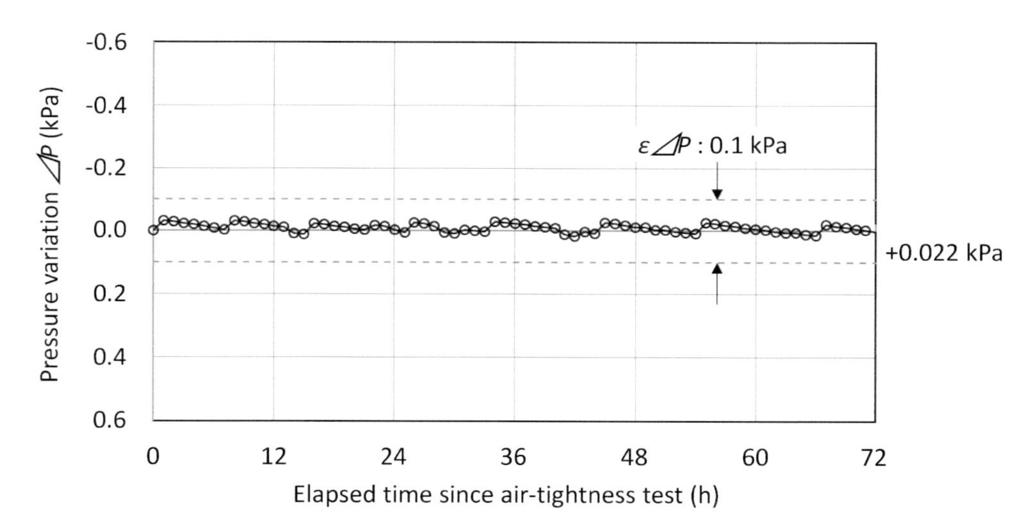

Figure 2.24 Result of calculation of variation of pressure ΔP.

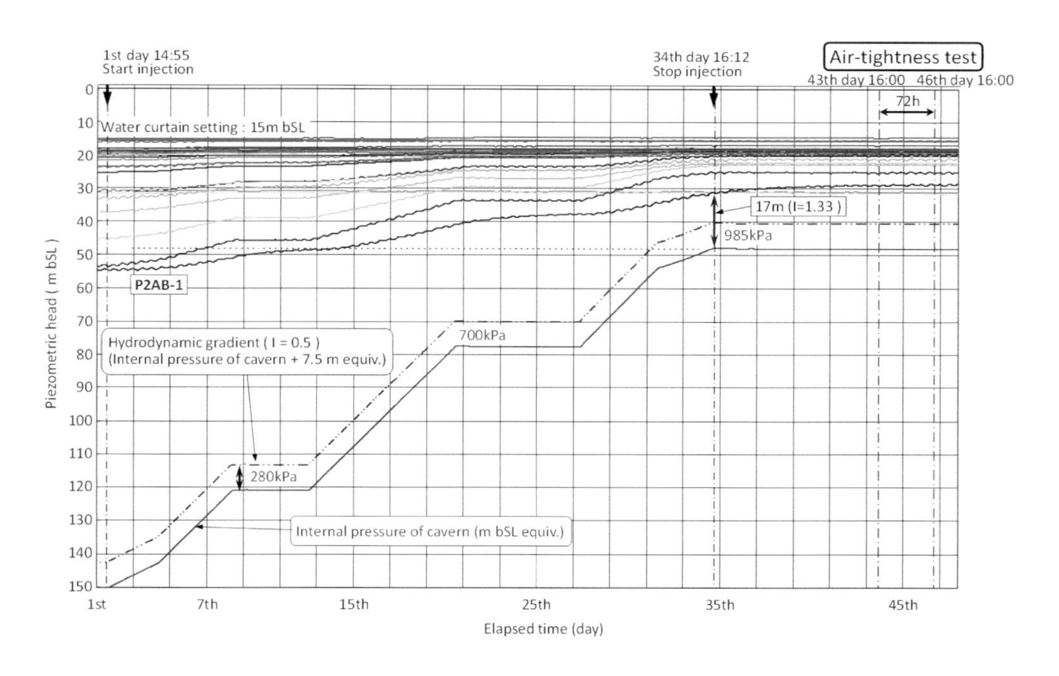

Figure 2.25 An example of temporal variation of pore water at pressurising and air-tightness test.

References and Further Readings

Aoki K., Yamamoto T., Shirasage S. (2003) Prediction of the rock properties ahead of the tunnel face in TBM tunnels by geostatistical simulation technique. *Proceedings of the First Kyoto International Symposium on Underground Environment (UE-KYOTO)*, Kyoto 247–252.

Bergman M. (1977) Storage in excavated rock caverns. *Proceedings of the First International Symposium*, Stockholm. P. 832.

Geological Survey of Japan, AIST (ed.) (2009) *Seamless Geological Map of Japan at a Scale of 1:200,000*. DVD edition. Geological Survey of Japan, National Institute of Advanced Industrial Science and Technology, Tsukuba.

Nilsen B., Olsen J. (1989) Storage of gases in rock caverns. *Proceedings of the International Conference on Storage of Gases in Rock Caverns*, Trondheim, p. 398.

Chapter 3

3-Dimensional Hydrogeological Model

A detailed hydrogeological survey was carried out at the early stage of excavation to understand the inhomogeneous behaviour (the distribution of the pore water pressure and its temporal variation) of the entire storage cavern area of the Kurashiki and Namikata sites. Based on results of the survey's, a 3-dimensional (3D) hydrogeological model was constructed. The distribution of the groundwater pressure was monitored during excavation and analysed by comparing it with the prediction of pore water pressure from the 3D hydrogeological model.

To model the complex geological structure of the sites of this project, the 3D distribution of the hydraulic conductivity of the rock was estimated around the cavern using a geostatistical technique. It analysed a large dataset of Lugeon tests from boreholes for scouting, water level measurements and water-tightness.

The validity of the initial 3D hydrological model was examined based on the newly acquired Lugeon test data at each stage of cavern excavation. The model prediction was updated for the next-stage prediction of the seepage flow rate and pore water pressure by comparing it with the actual measurements. These analyses and examinations confirmed the functionality of a water curtain and the effectiveness of grouting.

3.1 Hydrogeological Structure of the Construction Sites

As stated in Section 2.2.1 on the "geological characteristics of the construction sites", large amounts of data were collected from geological investigation at the time of excavation including core examination of the test boreholes and water curtain boreholes, observation by borehole televiewers and geological observation of the excavation face of the storage caverns and water curtain tunnels. Particularly in the granite around the cavern, detailed investigations were carried out including the orientation, density and characters of cracks and analysis of the minerals filling the cracks under a microscope. Based on these data, the granites were classified according to their geological structure and the process of forming them was estimated.

3.1.1 Kurashiki Facility Site

Figure 3.1 shows a geological map of the floor of a storage cavern arch (187.5 m below sea level) with the frequency distribution of the Lugeon value measured at the time of pre-grouting the bench. Figure 3.1 also shows the five geological zones (I–V) explained in Section 2.2.1 and the spatial distribution of microfractures. Zone V is affected by faults F2 and F3, and the areas have concentrations of highly permeable microfractures. The fracture density of Zone V is around two per metre, and its maximum Lugeon value is as high as

DOI: 10.1201/9780367822163-3

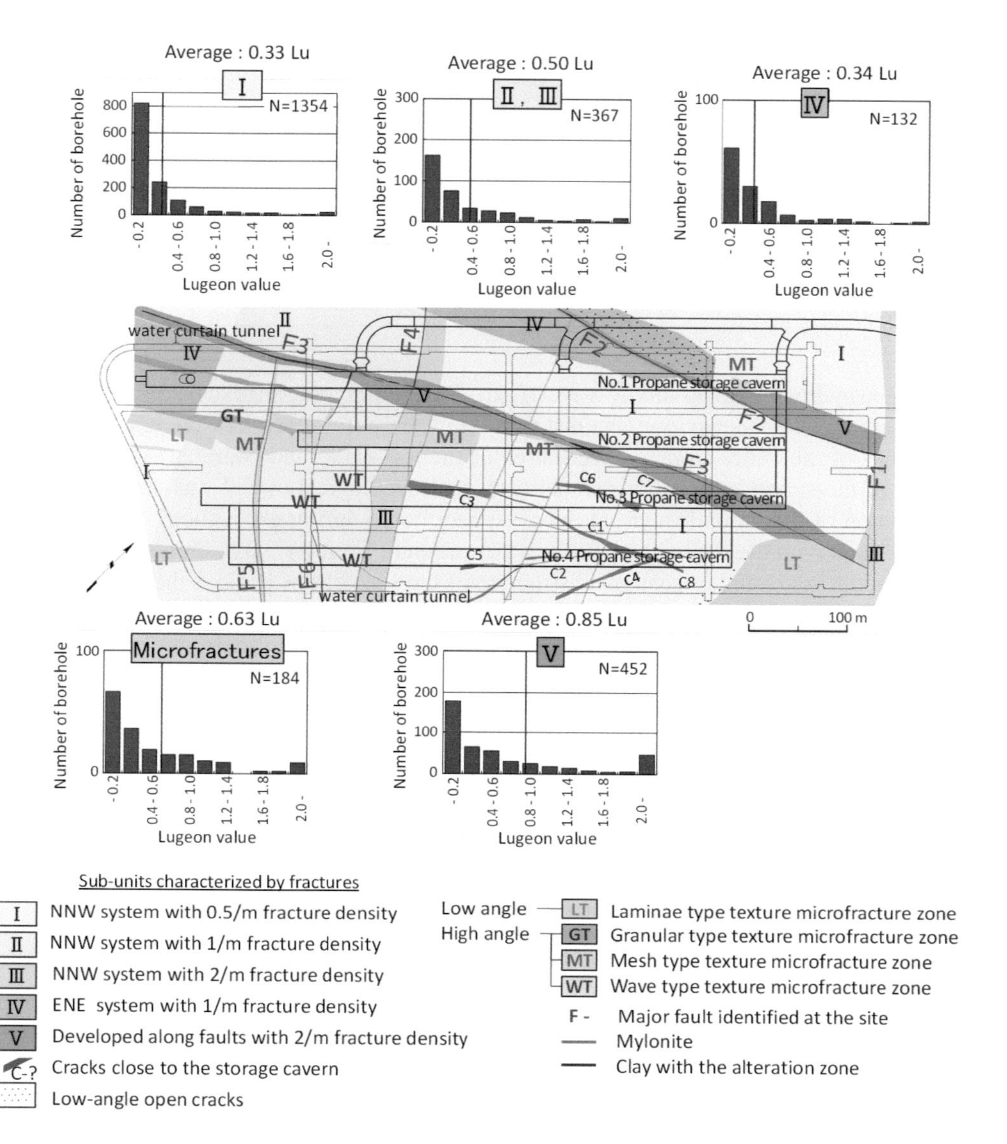

Figure 3.1 Frequency distribution of initial Lugeon values for each geological zone of the Kurashiki site (geology at 167 m bSL).

13 Lu (average 0.9 Lu). The maximum Lugeon value of the area with a concentration of microfractures is 3.3 Lu (average 0.6 Lu).

3.1.2 Namikata Facility Site

Figure 3.2 shows a geological map of the floor of a storage cavern arch (157.5 m below sea level) with pre-grouting Lugeon values of the arch which are the initial values. According

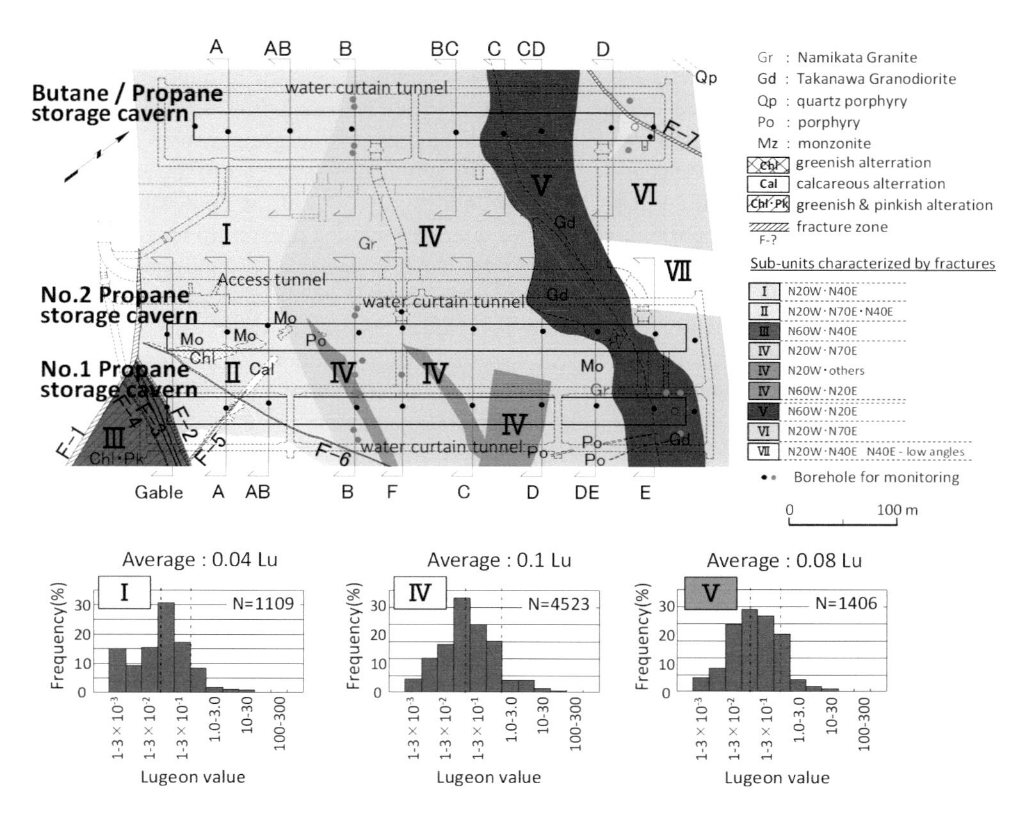

Figure 3.2 Frequency distribution of initial Lugeon values for each geological zone of the Namikata site (geology at 157.5 m bSL).

to fracture density, the Lugeon values of Zones I, IV and V are about 0.1 Lu. However, the frequency of the values over 1 Lu is different from zone to zone: about 4% in Zone 1, 10% in Zone IV and 6% in Zone V. Highly permeable areas were found near the propane storage caverns. Some of the highly permeable zones were considered to be near cracks in the orientation of the storage cavern, N70E, which crosses the structure of the N20W and N60W systems. When drilling grouting holes, encountered highly permeable cracks a large flow rate of seepage sometimes flowed into the grouting holes causing a sudden decrease of the pore water pressure near the cracks in the bedrock. This characterises the geology of the Namikata site, showing a tendency for the pore water pressure in the highly permeable cracks to be sensitive to seepage.

3.2 Groundwater Measurement System of Whole Storage Area

In constructing a large-scale underground storage, it is difficult to totally understand the hydrogeological characteristics and pore water distribution in their entirety. At the time of construction, therefore, the pore water pressure and kinetic behaviour (spatial and underground displacements) were closely monitored.

Understanding the distribution of the pore water pressure and the flow rate of seepage in the construction of an underground storage cavern under the condition of a heterogeneous bedrock, which is the concern of this book, is especially necessary through all the steps from excavation to pressurised storage. For an appropriate response, it is necessary to evaluate the water-tightness from an analysis of the behaviour of the groundwater using a 3D hydrogeological model which accurately reflects the actual measurements of the groundwater around the cavern. Therefore, a groundwater monitoring system of the pore water pressure around the cavern is very important. The gauges must be arranged differently for two different purposes: (1) to obtain 3D distribution of the pore water pressure all around the cavern and (2) to estimate the local variation of the pore water pressure due to geological heterogeneity.

To measure the groundwater level and the pore water pressure of the cavern area, gauges were distributed over a large area. For local variation, additional pore water pressure gauges were arranged around the cavern from the water curtain tunnel above the right and left walls of the cavern, the top and shoulders of the cavern and outside the grouting area on the floor. These gauges allowed the measurement of the pore water pressure in response to various construction activities: excavating, grouting, water-tightness tests, water-tightness enhancement and air-tightness tests. The arrangement of the gauges on the top, shoulders and floor of the cavern was considered with the knowledge of geological heterogeneity to ensure that the distribution of the pore water pressure around the cavern was captured (Figure 3.3).

3.2.1 The Kurashiki Facility

In the Kurashiki facility, water level gauges and pore water pressure gauges were arranged outside the water curtains to measure the groundwater level and to estimate the distribution of the pore water pressure in the large area around the cavern. Additional pore water pressure gauges were installed to cover the entire storage cavern inside the water curtains to understand the behaviour of the pore water pressure around the cavern. Figure 3.4 shows the arrangement of 46 water level gauges and 74 pore water gauges.

The pore water pressure gauges were systematically arranged on the top, shoulders and the floor of the cavern and around the cavern outside the grouting area. These were separated at 150–200 m intervals across the axis to form the main cross sections. Additional subordinate cross sections were drawn at locations critical for water-tightness: faults, areas

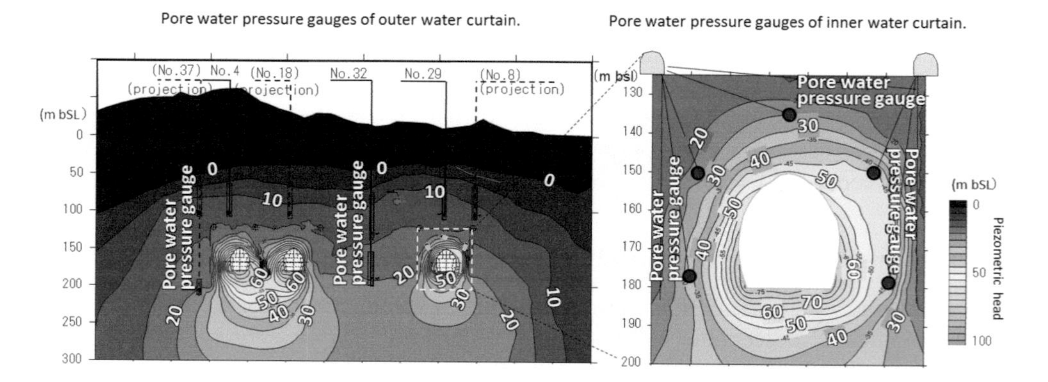

Figure 3.3 Concept for the arrangement of a pore water pressure gauge.

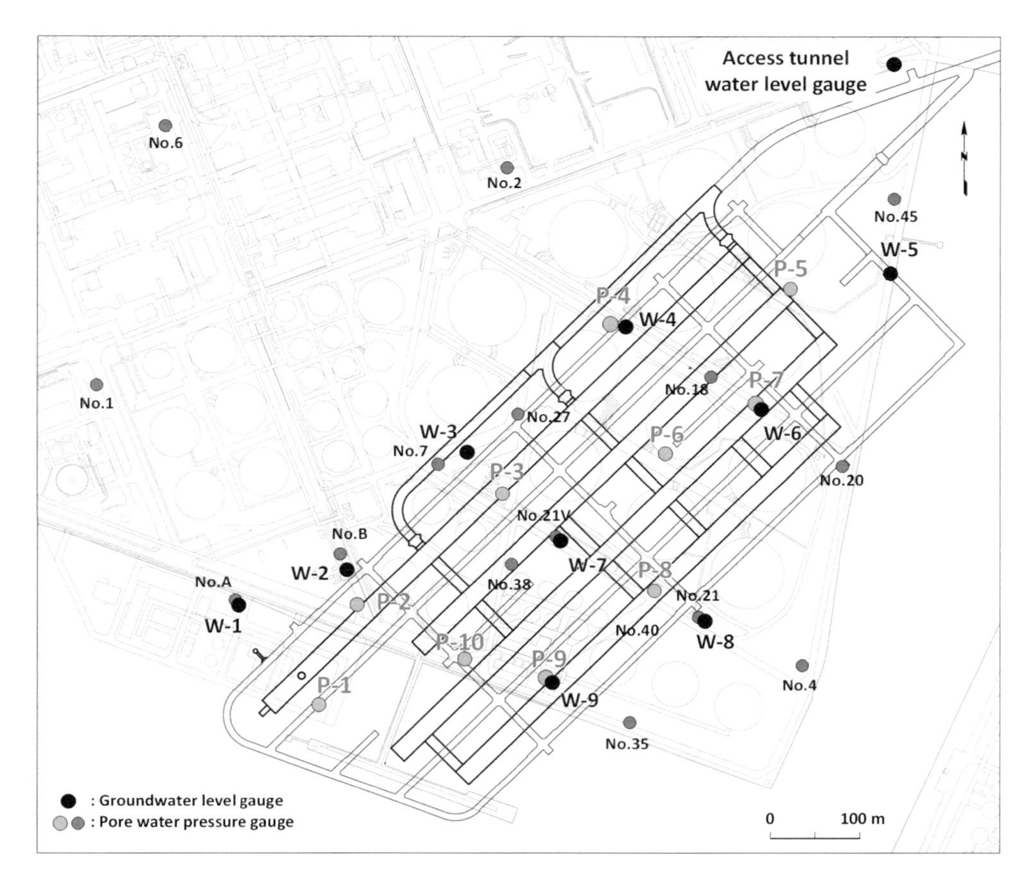

Figure 3.4 Arrangement of water level gauges and pore water pressure gauges outside the water curtains in the Kurashiki facility.

with highly permeable microfractures, less permeable rocks, around the piping shafts and the intersections between tunnels. The total number of pore water pressure gauges was 127. The pore water pressure gauges near the cavern were installed 12 m from the cavern wall where the air-tightness of the cavern could be monitored. Figure 3.5 shows the arrangement of the pore water pressure gauges in a plan view, and Figure 3.6 is an example of a cross section for the measurement. Additional measurements were taken of the water level and the flow rate of the injected sealing water at the piping and water injection shafts, the water level in the circulation pipes and the flow rate of the injected sealing water in the water curtain boreholes.

3.2.2 The Namikata Facility

In the Namikata facility, as in the Kurashiki facility, water level gauges and pore water pressure gauges were arranged outside of the water curtains to measure the groundwater level and estimate the pore water pressure distribution. Additional pore water pressure gauges were installed to cover the entire storage cavern between the water curtains and the cavern inside

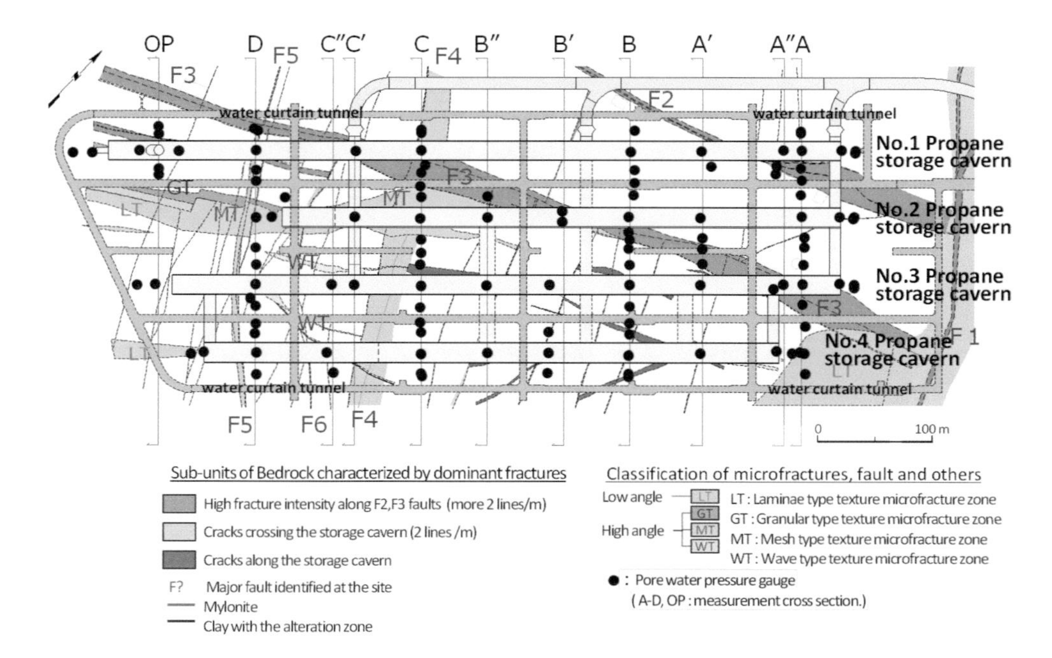

Figure 3.5 Arrangement of pore water pressure gauges at the top and shoulders inside the water curtains in the Kurashiki facility.

the water curtains to understand the behaviour of the pore water pressure around the cavern. Figure 3.7 shows the arrangement of 30 water level gauges and 82 pore water gauges.

The pore water pressure gauges were systematically arranged on the top, shoulders and floor of the cavern, and around the cavern outside of the grouting area. These were separated at 50–100 m intervals across the axis to form the main cross sections. Additional subordinate cross sections were drawn at locations critical for water-tightness: faults, areas with highly permeable microfractures, less permeable rocks, around the piping shafts and the intersections between tunnels. The total number of pore water pressure gauges was 135. The pore water pressure gauges near the cavern were installed 15 m from the cavern wall to monitor the air-tightness of the cavern. Figure 3.8 shows the arrangement of the pore water pressure in a plan view, and Figure 3.9 is an example of a cross section for the measurement. Additional measurements were taken for the water level and the flow rate of the injected sealing water at the piping and water injection shafts, the water level in the circulation pipes and the flow rate of the injected sealing water in the water curtain boreholes.

3.3 3D Modelling of Hydrogeological Structure

3.3.1 Application of Geostatistics

An accurate prediction of the distribution of the pore water pressure over the entire cavern is very important for constructing an underground rock storage, controlling the groundwater by water curtains and grouting. The distribution of the pore water pressure was mapped over the entire cavern area. For accurate mapping in this heterogeneous environment, the hydraulic

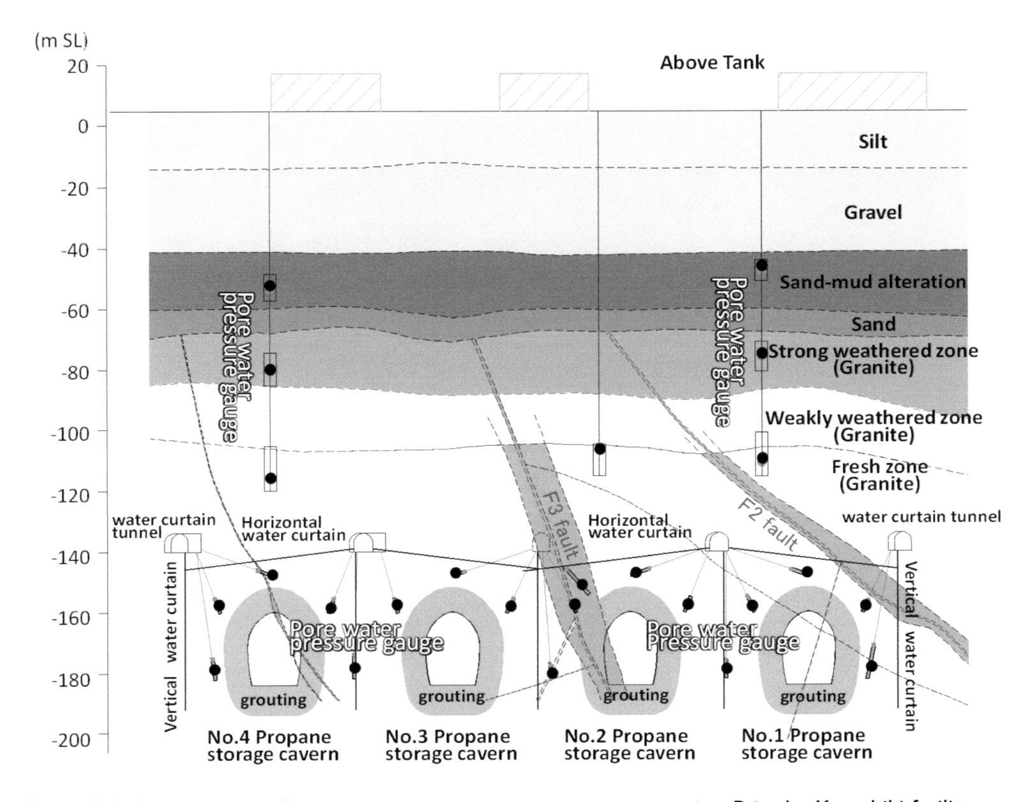

Figure 3.6 Arrangement of pore water pressure gauges on cross section B in the Kurashiki facility.

conductivity of the rocks needed to be accurately estimated from the measurements. In this project, the geostatistic technique was used to estimate the spatial distribution of hydrological characteristics of the rocks in an attempt to improve the accuracy of the estimation where actual measurement did not take place.

3.3.1.1 Geostatistics

Geostatistics is a field of applied statistics developed in the early 1950s by D. C. Krige, a mining engineer, and H. S. Sichel, a statistician, for estimating mineral resource reserves in the mining industry. This is a technique to statistically and probabilistically treat spatial and temporal variant data. One of its main purposes is to estimate the distribution of the parameters in the entire space from limited data samples. Its feature is that it probabilistically evaluates the referenced physical parameters as a conditional distribution.

3.3.1.2 Geostatistical Technique

Estimation by the geostatistical technique follows three steps:

1. Modelling spatial structure: To build a model as a random field from the observed values by describing the spatial distribution characteristics of the referenced physical parameters.

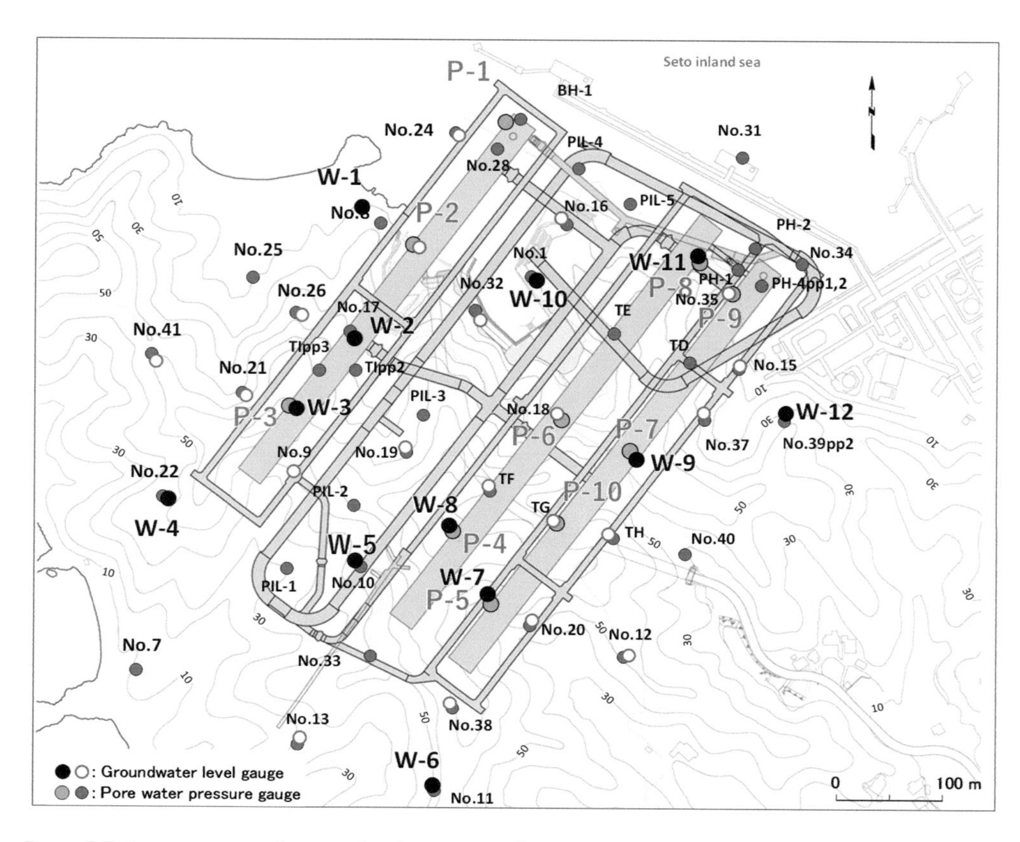

Figure 3.7 Arrangement of water level gauges and pore water pressure gauges outside the water curtains in the Namikata facility.

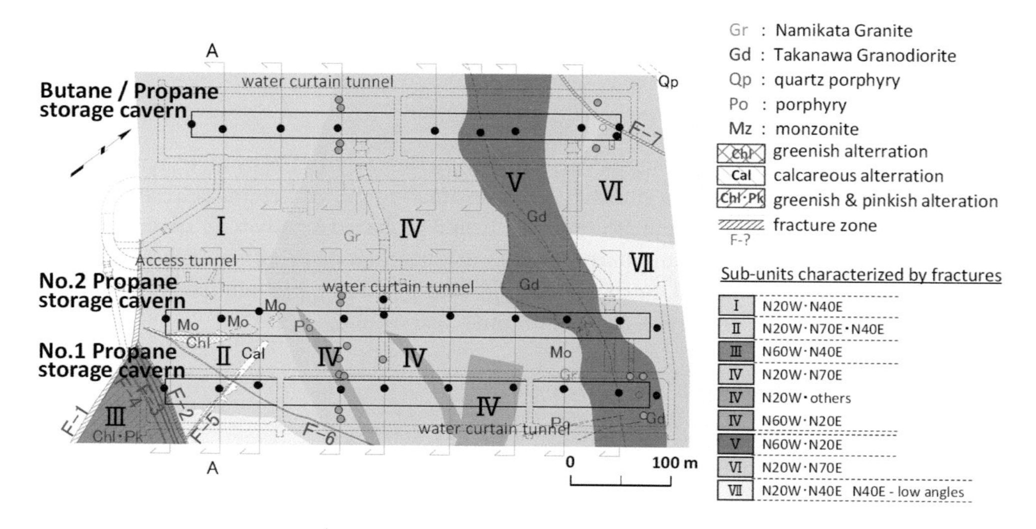

Figure 3.8 Arrangement of pore water pressure gauges at the top and shoulders inside the water curtains in the Namikata facility.

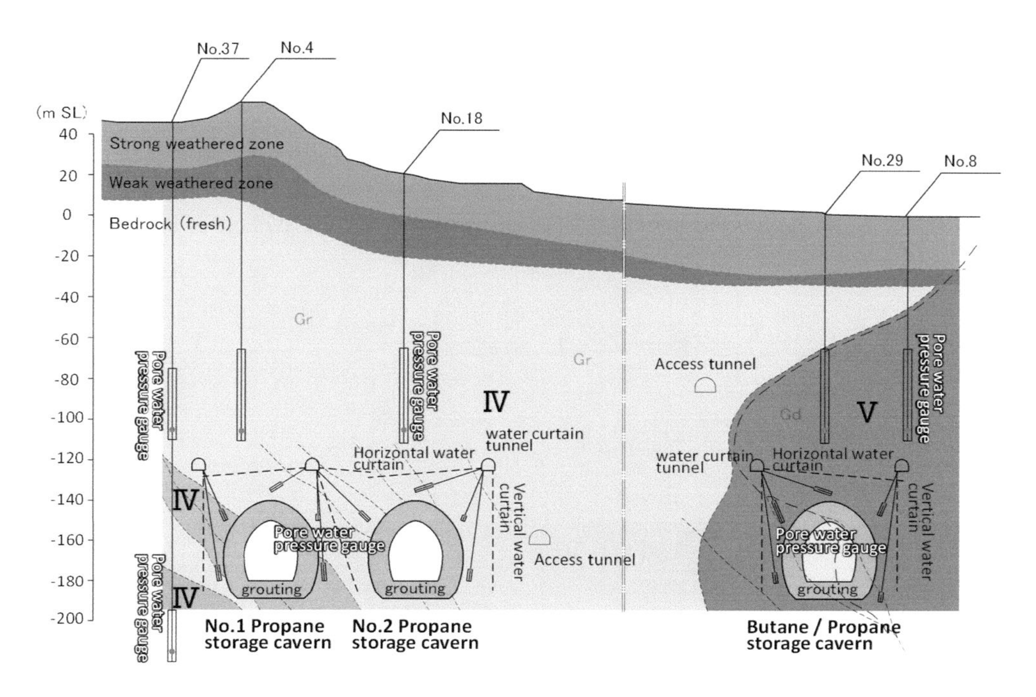

Figure 3.9 Arrangement of pore water pressure gauges on cross section C in the Namikata facility.

2. Estimation of the value in the space: Values at arbitrary points in the space are estimated by a technique called kriging.
3. Geostatistical simulation to suit the characteristics of the data: The physical parameters are sometimes estimated by probabilistic simulation.

The first stage estimates the value of covariance or variogram, which represents the average and spatial scatter of the physical property concerned, from observation. The second stage estimates the value of the locations where observation is missing as a linear sum of the observed values based on the model built in the first stage. (Depending on the nature of the observed values, non-linear "indicator kriging" [IK] may be used. This method treats the observed values as random variables, and accommodates the random variable at arbitrary locations to be outside of the upper or lower threshold.) The third step estimates the physical property concerned by the probabilistic simulation.

3.3.1.2.1 KRIGING

Kriging is a traditional method in geostatistics. The difference between kriging and classic linear regression is that kriging takes the probabilistic correlation between data values into account. Kriging is an estimation by linear regression and the estimated value $Z^*(x_0)$ at an arbitrary location x_0 is expressed as

$$Z^*(x_0)-\mu(x_0)=\sum_{i=1}^{n}\lambda_i\left[Z(x_i)-\mu(x_i)\right] \tag{3.1}$$

where $\mu(x_0)$ is the expected value of the random variable $Z^*(x_0)$, $\mathrm{E}\left[Z(x_0)\right]$. To calculate the estimated value using Eq. (3.1), weighting coefficients $(\lambda_1, \lambda_2, ..., \lambda_n)$ are necessary. The kriging equation estimates these coefficients fundamentally fulfilling two conditions: minimising the estimated error variance and the unbiased estimator.

The ordinary kriging method (OK method), which is generally used, is explained here. Eq. (3.2) describes the estimation error, the difference between the estimated value $Z^*(x_0)$ and the true value $Z(x_0)$. The condition of unbiasedness is that the average of this estimation error is 0.

$$Z^*(x_0) - Z(x_0) = \sum_{i=1}^{n} \lambda_i Z(x_i) - Z(x_0) \tag{3.2}$$

Suppose $\mathrm{E}\left[Z(x_0)\right] = \mathrm{E}\left[Z^*(x_0)\right] = \mu(x_0)$,

$$\begin{aligned} E\left[Z^*(x_0) - Z(x_0)\right] &= \sum_{i=1}^{n} \lambda_i \mu(x_0) - \mu(x_0) \\ &= \left(\sum_{i=1}^{n} \lambda_i - 1\right) \mu(x_0) \end{aligned} \tag{3.3}$$

To satisfy the unbiasedness of the estimated value at arbitrary $\mu(x_0)$, the condition is

$$\sum_{i=1}^{n} \lambda_i = 1 \tag{3.4}$$

Next, the estimated error variance $\sigma^2(x_0)$ is

$$\begin{aligned} \tilde{A}^2(x_0) &= E\left[\left(Z^*(x_0) - Z(x_0)\right)^2\right] \\ &= E\left[\left(\sum_{i=1}^{n} \lambda_i Z(x_i) - Z(x_0)\right)^2\right] \\ &= \sum_{i=1}^{m}\sum_{j=1} \lambda_i \lambda_j E\left[Z(x_i) \cdot Z(x_j)\right] - 2\sum_{i=1}^{n} \lambda_i E\left[Z(x_i) \cdot Z(x_0)\right] + E\left[\left(Z(x_0)\right)^2\right] \\ &= \sum_{i=1}^{m}\sum_{j=1} \lambda_i \lambda_j C\left(\left|x_i - x_j\right|\right) - 2\sum_{i=1}^{n} \lambda_i C\left(\left|x_i - x_0\right|\right) + C(0) \end{aligned} \tag{3.5}$$

Therefore, estimating the weighting coefficients is equivalent to minimising Eq. (3.5) with the condition given in Eq. (3.4). Using Lagrange's undetermined coefficient method, it is the solution λ and η for the simultaneous equation:

$$\begin{cases} \sum_{i=1}^{n} \lambda_i C\left(\left|x_j - x_i\right|\right) + \eta = C\left(\left|x_j - x_0\right|\right), \quad j = 1, 2, \ldots, n \\ \sum_{i=1}^{n} \lambda_i = 1 \end{cases} \tag{3.6}$$

Using the relationship between the variogram and the covariance function, the kriging equation can be expressed with variogram $\gamma(h)$ as in Eq. (3.8):

$$C(h) = C(0) - \gamma(h) \tag{3.7}$$

$$\begin{cases} \sum_{i=1}^{n} \lambda_i \gamma\left(\left|x_j - x_i\right|\right) + \eta = \gamma\left(\left|x_j - x_0\right|\right), \quad j = 1, 2, \ldots, n \\ \sum_{i=1}^{n} \lambda_i = 1 \end{cases} \tag{3.8}$$

A matrix expression of the solutions of this equation is

$$\begin{bmatrix} \lambda_1 \\ \lambda_2 \\ \vdots \\ \lambda_n \\ \eta \end{bmatrix} = \begin{bmatrix} \gamma(0) & \gamma\left(\left|x_1 - x_2\right|\right) & \ldots & \gamma\left(\left|x_1 - x_n\right|\right) & 1 \\ \gamma\left(\left|x_2 - x_1\right|\right) & \gamma(0) & \ldots & \gamma\left(\left|x_2 - x_n\right|\right) & 1 \\ \vdots & \vdots & \vdots & \vdots & \vdots \\ \gamma\left(\left|x_n - x_1\right|\right) & \gamma\left(\left|x_n - x_2\right|\right) & \ldots & \gamma(0) & 1 \\ 1 & 1 & \ldots & 1 & 0 \end{bmatrix}^{-1} \begin{bmatrix} \gamma\left(\left|x_1 - x_0\right|\right) \\ \gamma\left(\left|x_2 - x_0\right|\right) \\ \vdots \\ \gamma\left(\left|x_n - x_0\right|\right) \\ 1 \end{bmatrix} \tag{3.9}$$

Thus, the OK method is formulated.

3.3.1.2.2 INDICATOR KRIGING

In the cracked bedrock, the subject of this book, the hydraulic conductivity greatly varies between a highly permeable fault region and the surrounding less permeable rocks. The spatial distribution of the Lugeon values estimated by statistical methods may not sufficiently express a sudden hydrogeological variation. Indicator kriging is a method to estimate the spatial distribution in such a situation. The IK method, unlike the OK method, does not assume probabilistic distribution, but categorises the data values and reconstructs the cumulative probability distribution for each category (Figure 3.10 (1)). To analyse the spatial correlation,

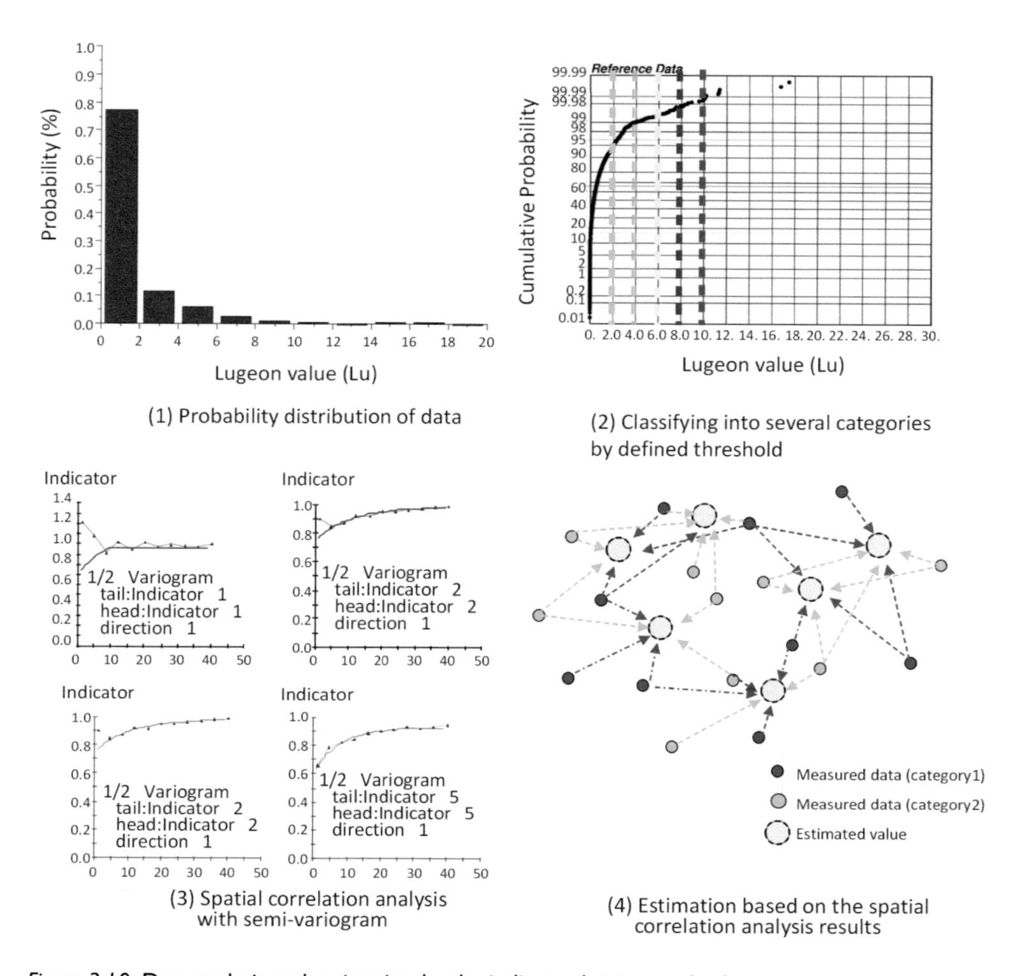

Figure 3.10 Data analysis and estimation by the indicator kriging method.

the data are categorised according to predefined thresholds (Figure 3.10 (2)). The spatial correlation of each category is then analysed (Figure 3.10 (3)). Using the result, the 3D spatial distribution is estimated from the cumulative probability distribution (Figure 3.10 (4)).

3.3.1.2.3 GEOSTATISTICAL SIMULATION

Geostatistical simulation considers the probability distribution based on estimated values from a probabilistic simulation, and evaluates the uncertainty of the physical parameter concerned. It is also called "conditional simulation". In this book, the successive method is used. In the successive method, the value at a location is estimated from the values in its vicinity and the probability distribution function is calculated from the previous simulation. This method uses a random sequence to successively select the grid points within the simulation area. The value of the selected point is estimated from the values (actual and simulated) in the vicinity.

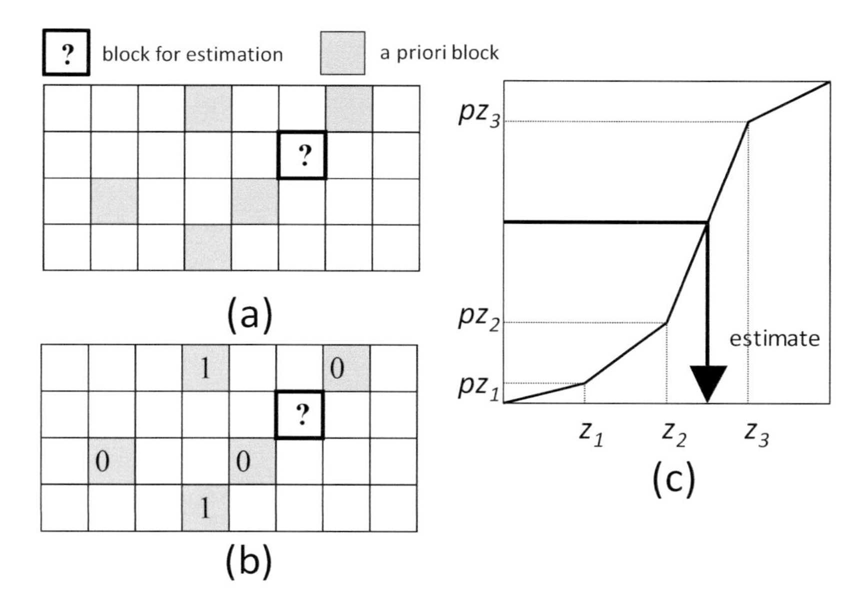

Figure 3.11 Procedure of successive indicator simulation (SIS).

The sequential indicator simulation (SIS) method calculates the local probability distribution using the indicator kriging method.

At the start of the estimation by the SIS method, a threshold value z_l is set for the variable to be estimated. Then, a point for estimation is randomly selected (Figure 3.11a) and an indicator value $i(x; z_l)$ is assigned to the point from the existing data (Figure 3.11b). The indicator value is defined as

$$i(x; z_l) = \begin{cases} 1 & \text{if } z(x) < z_l \\ 0 & \text{if not} \end{cases} \tag{3.10}$$

A variogram is then generated with respect to this threshold value.

$$\gamma_I(h) = \frac{1}{2} E\left\{\left[I(x; z_l) - I(x+h; z_l)\right]^2\right\} \tag{3.11}$$

Next, weight coefficients $(\lambda_1, \lambda_2, \ldots, \lambda\ )$ are calculated based on the obtained variogram from Eq. (3.9) to find the probability $P_{z_i}^*(x)$ to make the estimated value below the threshold at x_m, at the estimation point x_m.

$$P_{z_i}^*(x_m) = \sum_{i=1}^{n} \lambda_i(x_m) \cdot I(x_i; z_l) + \left[1 - \sum_{i=1}^{n} \lambda_i(x_m)\right] \cdot m_{z_i} \tag{3.12}$$

The probability for each threshold value is established and a cumulative distribution function is drawn based on the probability under the threshold. Selecting a random number between 0 and 1, an estimated value for this location is calculated based on the distribution function (Figure 3.11c). The estimated values thus calculated are now treated as known data, and the estimation is repeated for the next point randomly selected. Estimation is complete when the values for all the points are assigned.

Using this procedure, the result is not unique: several realisations may occur. By calculating the upper probability with a threshold z_c, a map capable of evaluating the uncertainty of a physical quantity, called a "value at risk (VaR) map", can be drawn, where $N(x|^*)$ is the number of realisations of x under the condition *:

$$P(z_c) = \frac{N_R\left(x \middle| Z(x) \rangle z_c\right)}{N_R(x)} \tag{3.13}$$

3.3.2 Constructing a Hydrogeological Model Using Actual Data at Excavation

3.3.2.1 Example of Constructing a Hydrogeological Model

This section describes an example of the construction of a hydrogeological model of bedrocks using the geostatistical technique on the actual Lugeon test data at the Namikata site. Here, hydraulic conductivity was estimated by the OK method and geostatistically simulated by the SIS method at each stage of excavation.

1. ESTIMATION OF THE DISTRIBUTION OF HYDRAULIC CONDUCTIVITY BY THE OK METHOD

A site-scale 3D model of hydraulic conductivity was built using the OK method based on the Lugeon test data of the borehole from the surface before underground excavation. The model was successively updated by integrating additional hydraulic conductivity data from subsequent excavation of the water curtain tunnels and drilling of the water curtain boreholes. Figure 3.12 shows the successive improvement of the distribution of hydraulic conductivity and its estimated error variance by additional data from the ongoing excavation (plan view at the level of the water curtain tunnel). These figures show the improvement of the accuracy of the estimation by the additional data integration. From the estimated error variance, the uniformity of the data coverage is examined. This estimate is effective in avoiding missing data and areas of poor coverage.

2. VAR LEVEL MAP OF HYDRAULIC CONDUCTIVITY BY THE SIS METHOD

The SIS method successively estimated the values for arbitrary points within the investigation area from the values initially estimated by the IK method. This distribution of prediction is called a "realisation map". The advantage of the SIS method is its ability to predict with higher accuracy than the OK method which may suffer from the smearing effect because it successively estimates values and it takes the spatial correlation between the estimated data into consideration. It also allows evaluating uncertainty as it can produce different results depending on the random sequence used. In this project, a distribution of reliability level was generated with a certain threshold from the estimated probability distribution of

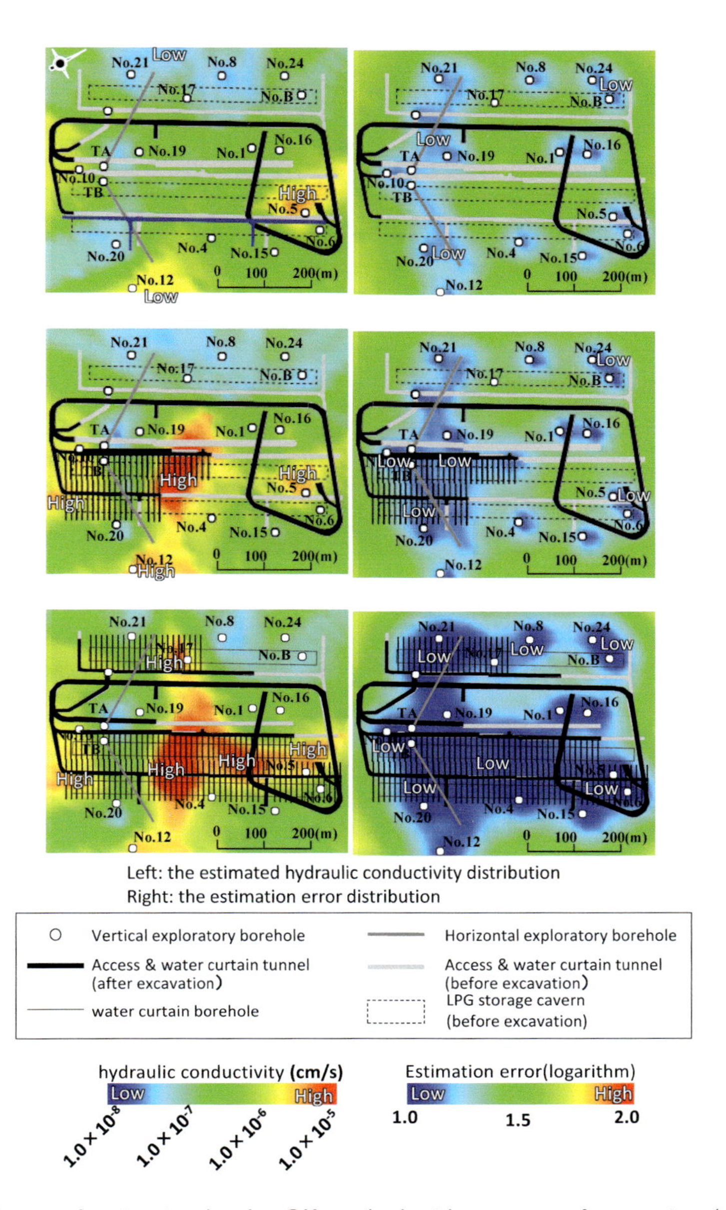

Figure 3.12 Change of estimation by the OK method with progress of excavation (plan view at 126 m bSL).

many possible values calculated by varying sequences randomly at each point in the space. Figure 3.13 shows an example.

The distribution map in Figure 3.13 shows the probability of the data value at the location being higher than the threshold. This can be used as the basic data for a risk analysis when the threshold of the hydraulic conductivity is set to the design value or improvement standard.

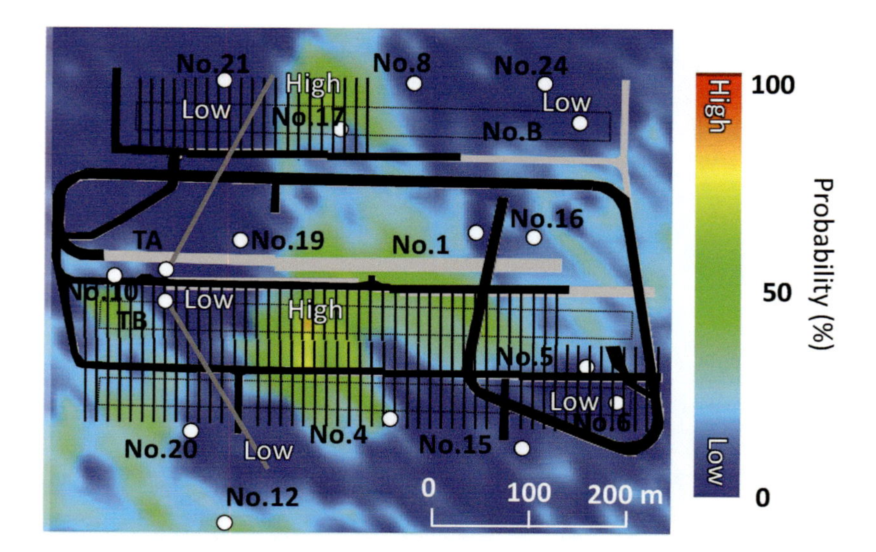

Figure 3.13 VaR level map by the SIS method (at VaR = 10^{-5} cm/s).

3. BUILDING A HIGH-PRECISION HYDROGEOLOGICAL MODEL BY SATURATED FLOW ANALYSIS

Using a saturated flow analysis on the 3D hydrogeological model constructed from the data on the excavation of water curtain tunnels, an attempt was made to build a high-precision hydrogeological model at the site scale including the subsequent excavation.

a. Methodology of 3D saturated flow analysis The geostatistical technique allows easy estimation of the data values at regular grid points. In this project, the 3D saturated flow analysis was applied to the structural grid using the finite flow rate method. An analytical model was constructed reflecting the heterogenic characteristics of rocks with a hydraulic conductivity value estimated by the geostatistical technique in the centre of a cube. On the top, bottom and sides, the natural water pressure was given as a fixed boundary condition of pressure. Under this condition, the distribution of the static pore water pressure was obtained first, and then a transient 3D saturated flow analysis was performed during excavation of the water curtain tunnels.

The analysis procedure is

1. Set up a boundary condition for the flow rate based on the measurement of the seepage flow rate during excavation of the water curtain tunnels.
2. Run a saturated flow analysis for the analysis period of 30 days.
3. Use the water level from the analysis of (2) for the initial condition of analysis for the next analysis period.
4. Repeat (1)–(3) up to six months.

b. Simulation using an estimation model by the OK method First, 3D distribution of the hydraulic conductivity (Figure 3.14) was estimated using the OK method. Inputting this distribution, a saturated flow model (Figure 3.15) was constructed. A simulation was run on the model.

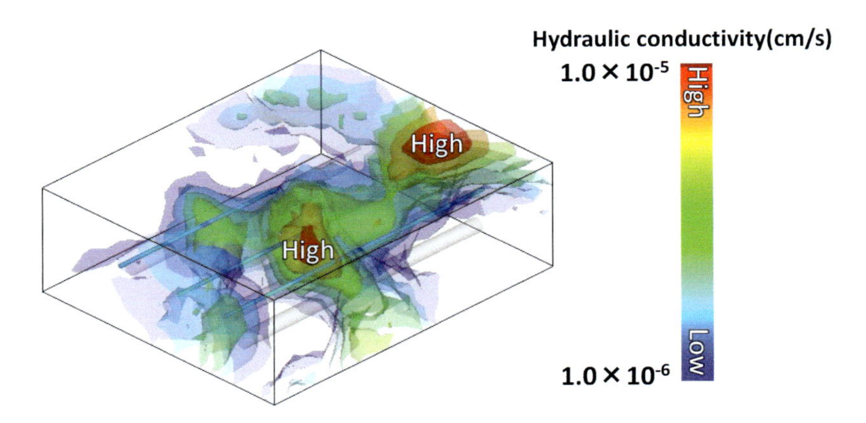

Figure 3.14 Distribution of hydraulic conductivity by the OK method.

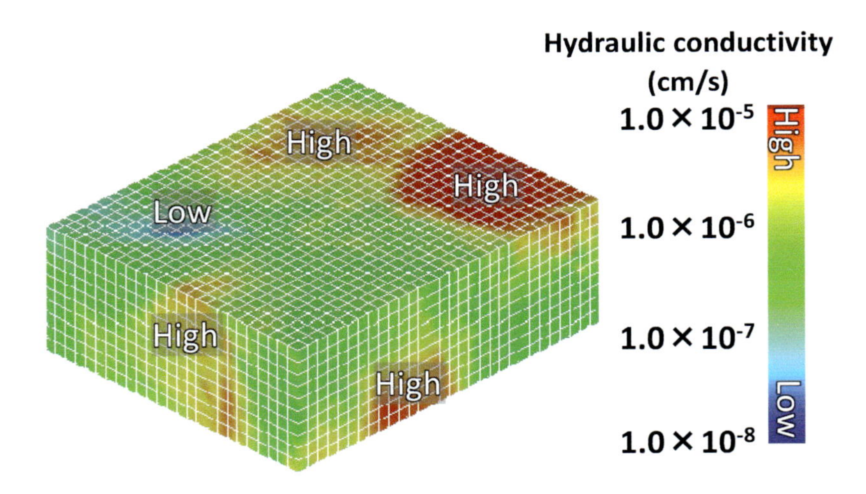

Figure 3.15 Analysis model by finite volume model.

Figure 3.16 shows an example of a comparison between the actual values and the predicted values along a temporal axis at two points (boreholes No. 15 and 16; see Figure 3.12) where the pore water pressure was repeatedly measured. This figure shows that the predicted values and the actual measurement had the same decreasing trend as excavation progressed, but the predicted values were slightly lower than the actual measurements. The reason for this may be that the OK method has a data smoothing effect resulting in the adjacent data points more similar than actual, and it overestimates the hydrogeological continuity in the highly permeable area.

4. SIMULATION BY THE SIS METHOD USING THE REALISATION MAP

Based on the discussion in the previous sections, a simulation was performed constructing a saturated flow analysis model based on the 30 realisation maps of the hydraulic conductivity

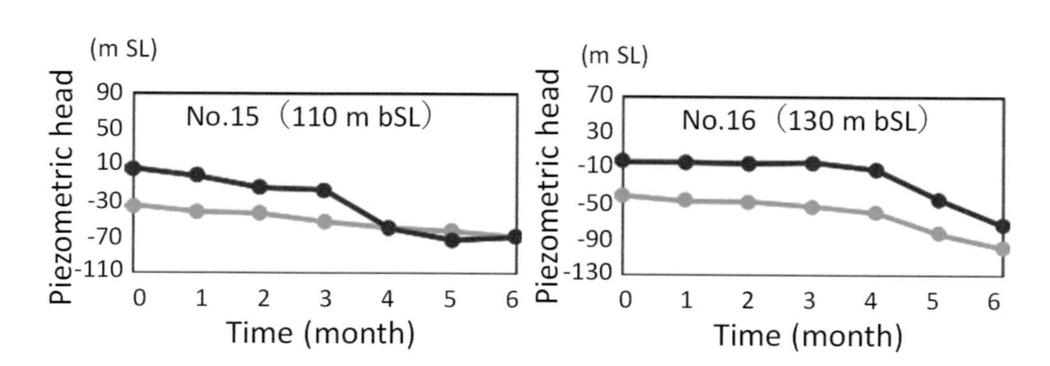

Figure 3.16 Comparison of pore water pressure between actual measurement and estimation by analysis.

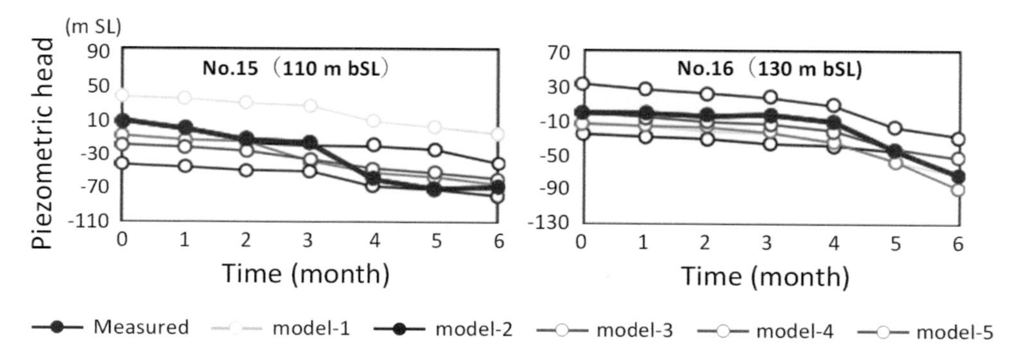

Figure 3.17 Comparison of pore water pressure between actual measurement and estimation by analysis (SIS).

predicted by the SIS method. As stated before, the realisation maps by the SIS method avoid the smoothing effect of the OK method as it reflects the spatial correlation between the estimated values. Figure 3.17 shows a comparison between the actual measurement of the pore water pressure and the predicted values of five models from five realisation maps along the time axis. In these graphs, all the modelled pore water pressures decrease over time. However, some scatters are noted between the models. These scatters are considered due to their non-uniqueness and it is reasonable to consider the model best fitted to the measurement as the optimum model.

Here, Model 3 was selected as the optimum model based on its smallest error against the actual measurement. Figure 3.18 shows the 3D distribution of the hydraulic conductivity expressed in contour surfaces.

Using this model, the pore water pressure at two other points (boreholes No. 5 and 20), where the actual pore water pressure was measured over time, was predicted and the result is shown in Figure 3.19. The results of prediction by the OK method are also plotted in this figure. It shows good agreement between the analysis results and the actual measurements,

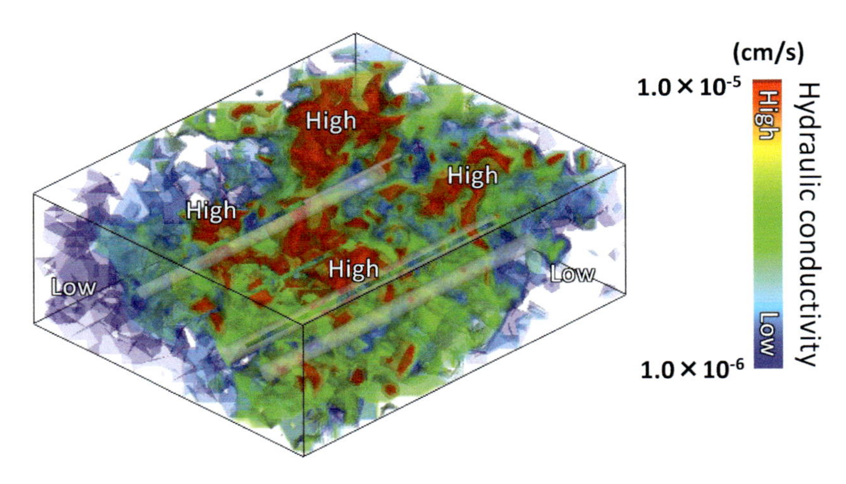

Figure 3.18 Distribution of hydraulic conductivity based on the SIS method and saturated flow analysis.

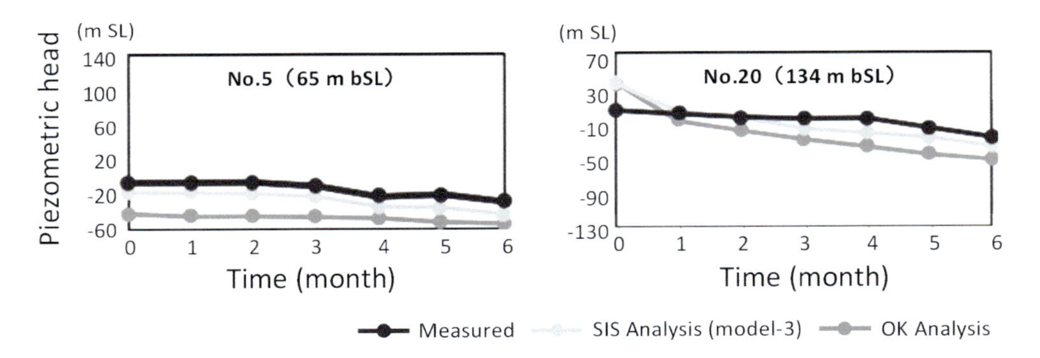

Figure 3.19 Comparison of pore water pressure between actual measurement and estimation by analyses (SIS and OK).

ensuring the reliability of the model. The pore water pressure values predicted by the OK method generally present lower values than results by the SIS method.

3.3.2.2 Application of the Hydrogeological Model

This section illustrates examples of building a hydrogeological model for groundwater analysis. More details of the evaluation of the water-tightness function and subsequent measures for improvement at each stage are described in Chapter 4.

3.3.2.2.1 THE KURASHIKI FACILITY

3.3.2.2.1.1 Determining the Area for Hydrogeological Modelling The 3D unsaturated groundwater flow analysis technique was used for groundwater analysis from excavation to the present. Considering the geological formation at the ground surface, the area for analytical modelling was set to 1900 × 1300 m with a depth of 400 m below ground level. The extent of the estimation of the 3D distribution of the hydraulic conductivity by a heterogeneous hydrogeological model was set to 900 m (length) × 400 m (width) × 150 m (vertical: from 100 to 250 m below sea level) which included the propane storage caverns. Grid points to estimate the hydraulic conductivity were set at 5 m intervals (Figure 3.20).

The construction of a heterogeneous hydrogeological model is discussed below.

3.3.2.2.1.2 Result of Estimation of Distribution of Hydraulic Conductivity on Completion of Excavation of Arch The results of a hydrogeological analysis on the flow rate of seepage in the cavern and the pore water pressure on excavation of the arches showed that faults F2 and F3 were filled with continuous clay and had high water-tightness. High hydraulic conductivity was seen in the localised microfractures and low angle cracks. Accordingly, faults F2 and F3 and the microfractures were modelled where they were postulated and their hydraulic conductivity was set from the results of the tests. The distribution of hydraulic conductivity in the 3D space at other locations was estimated from various survey data using the OK method (Figure 3.21).

The Lugeon test data through to completion of the excavation included Lugeon tests during drilling of the groundwater observation boreholes, Lugeon values at the reconnaissance excavation of the water curtain tunnels, Lugeon values from the boreholes for pore water pressure gauges around the caverns, Lugeon values in the water curtain boreholes and Lugeon tests in the grouting holes. As the hydraulic conductivity in a grouting hole is affected

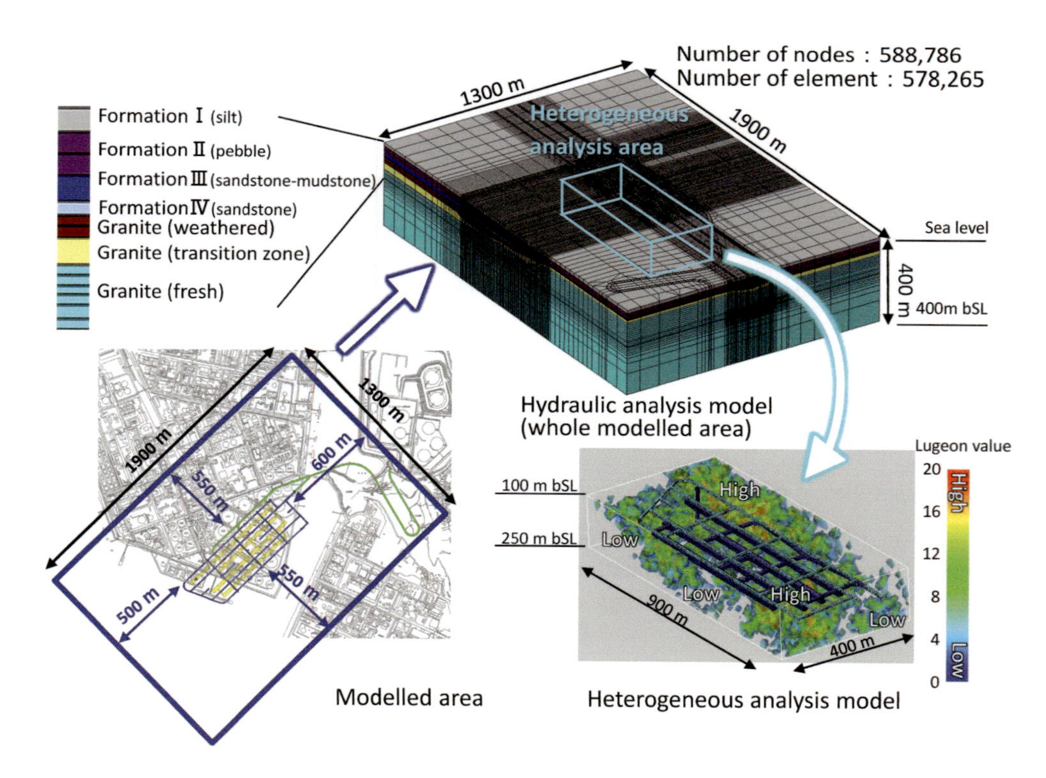

Figure 3.20 Outline of 3D hydrological analysis model.

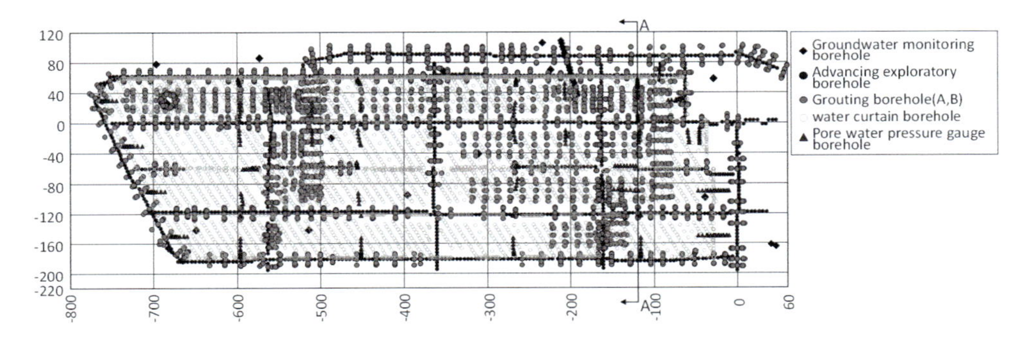

Figure 3.21 Locations of the data points used for the OK method.

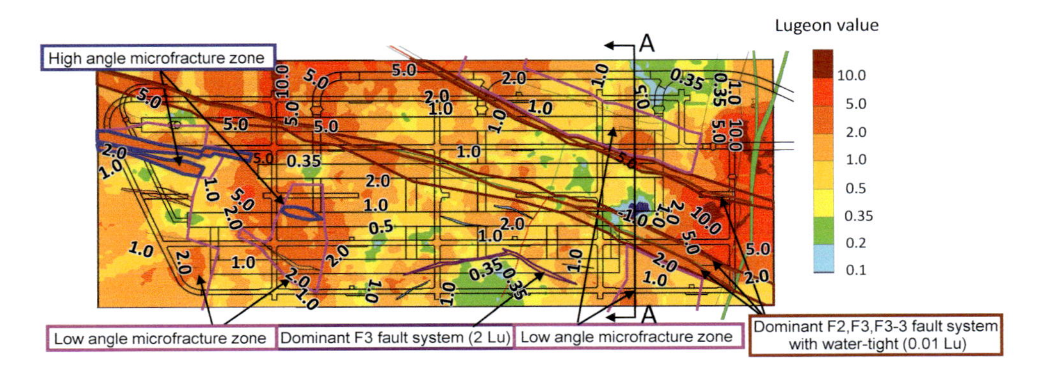

Figure 3.22 Distribution of the Lugeon value by the OK method and modelled geological structure.

by the adjacent grouting borehole, the values from the primary boreholes that represent the hydraulic conductivity of the bedrock were used to represent the pre-grouting permeability. The result of the Lugeon tests at the final stage was deemed to represent the post-grouting permeability and it was used for estimating and constructing the hydrogeological model.

Figure 3.22 shows the distribution of hydraulic conductivity at the floor of a cavern arch (167.5 m below sea level), as an example of the distribution of hydraulic conductivity around the cavern estimated by the OK method. This planar distribution of the hydraulic conductivity shows that the area around faults F1 and F2 and the region microfractures developed in the southwest of the cavern and were highly permeable. On the other hand, the local area between faults F2 and F3 was confirmed to be less permeable.

The hydrogeological model along cross section A–A is shown in Figure 3.23. It confirms that hydrogeological heterogeneity is present in less permeable regions between faults F2 and F3. The grout at the arch was modelled according to improvements at the inner and outer rings from an analysis of the performance of pre-grouting. The hydraulic conductivity at the time of bench excavation was predicted and modelled using its target value by grouting.

3.3.2.2.1.3 Result of Estimation of Distribution of Hydraulic Conductivity on Completion of Excavation of Bench The hydrogeological structure was analysed using data from the pore

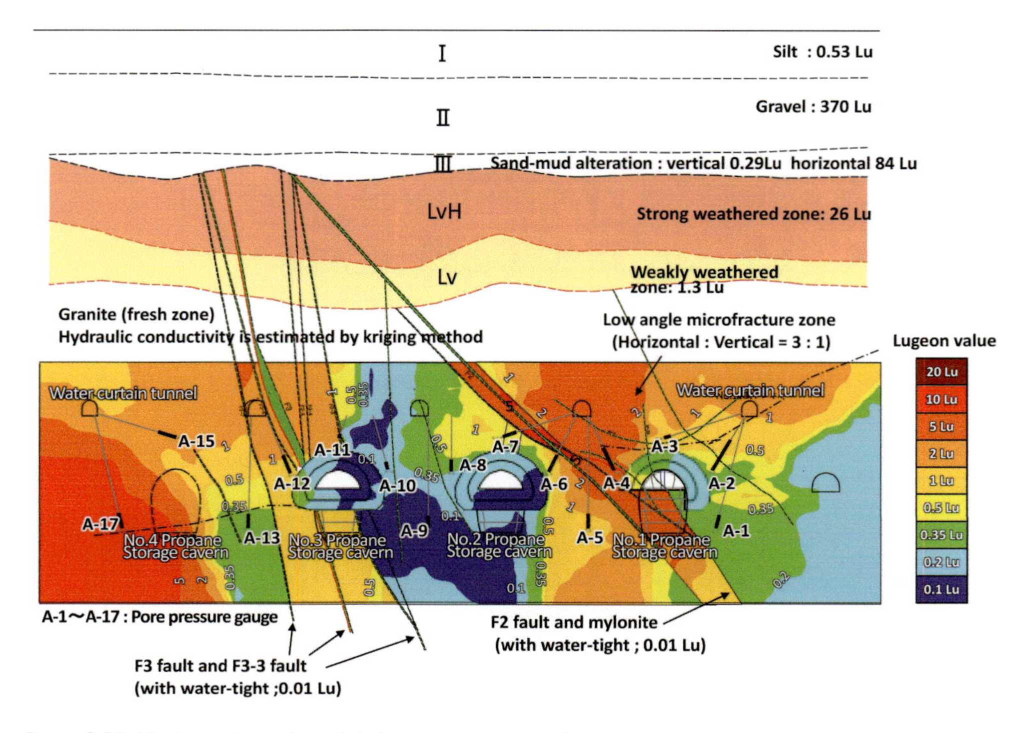

Figure 3.23 Hydrogeological model along cross section A.

water pressure behaviour on excavation of the benches and the test data from grouting including the Lugeon test. The result clarified the regions affected by faults F2 and F3 as postulated at the time of the excavation of the arches as well as regions of microfractures.

Figure 3.24 shows the probability distribution of hydraulic conductivity from Lugeon tests at geological divisions, regions affected by faults F2 and F3 and the microfracture zones WT, MT and GT. It shows that the regions affected by the faults and microfractures are extensive and highly permeable unlike the general surroundings. On the other hand, the Lugeon value is relatively low at around 0.7–1.3 Lu on average in other areas (geological subdivision I–IV). The hydraulic conductivity values in these regions are continuous and their distribution is not normal but exponential. Accordingly, spatial distribution of the hydraulic conductivity was estimated using the IK method which does not assume random distribution of the data.

Used as the permeability data were Lugeon values from reconnaissance boreholes, groundwater observation boreholes, water curtain boreholes, pore water pressure measurement boreholes and primary cavern grouting outer ring boreholes. The spatial distribution of the hydraulic conductivity was estimated for each geological division (I–V), regions affected by the faults and microfractures, using the IK method. The IK method was also applied to regions improved by grouting within an 8 m radius of the caverns using the Lugeon values from several final grouting boreholes. Figure 3.25 shows the final hydrogeological model of the Kurashiki facility on completion of post-grouting.

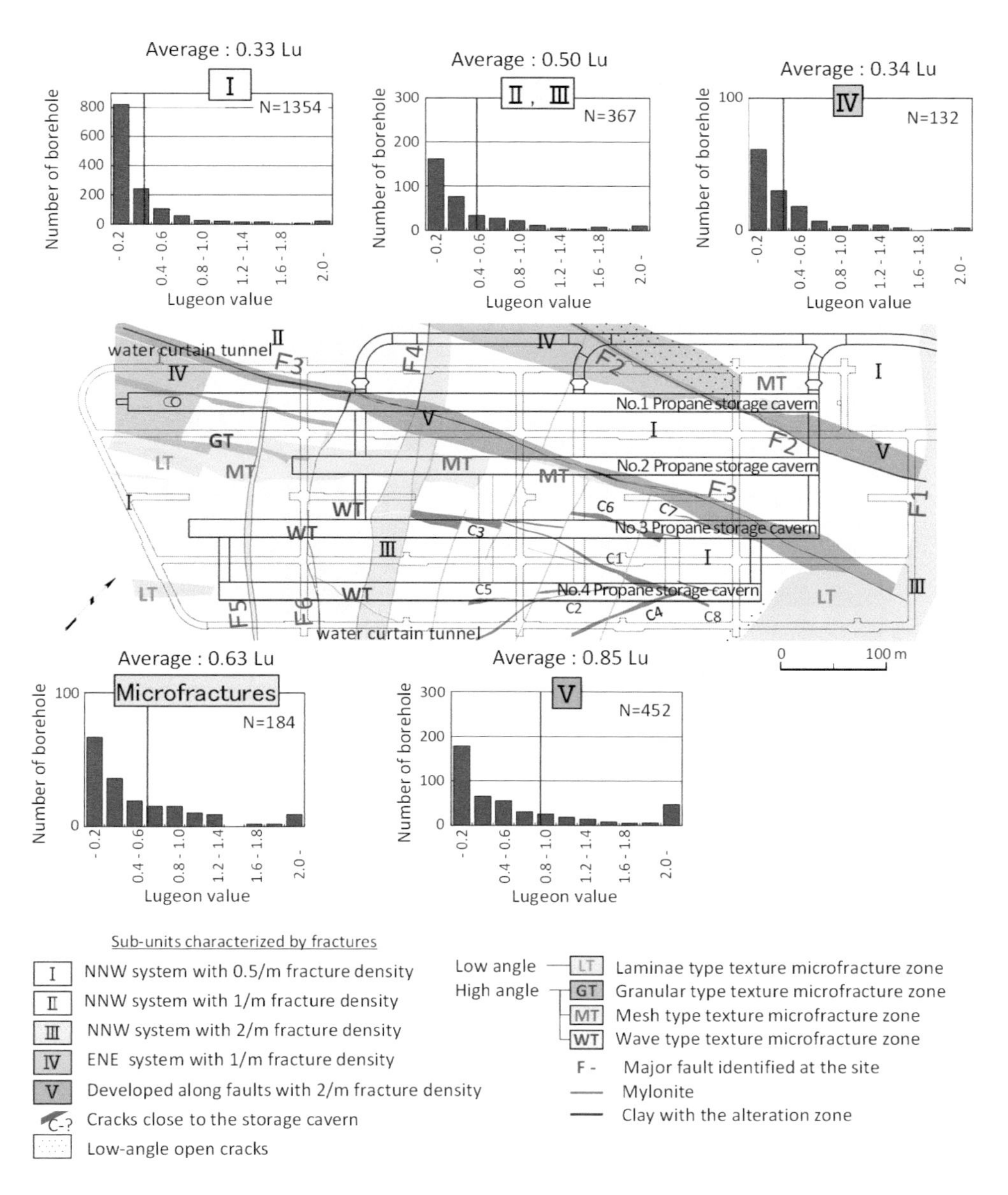

Figure 3.24 Random distribution of hydraulic conductivity by geological division, areas affected by the faults and microfractures.

Maps of the distribution of the Lugeon value show

- Areas along faults F1 and F2 are relatively permeable, especially to the southwest of fault F4.
- In the region with the development of mylonite near fault F5, a highly permeable trend across the orientation of the cavern is present.

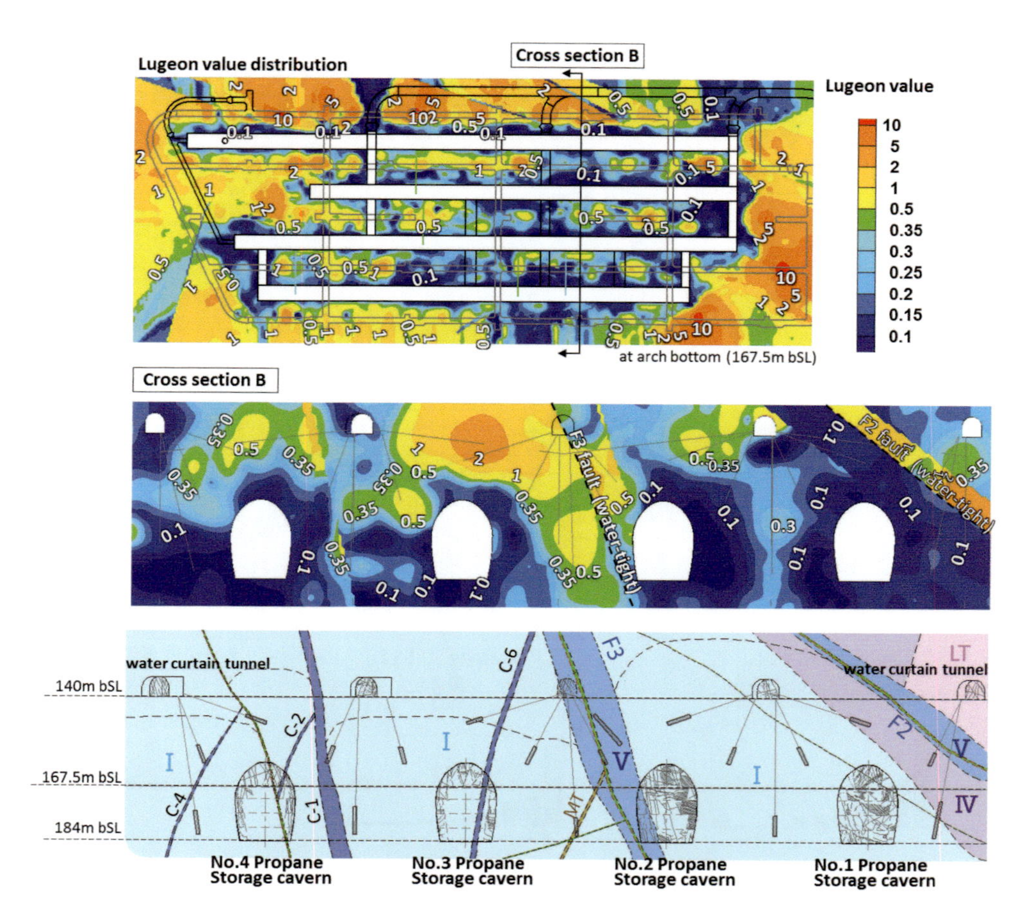

Figure 3.25 Distribution of the Lugeon value on completion of the bench excavation at the Kurashiki facility.

- Less permeable areas are distributed in the region between faults F2 and F3 in the north-eastern part of the site, in the vicinity of piping shaft 1PC, near the central part of 4PC and to the southwest of 3PC and 4PC.
- The Lugeon value around the cavern is lowered by pre- and post-grouting improvements.

3.3.2.2.1.4 Example of Estimation of Behaviour of Groundwater Among the behaviours of the groundwater on completion of the excavation of the arches, the measured and predicted flow rates of seepage in storage cavern No. 1 are shown in Figure 3.26: the measured values up to completion of bench grouting after the excavation of the arches, and the predicted values to conclusion of cavern construction. The predictions in this figure are based on three models: the "homogeneous model" where the Lugeon value of the fresh granite was set to a constant 0.35 Lu; "Model 1" where the Lugeon value of the grouting region was set to the target value of improvement; and "Model 2" where the Lugeon value of the grouting region was set to the value at actual grouting.

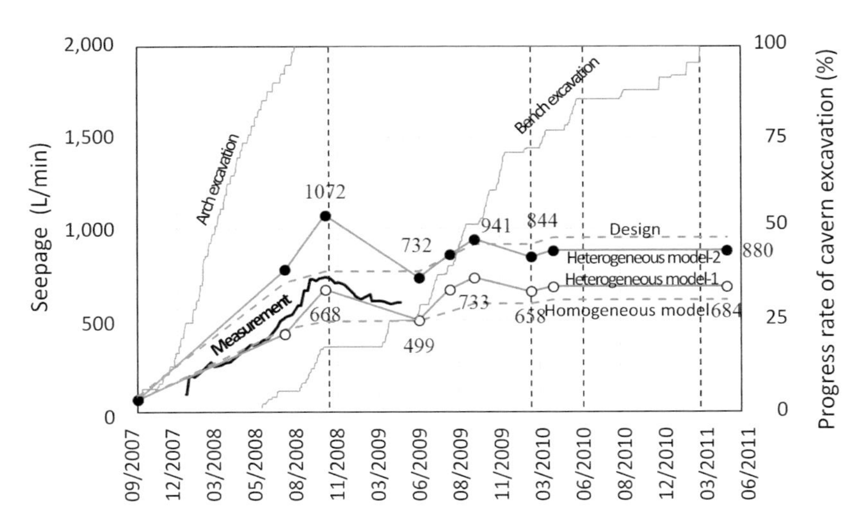

Figure 3.26 Actual and predicted volume of seepage on completion of arch excavation (storage cavern No. 1).

The actual measured values during the excavation of the arches showed good agreement with Model 1 and the homogeneous model. Then, the actual flow rate of the seepage after commencing excavation of the benches exceeded the prediction of the homogeneous model, while it accurately agreed with the prediction of Model 1. Model 1 also accurately predicted the decrease in seepage at post-grouting and bench grouting after excavation of the arches. The Kurashiki facility was constructed with less seepage than expected by the design due to appropriate post-grouting. The actual value on conclusion of the bench grouting was almost the same as predicted by Model 1. This confirmed that the Lugeon value of the grouting region around the cavern was reduced to the target value.

Figure 3.27 shows the results of a predictive analysis of the flow rate of water and the behaviour of the pore water pressure using the hydrogeological model updated on completion of excavation. It also shows the flow rate predicted assuming a uniform Lugeon value of the fresh granite at 0.35 Lu. The actual measurement showed the flow rate less than that predicted by the heterogeneous model, which does not significantly differ from the prediction with the target value of improvement of the Lugeon value in the grouting area on completion of the arch excavation.

The measured values agreed with the predictions using the homogeneous model with 0.35 Lu, and confirmed that the Lugeon value of the grouted area around the storage cavern was improved to the target value.

The measured flow rate of seepage on completion of excavation was 727 L/min compared with 880 L/min by the model from the grouting results on completion of the arch excavation and 845 L/min by the model on completion of the bench excavation. This confirmed that the model constructed during arch excavation accurately predicted the seepage flow rate during bench excavation. The flow rate during the excavation of the storage cavern was almost equal to the value predicted by the homogeneous model with 0.35 Lu, which confirmed that the Lugeon value of the grouted area around the storage cavern was improved to the target value.

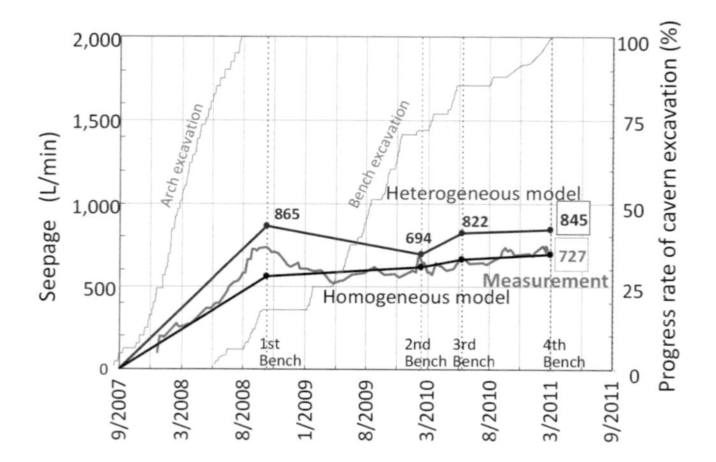

Figure 3.27 Actual and predicted flow rate of seepage on completion of bench excavation (storage cavern No. 1).

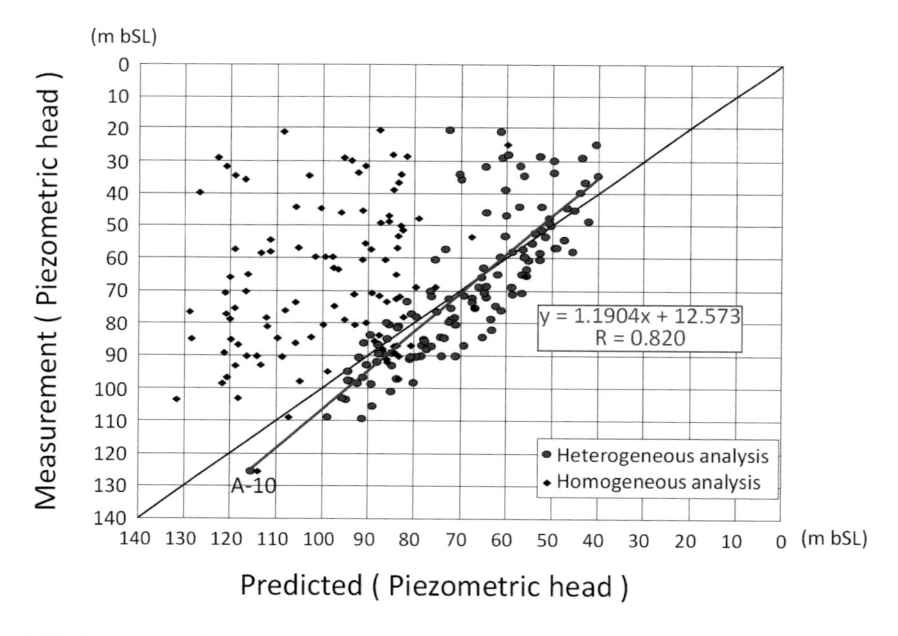

Figure 3.28 Comparison between pore water pressure around the storage cavern on completion of bench excavation.

Figure 3.28 shows a comparison between the pore water pressures around the storage cavern, actual and predicted by the homogeneous and heterogeneous models, as a cross plot. The heterogeneous model was constructed considering the geological structure and the effect of grouting around the cavern, and a 3D distribution of the hydraulic conductivity using the

IK method was estimated. While the piezometric head predicted by the homogeneous model was almost always higher than the actual values by about 20 m, errors in the heterogeneous model were within ±15 m. This means that the heterogeneous analysis model was adequately applicable for predicting the flow rate of seepage in the cavern and the pore water pressure.

3.3.2.2.1.2 THE NAMIKATA FACILITY

3.3.2.2.1.2.1 Determining the Area for Hydrogeological Modelling The 3D unsaturated groundwater flow analysis technique was used for groundwater analysis from excavation to present. Considering the geological formation of the ground surface, the area for analytical modelling was set to 1260 m (N-S) × 1340 m (E-W) with a depth of 500 m below ground level. The extent of the estimation of the 3D distribution of the hydraulic conductivity by a heterogeneous hydrogeological model was set to 900 m (length) × 400 m (width) × 150 m (vertical: from 100 to 250 m below sea level), which included the propane storage and the combined butane/propane storage caverns. Grid size to estimate the hydraulic conductivity were set at 5 m (Figure 3.29).

The results of the Lugeon tests were successively added from the survey stage through progress of the excavation of the cavern. This book explains the building and updating of

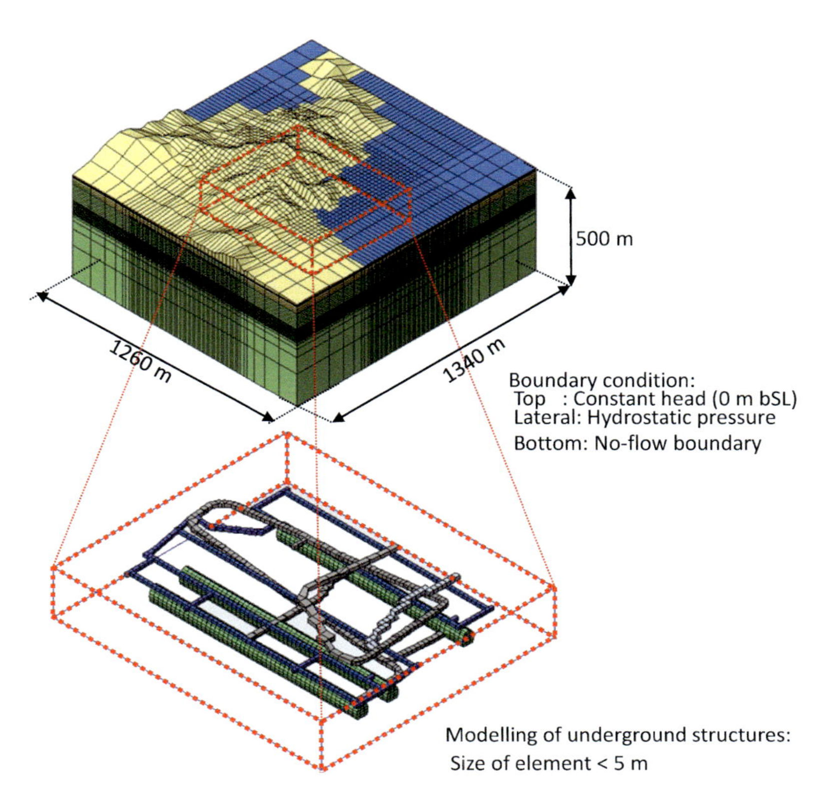

Figure 3.29 Outline of 3D hydrological analysis model and extent of estimation of the distribution of hydraulic conductivity.

the hydrogeological model on completion of the excavation of the arch and the bench (post-grouting and after additional water curtain boreholes).

3.3.2.2.1.2.2 Result of Estimation of Distribution of Hydraulic Conductivity on Completion of Excavation of Arch Figure 3.30 shows the results of the Lugeon tests through to completion of the excavation of the arches including reconnaissance survey boreholes, water curtain boreholes, measurement boreholes and primary grouting boreholes. Histograms of the hydraulic conductivity from the test results are shown in Figure 3.31. These histograms show that the curve is closer to normal distribution when the logarithmic scale, not the real scale, is chosen for the horizontal axis to plot. Although the hydraulic conductivity in the Namikata site was generally low, parts along continuous cracks had high hydraulic conductivity. Therefore, hydraulic conductivity was highly variable depending on the orientation of the cracks. Considering this variability, the lognormal distribution of the hydraulic conductivity is plausible. Accordingly, the 3D distribution of the hydraulic conductivity was estimated by the OK and SIS methods, the latter of which can estimate extreme values in the hydraulic conductivity distribution.

To estimate the 3D distribution of hydraulic conductivity by the SIS method, 30 maps of the realisation value of hydraulic conductivity were drawn, and occurrence probability maps (VaR level maps) for several values of hydraulic conductivity (VaR = 10^{-4} cm/s, VaR = $10^{-4.2}$ cm/s· · ·VaR = $10^{7.8}$ cm/s, VaR = 10^{-8} cm/s) were obtained. One distribution of the hydraulic conductivity was estimated by superimposing an arbitrary probability of occurrence (50% was used here) at several distributions of the arbitrary probability of occurrence.

Figure 3.32 shows the heterogeneous 3D distribution of the hydraulic conductivity estimated by the OK and SIS methods, and its horizontal distribution at the level of the arch floor (157.5 m below sea level) is shown in Figure 3.33. In comparison, the distribution of hydraulic conductivity estimated by the SIS method had a stronger contrast than that by the OK method. On the other hand, the hydraulic conductivity estimation by the OK method tended to over-smooth the local heterogeneity. This result suggested that the SIS method was

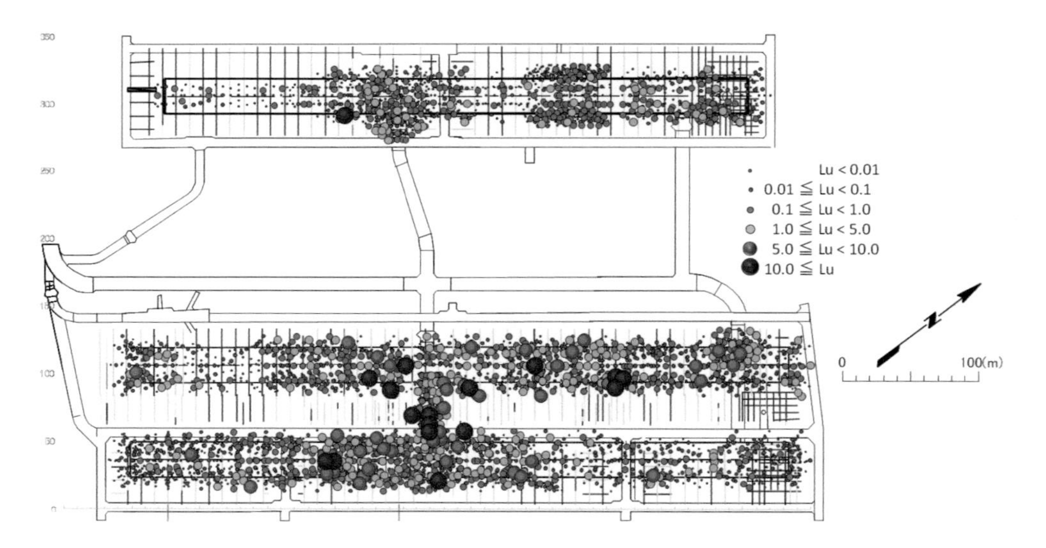

Figure 3.30 Lugeon value map before improvement by cavern grouting on completion of arch excavation.

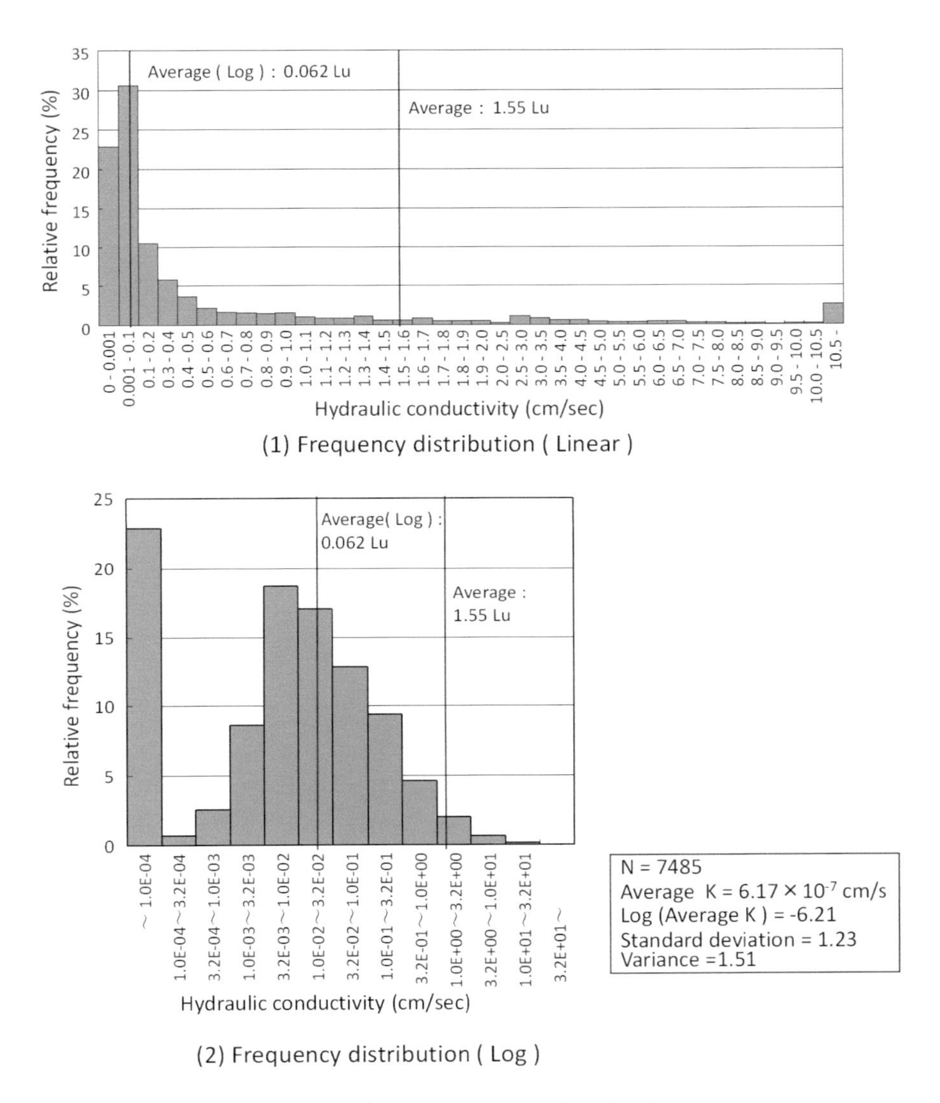

Figure 3.31 Histogram of hydraulic conductivity on completion of arch excavation.

more suitable for estimating the distribution of hydraulic conductivity in the Namikata site where the hydraulic conductivity was generally low with local high hydraulic conductivity anomalies.

3.3.2.2.1.2.3 Result of Estimation of Distribution of Hydraulic Conductivity on Completion of Excavation of Bench Analysis during the excavation of the benches included the flow rate of seepage, the behaviour of the pore water pressure and test data including a Lugeon test at the grouting and hydrogeological structure. The results showed that the different geological zones have different hydraulic conductivity values and zones with anisotropic hydraulic

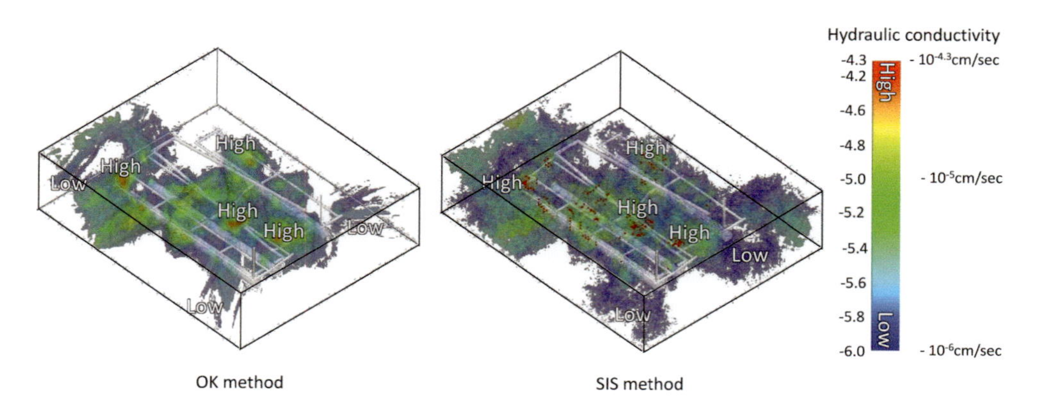

Figure 3.32 Estimated 3D distribution of hydraulic conductivity.

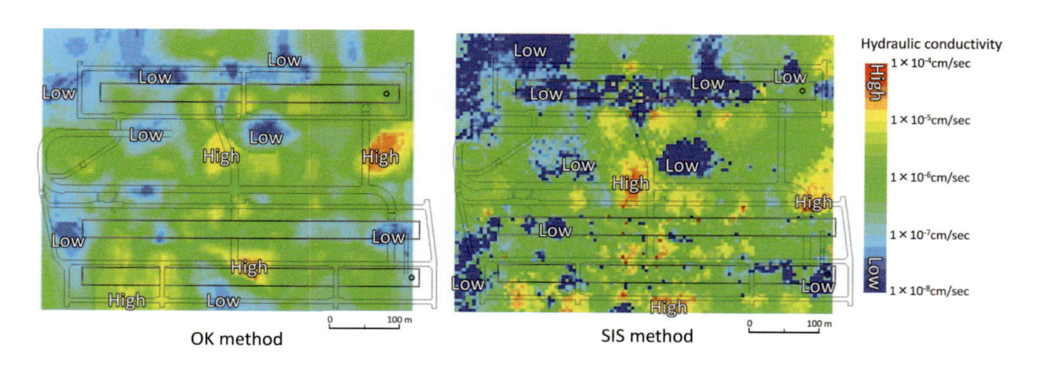

Figure 3.33 Estimated distribution of hydraulic conductivity at 157.5 m below sea level.

conductivity were identified. These elements were integrated in the hydrogeological model using the SIS method.

The heterogeneity of hydraulic conductivity was examined based on the Lugeon tests in the additional water curtain boreholes. Heterogeneous hydraulic conductivity was considered for the zones where heterogeneity was observed.

Figure 3.34 shows a comparison of Lugeon tests in boreholes of different directions in the same vicinity. In these figures, high hydraulic conductivity zones were not seen in the vertical boreholes while the oblique boreholes had several high hydraulic conductivity zones. This suggests the existence of cracks with high hydraulic conductivity across the cavern. Therefore, the hydraulic conductivity was thought to be different along and across the cavern. This was incorporated in the heterogeneous hydrogeological model.

The distribution of hydraulic conductivity was estimated by the SIS method using all the test data. Figure 3.35 shows the results of Lugeon tests and an example of the distribution of the hydraulic conductivity along cross section EE of propane storage No. 2 estimated by the SIS method. Areas with low hydraulic conductivity were concentrated to the east and under the floor of the cavern while areas with high hydraulic conductivity were in the vicinity of the water curtain boreholes.

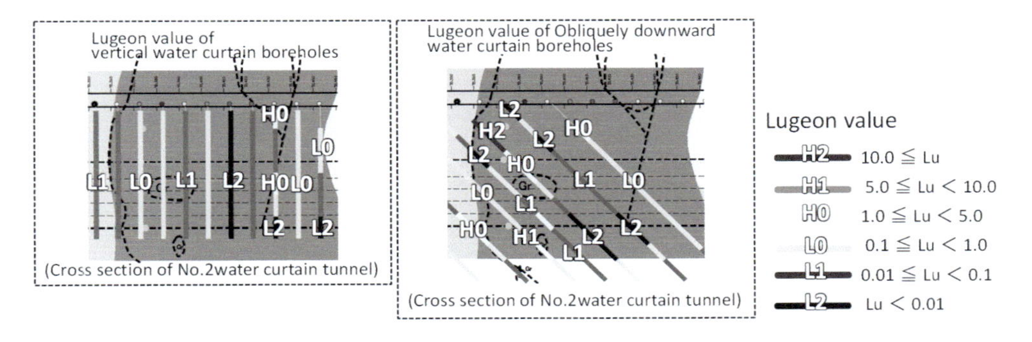

Figure 3.34 Results of Lugeon tests in vertical and oblique boreholes (cross section C of propane storage No. 2).

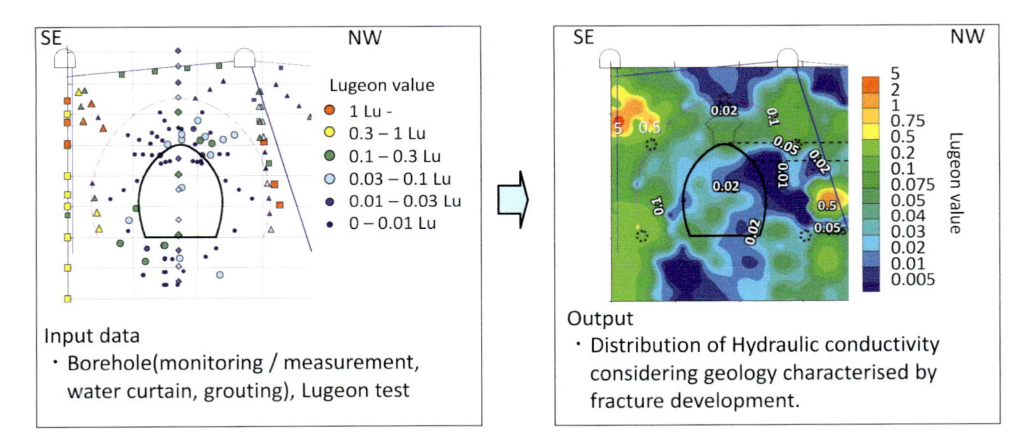

Figure 3.35 An example of the distribution of hydraulic conductivity by the SIS method.

Figure 3.36 shows an example of the distribution of hydraulic conductivity using the results of Lugeon tests in the vertical and oblique water curtain boreholes along the vertical curtain row. Cracks across the cavern were dominant in the Takanawa granite of Zone V. The histogram of the hydraulic conductivity of a water curtain borehole shows that the Lugeon values for oblique boreholes are one order of magnitude higher than the vertical water curtain boreholes, indicating anisotropy in hydraulic conductivity. Accordingly, anisotropy in hydraulic conductivity (high across and low along the cavern) was assigned to the hydrological characteristics of this zone.

Figures 3.37 and 3.38 show plan view maps of the estimated distribution of Lugeon values of the propane and butane storage caverns, respectively, at 150

m below sea level. In these figures, Lugeon values along and across the caverns are shown separately.

Next, the effect of post-grouting on the improvement in hydraulic conductivity was examined. The histograms in Figure 3.39 are the result of Lugeon tests up to completion of the excavation of the benches in the reconnaissance boreholes, water curtain boreholes,

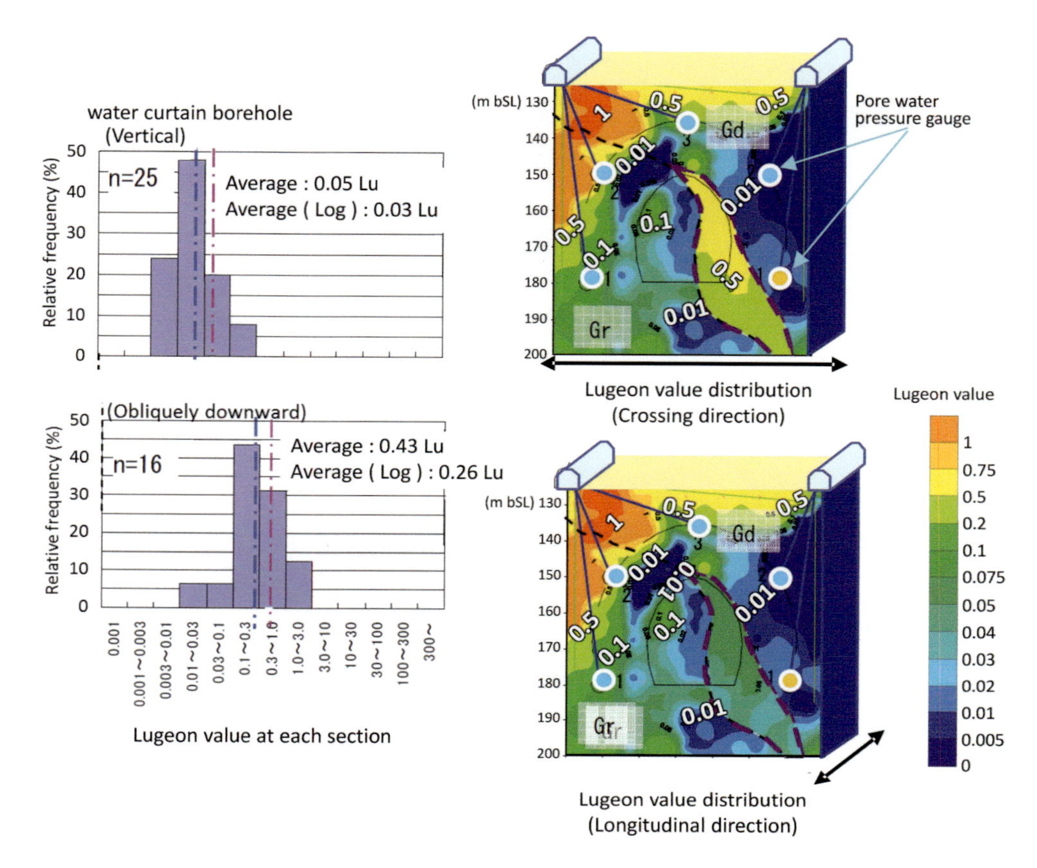

Figure 3.36 Histogram of the result of a Lugeon test (left). Example of the estimated distribution of Lugeon values with anisotropy (C section of propane storage No. 2).

measurement boreholes and grouting boreholes. The average Lugeon value before grouting was 0.093 Lu, which improved to 0.048 Lu.

Figure 3.40 shows the estimated distribution of hydraulic conductivity before and after grouting. The plan view is at 157.5 m below sea level, the level of the ceiling of the cavern and the cross section is Section AA of the propane storage. From this 3D distribution of the hydraulic conductivity, improvement was confirmed around the storage cavern from high to low hydraulic conductivity.

3.3.2.2.1.2.4 Example of Estimation of Behaviour of Groundwater Figure 3.41 shows the distribution of the difference between the predicted actual piezometric head values and the change in pore water pressure in cross section C of the butane storage cavern, where measurements were taken at each stage of the excavation.

Figure 3.42 is another example from cross section A. Here, the difference is expressed as prediction minus actual, and the distribution was generated from the five pore water pressure gauges around the cavern. These figures also show the predicted values of the pore water pressure assuming a 0.1 Lu value for the bedrock. The piezometric head values predicted by

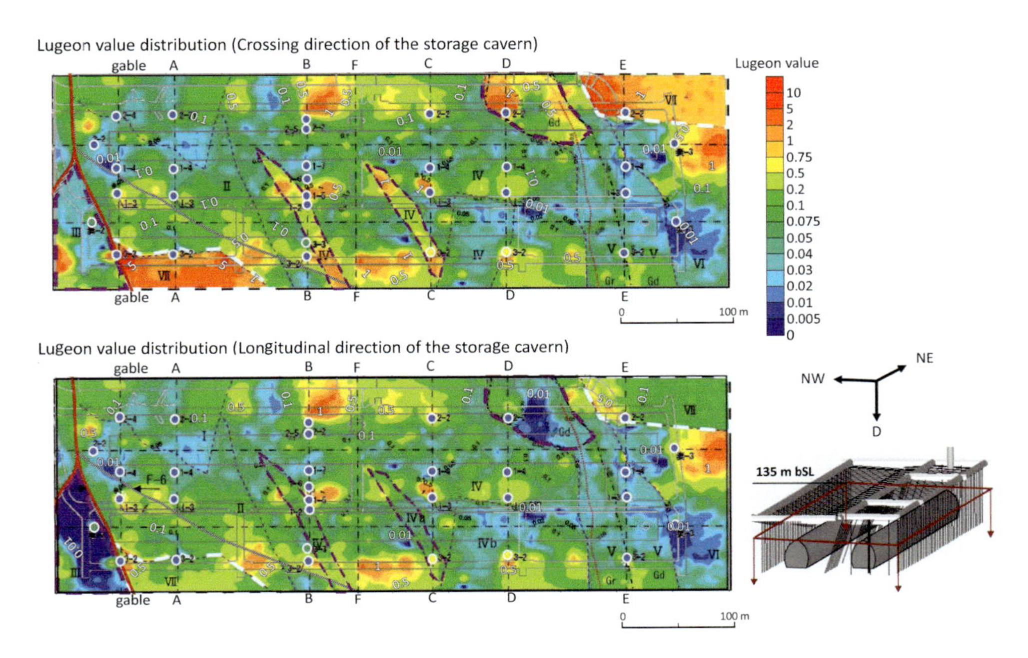

Figure 3.37 Estimated distribution of Lugeon values around the propane storage cavern (depth slice at 135 m bSL).

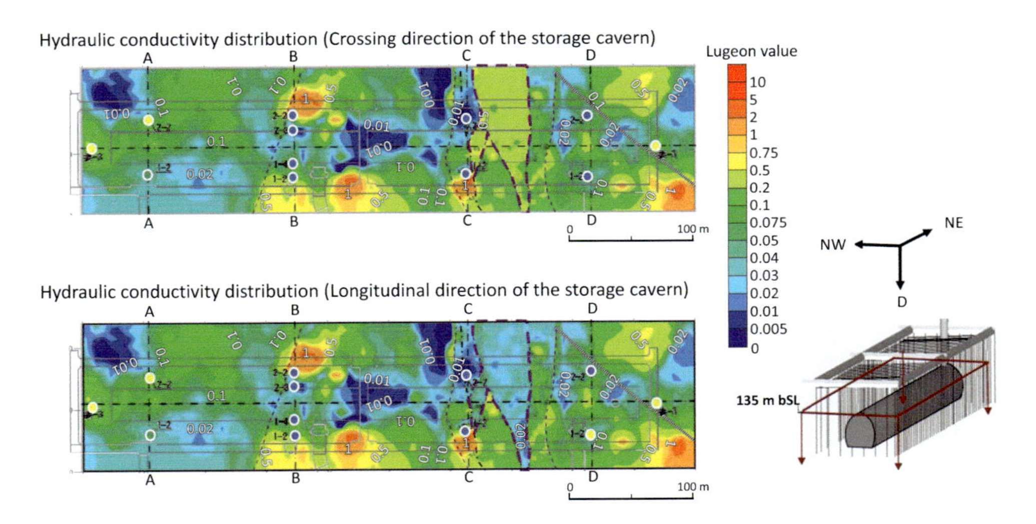

Figure 3.38 Estimated distribution of a Lugeon value around the butane storage cavern (depth slice at 135 m bSL).

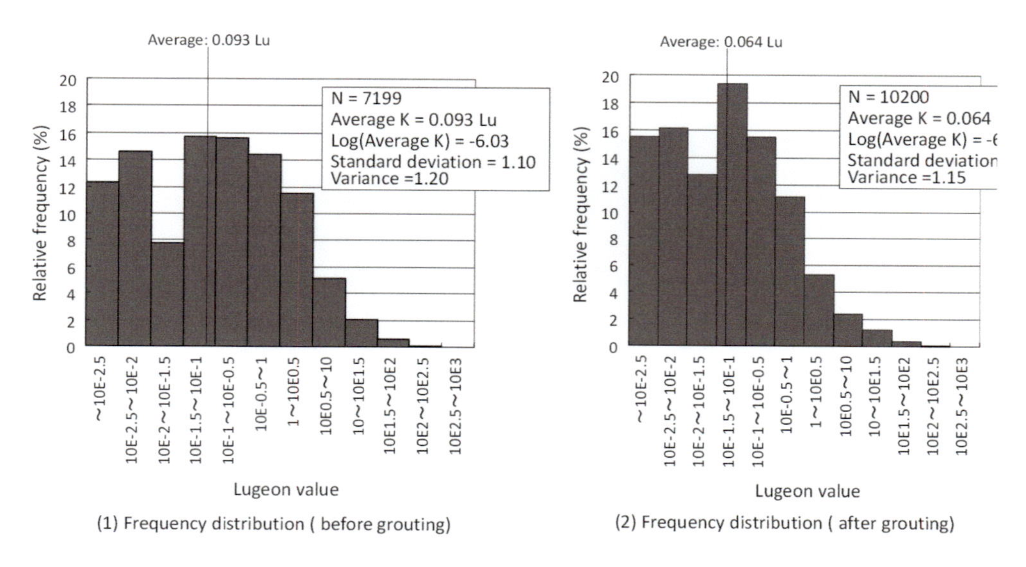

Figure 3.39 Histogram of hydraulic conductivity on completion of bench excavation.

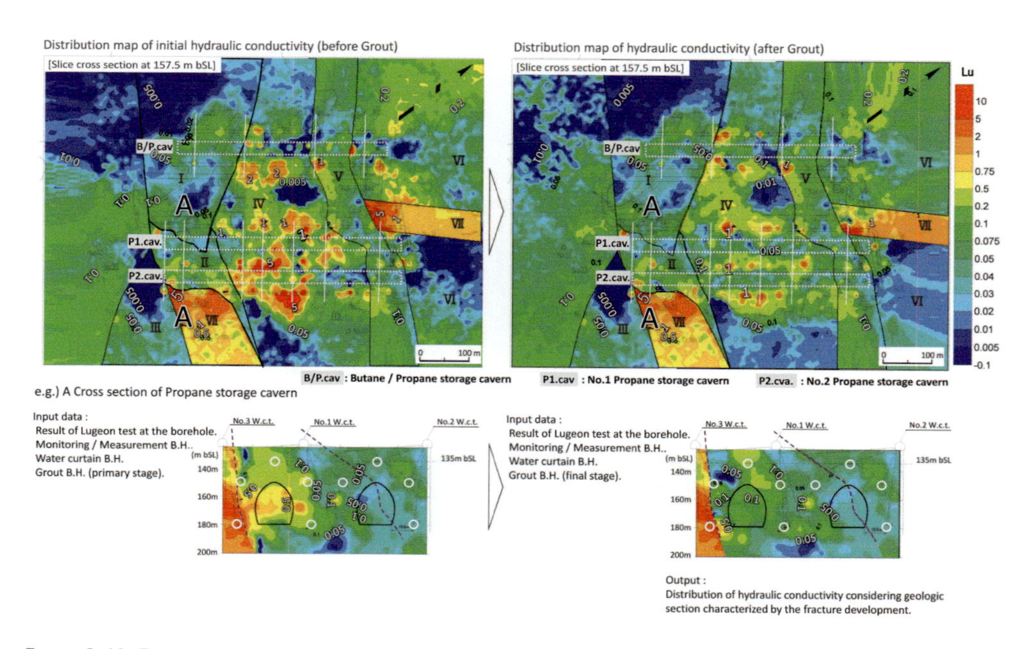

Figure 3.40 Estimated distribution of hydraulic conductivity (plan view at 157.5 m bSL).

the 3D heterogeneous hydrogeological model agreed with the actual measured values from the start of excavation of the arches through to completion of excavation. On the other hand, although the values predicted by the homogeneous hydrogeological models did not differ from the actual measurements, their difference grew larger as the excavation progressed.

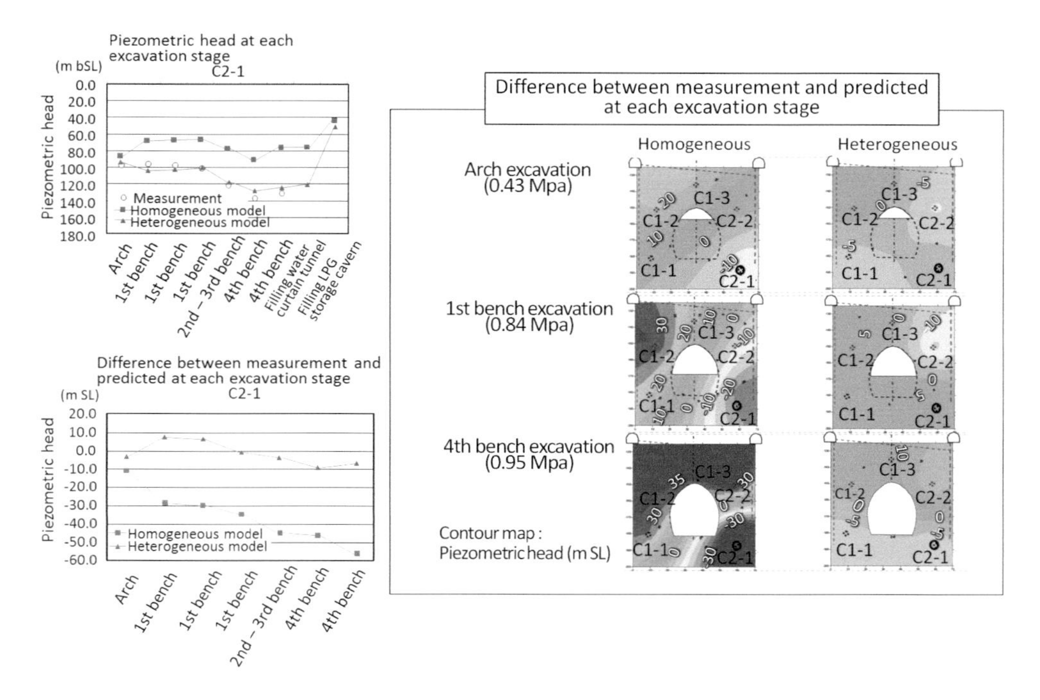

Figure 3.41 Change and distribution of pore water pressure through excavation steps in cross section C of butane storage cavern.

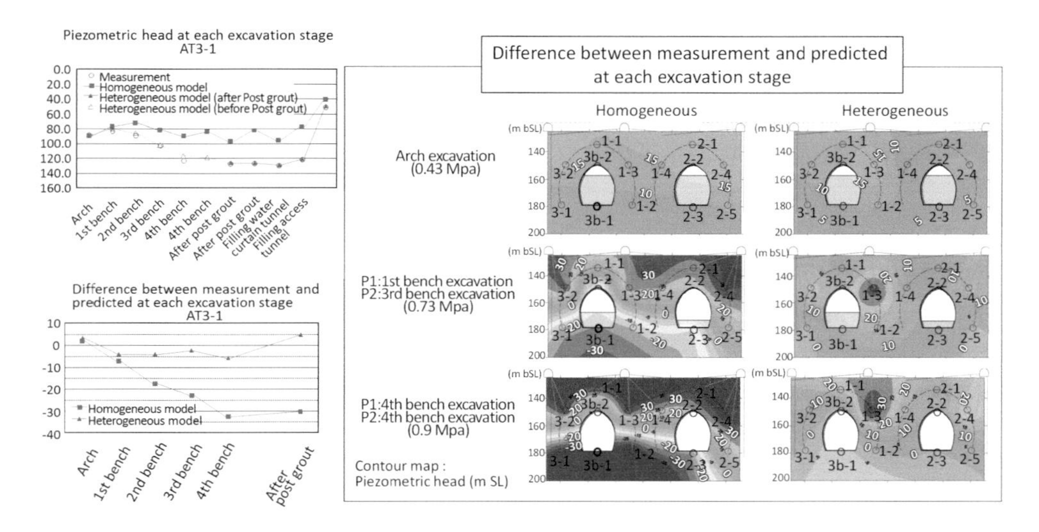

Figure 3.42 Change and distribution of pore water pressure through excavation steps in cross section A of propane storage cavern.

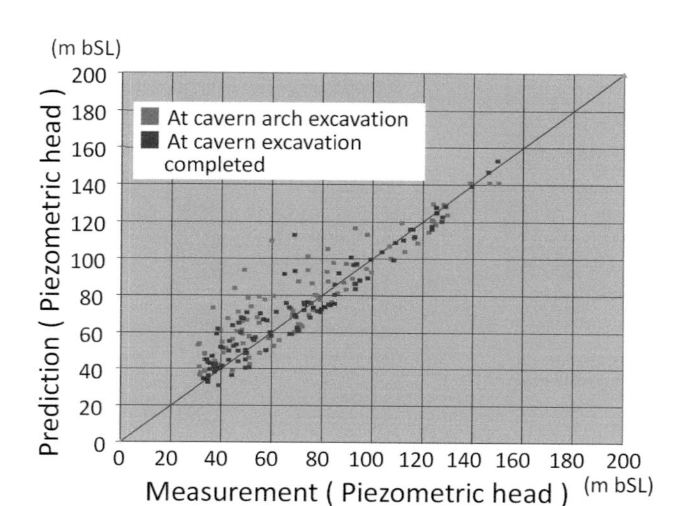

Figure 3.43 Comparison between measured pore water pressure and prediction on completion of arch and bench excavation.

Figure 3.43 shows a comparison between the measured pore water pressure and prediction on completion of the arch and bench excavation. The piezometric head values predicted by the hydrogeological model on completion of the cavern excavation generally fell within ±10 m of measured values. The scatter of the difference was smaller on completion of the arch excavation. This is because of the inclusion of a larger number of Lugeon test data in the grouting boreholes and additional water curtain boreholes, and updated geological structure knowledge with the orientation and density of cracks and anisotropy in hydraulic conductivity into the hydrogeological model. The hydrogeological model made on completion of cavern excavation was adequately applicable to the air-tightness tests after the excavation and prediction of the flow rate of seepage and pore water pressure behaviour through commissioning and during operation.

3.4 Prediction of Behaviour of Groundwater by the Hydrogeological Model During Excavation

3.4.1 The Kurashiki Facility

The excavation of the cavern progressed comparing the predicted ("management standard" hereafter) and the ongoing measurement of the flow rate of seepage and pore water pressure. Figure 3.44 shows a comparison between the predicted and measured flow rate of seepage at each stage of excavation. The actual flow rate of the seepage was generally lower than the flow rate predicted by the 3D hydrogeological model, and they had a similar trend. This indicated that the grouting was effective.

The distribution of the pore water pressure was predicted using the 3D hydrogeological model and was compared with the actual measurements. Figure 3.45 shows comparisons in three aspects around storage No. 1 outside of the water curtain system: distribution along the depth across the storage; distribution in a longitudinal cross section; and plan view under

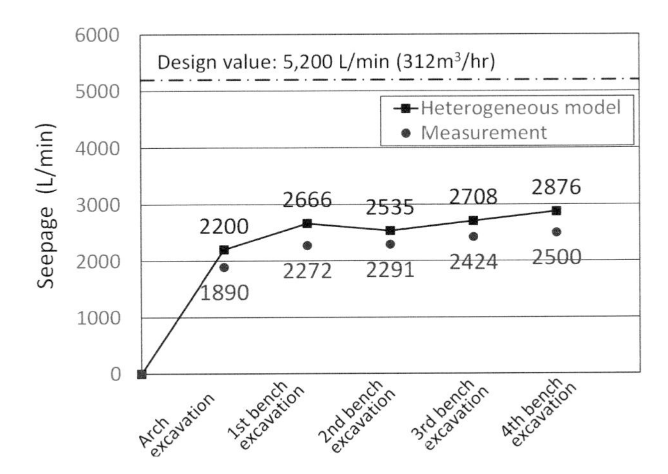

Figure 3.44 Comparison between predicted and measured flow rate of seepage at each stage of excavation.

the water curtain. As seen in Figure 3.45, the measured pore water pressure at ten points in four boreholes in the upper part of the water curtain system agreed well with the values predicted by the model. On the other hand, the horizontal distribution of the pore water pressure was not as heterogenic as predicted by the model. The difference between the predicted and measured values was mostly within 15 m expressed in piezometric head. From this result, the model is regarded as representing the actual heterogeneity of the bedrock.

Near cross section A, the hydraulic conductivity was low around the storage cavern and a sealing fault prevented the water supply from formations above. Figure 3.46 shows the distribution of hydraulic conductivity and pore water pressure along cross section A. The pore water pressure along "A" dropped during excavation of the water curtain tunnel and storage cavern. It was maintained high by enhanced water injection through the additional water curtain boreholes.

3.4.2 The Namikata Facility

The excavation of the cavern progressed comparing the predicted and ongoing measurement of the flow rate of the seepage and pore water pressure. Figure 3.47 shows a comparison between the predicted and measured flow rate of the seepage at each stage of excavation. The actual flow rate of the seepage was generally lower than the flow rate predicted by the 3D hydrogeological model, and they had a similar trend. This indicated that the grouting was effective.

Figure 3.48 shows the distribution of the pore water pressure on completion of cavern excavation along cross section A, a representative section of the storage area. The pressure around the storage cavern and in the access tunnel was the atmospheric pressure. It causes a drop in the pore water pressure in the vicinity. This atmospheric pressure in the cavern had no influence outside of the water curtain system because the water curtain system including tunnels and boreholes was in operation around the cavern. This confirmed the effectiveness of the water curtain system.

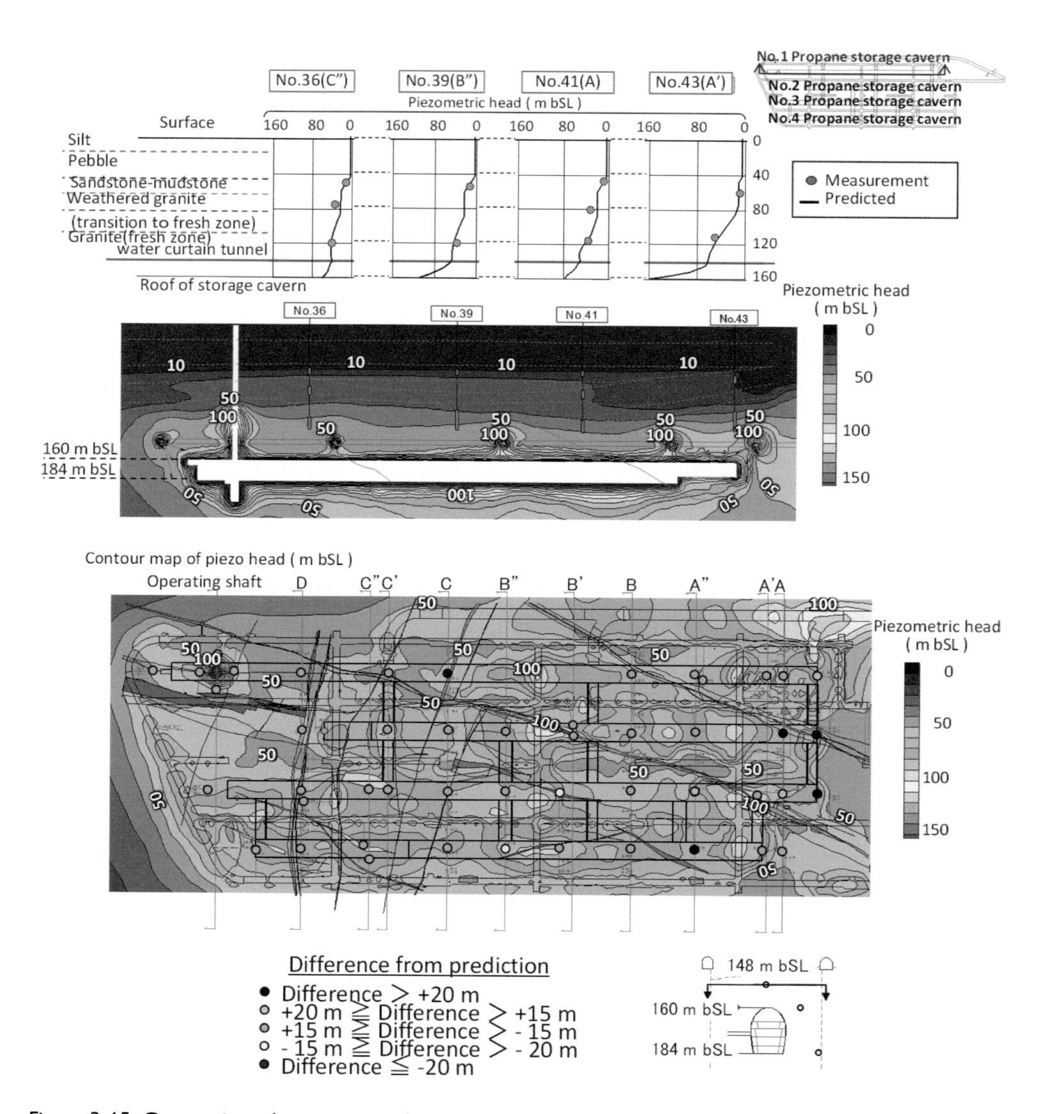

Figure 3.45 Comparison between predicted and measured pore water pressure on completion of bench excavation (Kurashiki).

Figure 3.49 shows the variation in pore water pressure with cavern excavation, while Figure 3.50 is a comparison with the actual measurement on completion. The pore water pressure in the fresh granite in the upper part of the water curtain (~105 and ~120 m below sea level) dropped when the water curtain tunnels were excavated. However, it did not drop significantly during the cavern excavation after starting the operation of the water curtain boreholes. This confirmed the effectiveness of the water curtains. Although the pore water pressure at the top (135 m below sea level) and upper part (148 m below sea level) of the cavern – both under the water curtains – dropped during cavern excavation, its magnitude was within the management standard. This indicates that the grouting was effective.

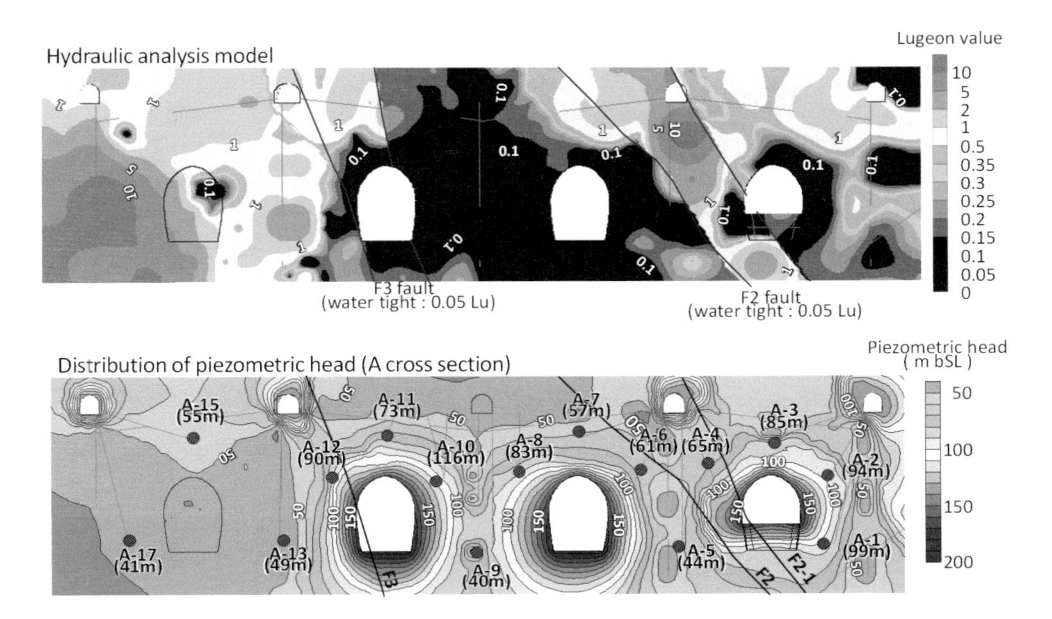

Figure 3.46 Distribution of Lugeon value (top) and pore water pressure (bottom) on completion of post-grouting.

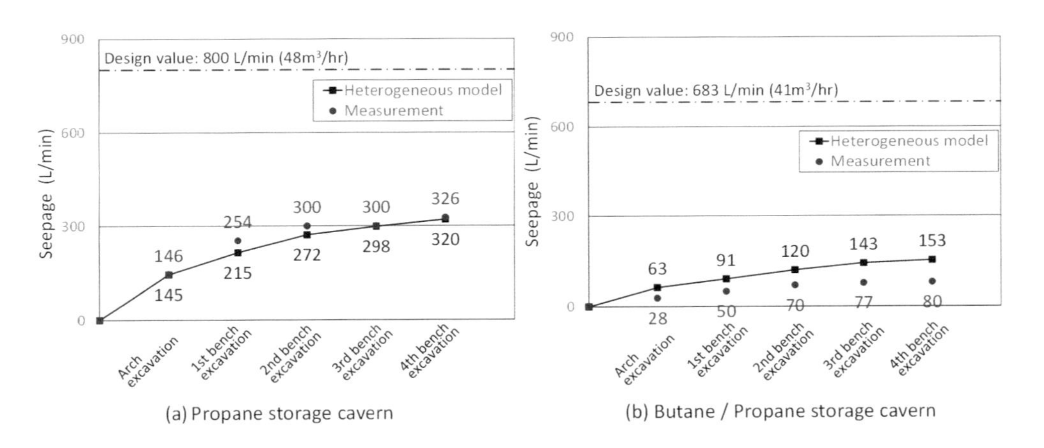

Figure 3.47 Comparison between predicted and measured flow rates of seepage at each stage of excavation.

Figure 3.51 is an example of the improvement by grouting a highly permeable area around the cavern: Lugeon value along cross section B of the combined butane-propane storage and pore water pressure (total piezometric head) by permeability analysis. This figure shows that the area improved by pre- and post-grouting maintained high pore water pressure compared with the areas not grouted.

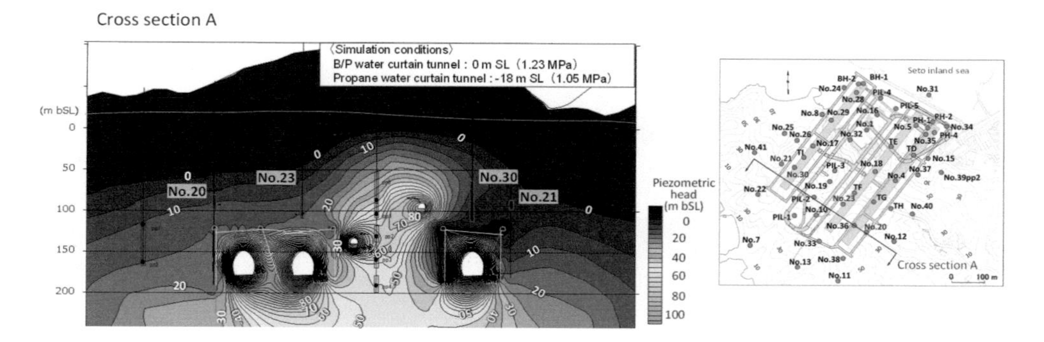

Figure 3.48 Distribution of pore water pressure on completion of post-grouting (cross section A).

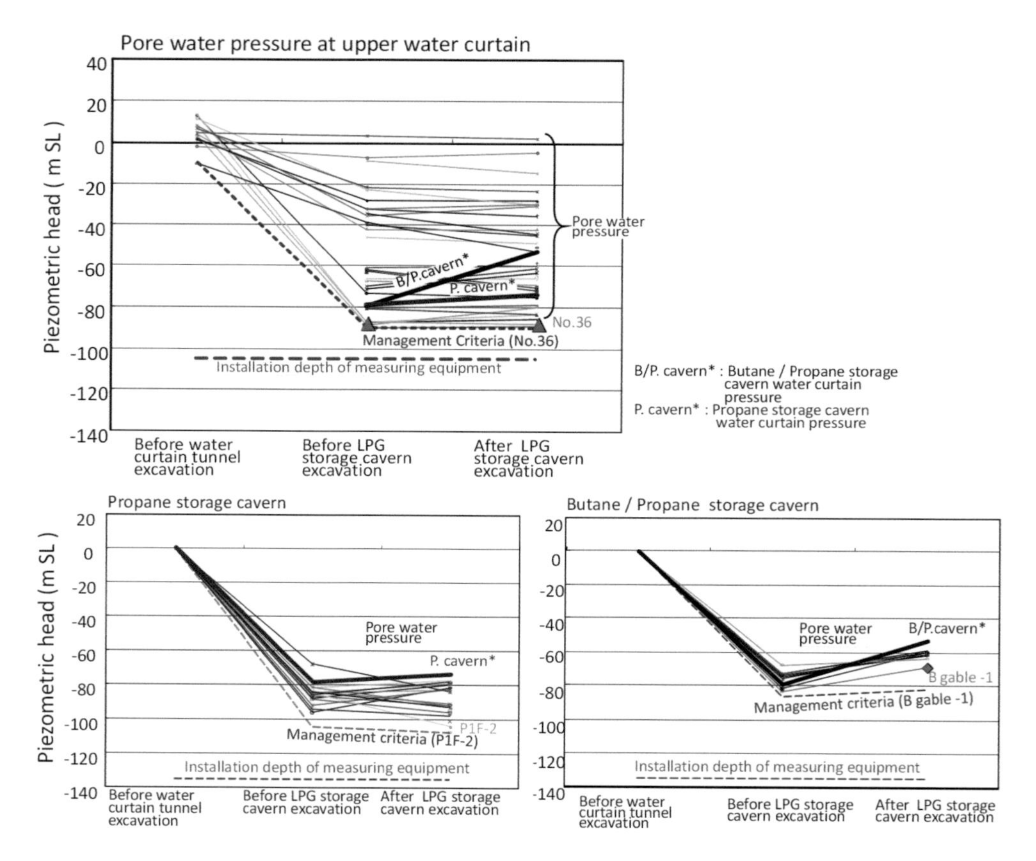

Figure 3.49 Variation of pore water pressure at upper part (105 m below sea level) and inside (135 m below sea level) water curtains.

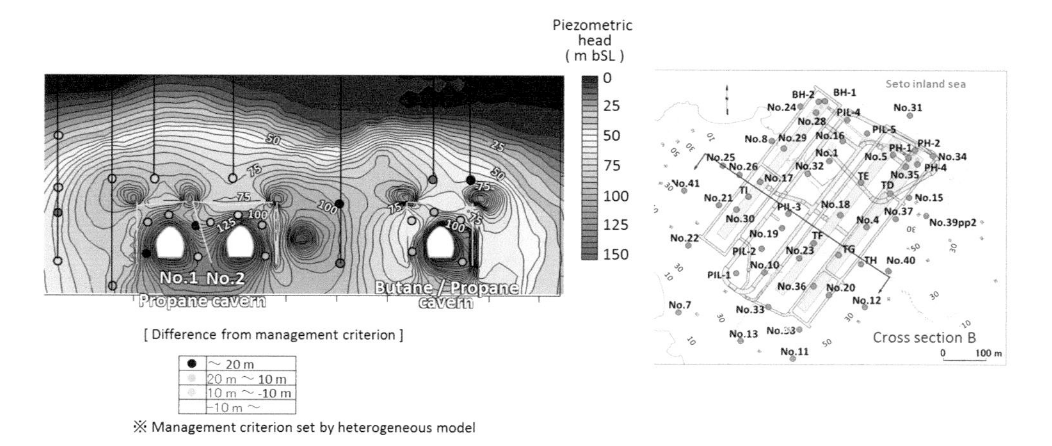

Figure 3.50 Comparison between predicted and actual pore water pressure during cavern excavation (cross section B).

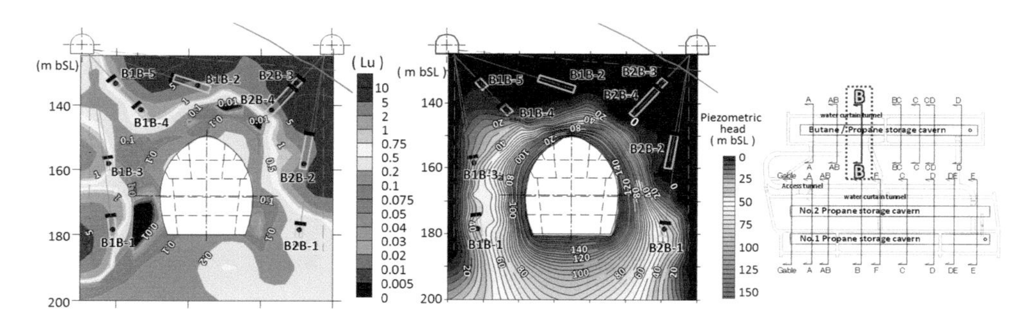

Figure 3.51 Distribution of Lugeon value (left) and pore water pressure (right) after post-grouting (cross section BB of combined butane-propane storage cavern).

A comparison between the predicted and actual pore water pressure at the time of water pressurising is shown in Figure 3.52. It shows a general agreement between the predicted and actual values of the pore water pressure response to the increase in the water sealing pressure as well as the piezometric head from the pore water pressure.

Figure 3.53 shows Lugeon values along cross section BC of the combined butane-propane storage cavern and the distribution of the pore water pressure (total piezometric head) from the analysis as an example of the situation of homogeneous low Lugeon values around the storage cavern where improvement by grouting was not needed. These figures show that the Lugeon value was low and the variation in pore water pressure was uniform from the cavern wall to the water curtain boreholes and that high pore water pressure was maintained around the cavern.

A comparison between the predicted and actual pore water pressure at the time of pressurising the sealing water is shown in Figure 3.54. The predicted values generally agreed with the actual values: they precisely predicted the piezometric head value of the pore water pressure and the response of the pore water pressure to the water pressurisation at the same time.

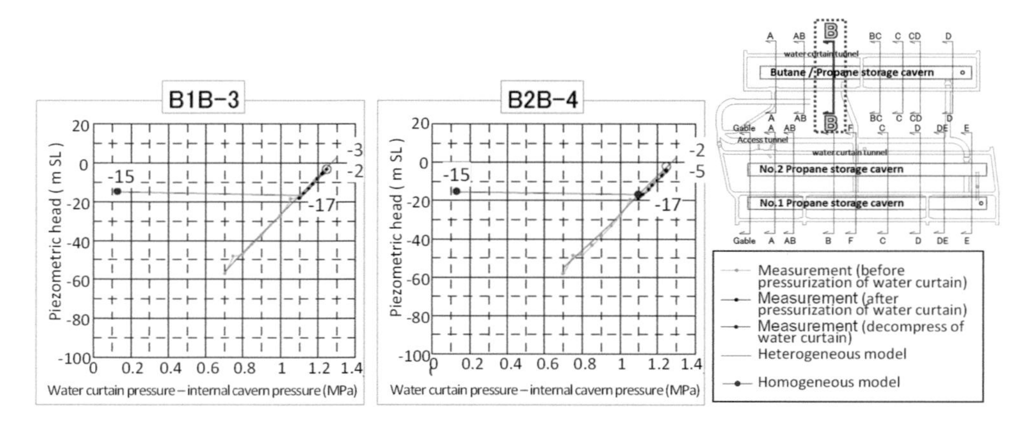

Figure 3.52 Variation of pore water pressure along cross section BB of the combined butane-propane storage cavern at time of pressurising the sealing water.

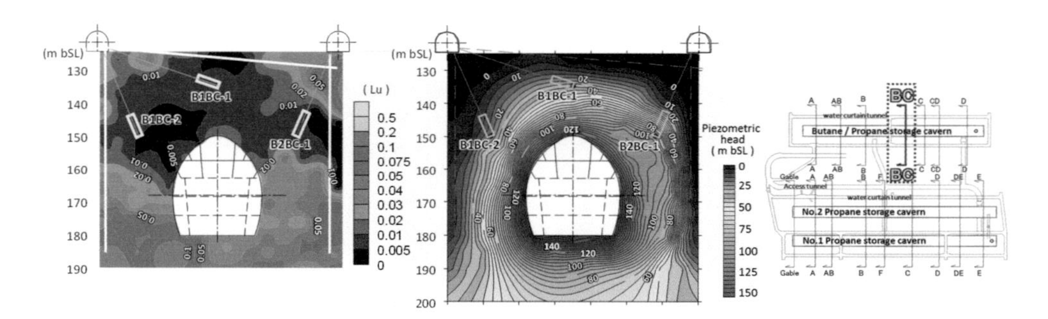

Figure 3.53 Distribution of Lugeon values and pore water pressure (cross section BC around the combined butane-propane storage cavern).

A 3D hydrogeological model was constructed based on detailed hydrogeological surveys from reconnaissance to excavation and a detailed pore water pressure measurement over the entire cavern areas with a heterogeneous hydrogeological structure. This 3D hydrogeological model was built applying geostatistics, considering the geological structure including the density and orientation of cracks and faults. The Kurashiki facility site adopted the IK method, while Namikata used the SIS method of geostatistical simulation. As seen, the results of geological surveys and hydrogeological tests were analysed and evaluated, and the methods of constructing the 3D hydrogeological models were selected appropriate to the local geological characteristics. Their accuracy was improved by updating the geological information and additional test results during excavation. The 3D hydrogeological models thus constructed were assessed by comparing the predicted and measured distributions of the pore water pressure around the cavern and the flow rates of seepage in the cavern through cavern excavation to completion of post-grouting. This result clarified that the system is applicable to prediction of the pore water pressure and flow rate of seepage in caverns in the future.

The new model building method, proposed in this book, combines the geostatistical technique and seepage flow analysis. This method is applicable to the behaviour of the

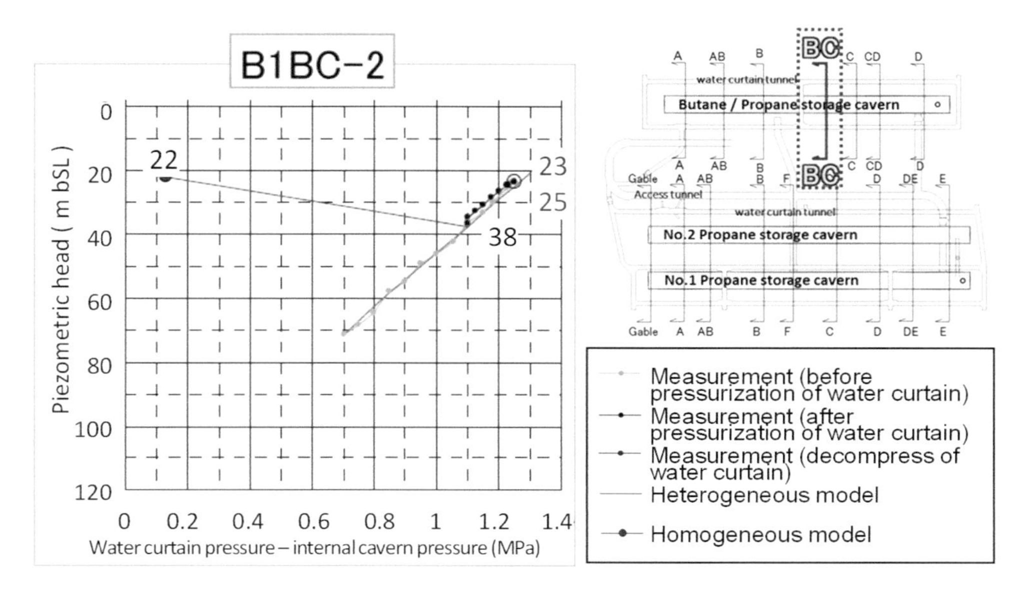

Figure 3.54 Comparison between predicted and actual pore water pressure at the time of pressurising the sealing water (cross section BC around the combined butane-propane storage cavern).

groundwater level, variation in the pore water pressure and the statistical relationship (spatial correlation) of hydraulic conductivity. This method was applied to hydrogeological modelling in the two storage construction sites. The application predicted with high accuracy the flow rate of seepage, the pore water pressure behaviour, variation in the pore water pressure and the hydrological characteristics. These results show that the modelling scheme proposed here is effective in building high-precision hydrogeological models necessary for evaluating the water-tightness and subsequent control of groundwater.

References and Further Readings

Aoki K., Maejima T., Morioka H., Mori T., Tanaka M., Kanagawa T. (2003) Estimation of rock stress around cavern by CCBO and AE method. *Proceedings of the Third International Symposium of Rock Stress*, Kumamoto, 203–209.

Aoki K., Mito Y. (2006) Hydrogeological modeling of rock mass by MDS-IDW technique. *Proceedings of the International Symposium of the International Society for Rock Mechanics, Eurock 2006*, Liege, 677–682.

Aoki K., Mito Y., Chang C.S. (2007) The geostatistical prediction technique of geological conditions ahead of the TBM driving tunnel face. *Proceedings of the 7th International Workshop on the Application of Geophysics to Rock Engineering*, Lisbon, 57–62.

Aoki K., Mito Y., Chang C.S., Tasaka Y., Maejima T. (2007) Hydro mechanical coupled discrete modelling for the assessment of air-tightness of unlined large rock cavern. *11th Congress of the International Society for Rock Mechanics*, Lisbon, 925–929.

Aoki K., Mito Y., Matsuoka T., Kondoh D. (2004) Design of gas storage rock cavern by the hydro-mechanical coupled discrete model. *Proceedings of the 2nd International PFC Symposium*, Kyoto, 289–299.

Aoki K., Shiogama Y., Kobuchi T. (1990) Groundwater behavior in jointed rock mass during excavation of underground crude oil storage caverns. *International Symposium on Rock Joints*, Loen, 363–368.

Bergman M. (1977) Storage in excavated rock caverns. *Proceedings of the First International Symposium*, Stockholm, 832.

Lindblom U.E. (1989) The development of hydrocarbon storage in hard rock caverns. *Proceedings of the International Conference on Storage of Gases in Rock Caverns*, Trondheim, 15–30.

Maejima T., Uno H., Mito Y., Chang C.S., Aoki K. (2007) Three-dimensional hydrogeological modelling around the large rock cavern for the LPG storage project. *11th Congress of the International Society for Rock Mechanics*, Lisbon, 1273–1277.

Miyashita K., Aoki K., Hnamura T., Kashiwagi N. (1983) An investigation of geomechanics and hydraulics around an underground crude oil storage cavern. *International Symposium on Field Measurement in Geomechanics*, Zurich, 1117–1126.

Nilsen B., Olsen J. (1989) Storage of gases in rock caverns. *Proceedings of the International Conference on Storage of Gases in Rock Caverns*, Trondheim, 398.

Shirasagi S., Yamamoto T., Murakami K., Ogura E., Mito Y., Aoki K. (2005) Evaluation system of tunnel excavation by geostatistics applying for seismic reflective surveys and TBM. S driving date. *Proceedings of the International Symposium on Design, Construction and Operation of Long Tunnels*, Taipei, 1255–1262.

Chapter 4

Groundwater Controlling System

The bedrock of the Namikata site is generally of low permeability, and contains cracks with hydrological continuity. The complex geology of the Kurashiki site includes highly permeable cracks of faults and joints, layers of impermeable fault clay and a local zone of extremely low permeability. For these complex bedrocks, maintaining a stable groundwater level is necessary from construction through to operation. For this reason, water curtain systems were constructed, whereby groundwater is artificially supplied through boreholes designed on the top and in the sides of caverns. In addition, the highly permeable zones around the caverns were improved by grouting to reduce seepage and the flow of groundwater (this system of water curtains and grouting is referred to as the "groundwater control system").

The water curtains supply groundwater through the water curtain boreholes previously drilled on the top and in the sides of the cavern from construction to operation. The water curtain boreholes were arranged in accordance with the prescribed basic pattern. However, more water curtain boreholes were added according to the variation in the pore water pressure and the seepage rate and hydrogeological tests analysed the hydrogeological characteristics.

The bedrocks around the caverns were improved by pre-grouting before excavation to reduce seepage. The effect of pre-grouting was judged after excavating the pre-grouted zones by measuring the seepage rate and the pore water pressure around the cavern. Accordingly, post-grouting was added where deemed necessary. The effect of the groundwater control system at each stage of construction was monitored by comparing the actual seepage rate and the pore water pressure around the caverns with the values predicted by the 3-dimensional (3D) hydrogeological model described in Chapter 3.

On completion of the excavation of the caverns, it is necessary to check and confirm the water-tightness over the entire cavern area. For this, the water curtain tunnels and water curtain boreholes were filled with water and pressurised up to the injection pressure of the operation stage. Under this condition, the water level, pore water pressure and seepage rate in the caverns were analysed globally and locally, and the final water-tightness functionality was confirmed. The use of 3D hydrogeological models built during construction enabled a high-precision prediction and evaluation of the distribution and flow of the groundwater at the pressurising stage.

DOI: 10.1201/9780367822163-4

4.1 Construction of Groundwater Control System (Water Curtains)

4.1.1 General Policy in Constructing Water Curtains

A water curtain is an artificial water sealing screen surrounding storage caverns by injecting pressurised groundwater from a water curtain borehole drilled horizontally and vertically from the water curtain tunnel on top of the caverns. Figure 4.1 shows the construction procedure for the water curtains. First, water curtain boreholes are drilled vertically and horizontally at equal distance from the water curtain tunnels which have already been excavated. Then, the water curtains are formed by pressurised water injected from the water curtain boreholes. Next, the excavation of the storage caverns starts from the arch of the cavern and the benches, one after another. During excavation of the storage caverns, the seepage rate and the pore water pressure are constantly monitored for heterogeneous hydrogeological characteristics such as faults and cracks. These measurements are compared with the values predicted by 3D hydrogeological modelling. In order to monitor the functionality of the water curtains, the function of the water injection boreholes is tested and the effect of water injection is judged. Depending on the judgement, additional water curtain boreholes are drilled if

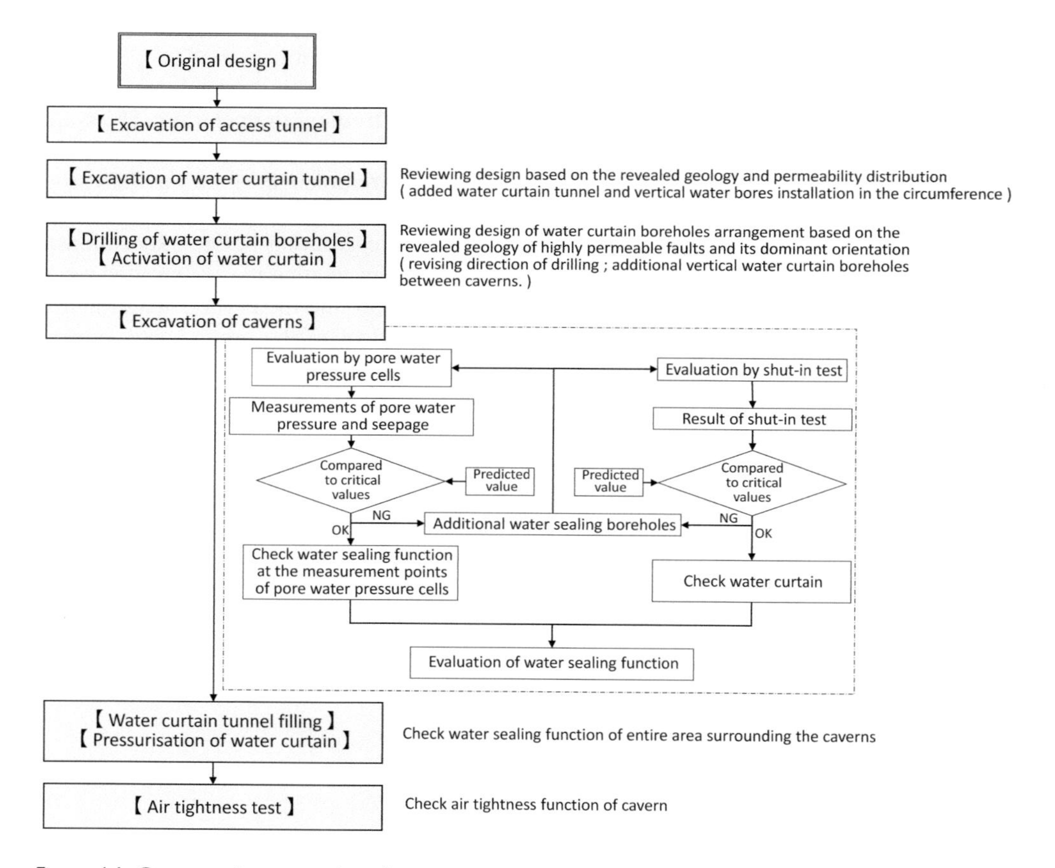

Figure 4.1 Construction procedure for water curtains.

necessary to form water curtains. The position, direction and length of the additional water curtain boreholes are determined from the analysis of a comparison between the measured and predicted pore water pressure and seepage rate, referring to the results of the hydrogeological characteristics survey and the hydrological data at grouting.

Once the initial construction is completed, it is very difficult to add any construction work because all the tunnels leading to the storage caverns are immersed in water. To avoid the necessity of additional work, water-tightness is tested and confirmed by pressurising the completed storage caverns before sealing. In this test, water curtain tunnels are immersed and the injection pressure of the water curtain boreholes is increased while the storage cavern is kept at atmospheric pressure. The water-tightness is evaluated from the recovery of the pore water pressure around the cavern and the seepage rate. Comparing the measured and predicted pore water pressure, the validity of the 3D hydrogeological model is confirmed. The 3D hydrogeological model is used to predict the pore water pressure and the seepage rate at the final air-tightness test of the caverns.

The basic design of the water curtain system is similar to other cases of water curtains at existing facilities. After drilling the water curtain boreholes, a packer is installed at the start of the borehole and sealing water is supplied through temporary piping during the excavation of the caverns. These temporary pipes are removed after topping up the water curtain tunnels on completion of cavern excavation. Then, the water curtain tunnels are quickly immersed and water is injected from the tunnels through the water curtain boreholes.

For the long-term maintenance of the water curtain system, the issue of water circulation must be addressed. The water that is stored in the water curtain tunnel and supplied to the water curtain boreholes contains minute rock particles from the tunnel walls. These may clog the water curtain boreholes. Similar issues may be found with fault clays and calcium precipitation in the groundwater.

In response to these issues, a continuous water supply system was adopted in both the Kurashiki and Namikata facilities. Firstly, the water was supplied to the water curtain boreholes through the water supply pipe connected to the ground surface from cavern excavation through to operation, so that no change in the supply system, hence no downtime, was incurred. In the Kurashiki facility, the water used for sealing was industrial water from a nearby river which was filtered with a membrane. As there is no river near the Namikata facility, the sea water used was desalinated by membrane filters. Thus, clogging of the water curtain boreholes by calcium scaling and organic slime was prevented.

A water circulation function was added to the supply system. This enables washing of the water curtain boreholes if clogging of the boreholes becomes an issue in the future. The water supply pipe is designed in two layers: the structure allows the pipes and boreholes to be washed by the packers to control the injection through the inner pipe while draining through the outer pipe. Figure 4.2 shows this water circulation system with the water supply and return pipes from the surface connection to the water curtain boreholes. The structure of the supply and return pipes and the recirculation packer is shown in Figure 4.3.

4.1.2 Construction Method of the Water Curtain System

4.1.2.1 Hydrological Test During Drilling Water Curtain Borehole

During the drilling period of the water curtain boreholes, various surveys and tests were carried out to collect the hydrogeological data used for constructing the 3D hydrogeological

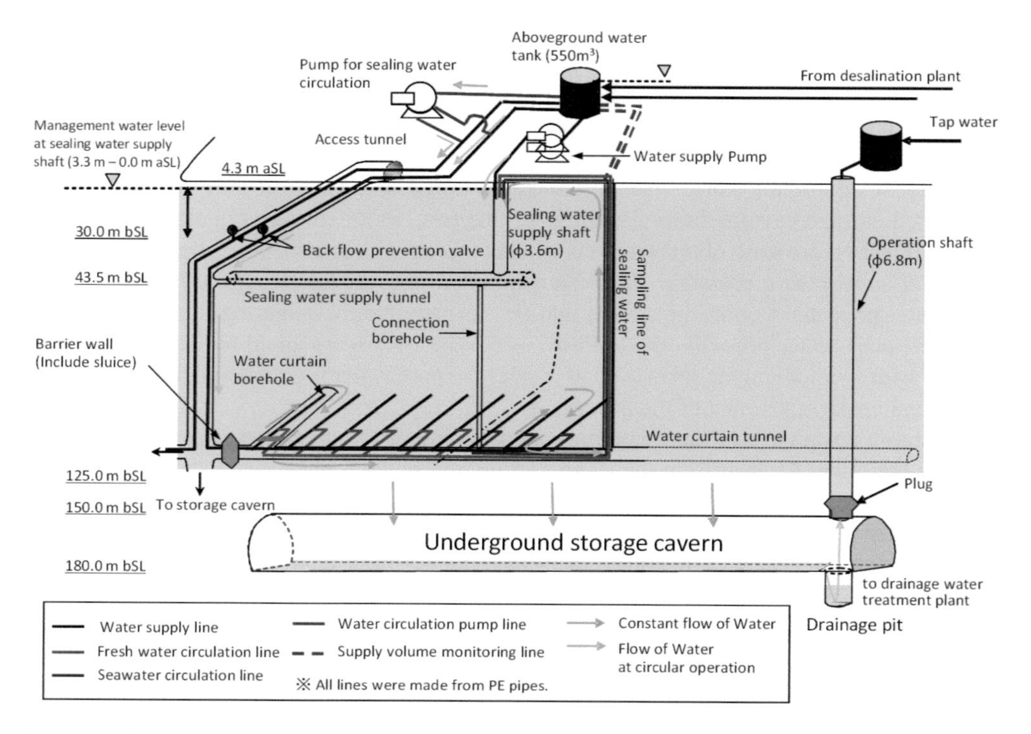

Figure 4.2 Conceptual diagram of water supply and circulation system.

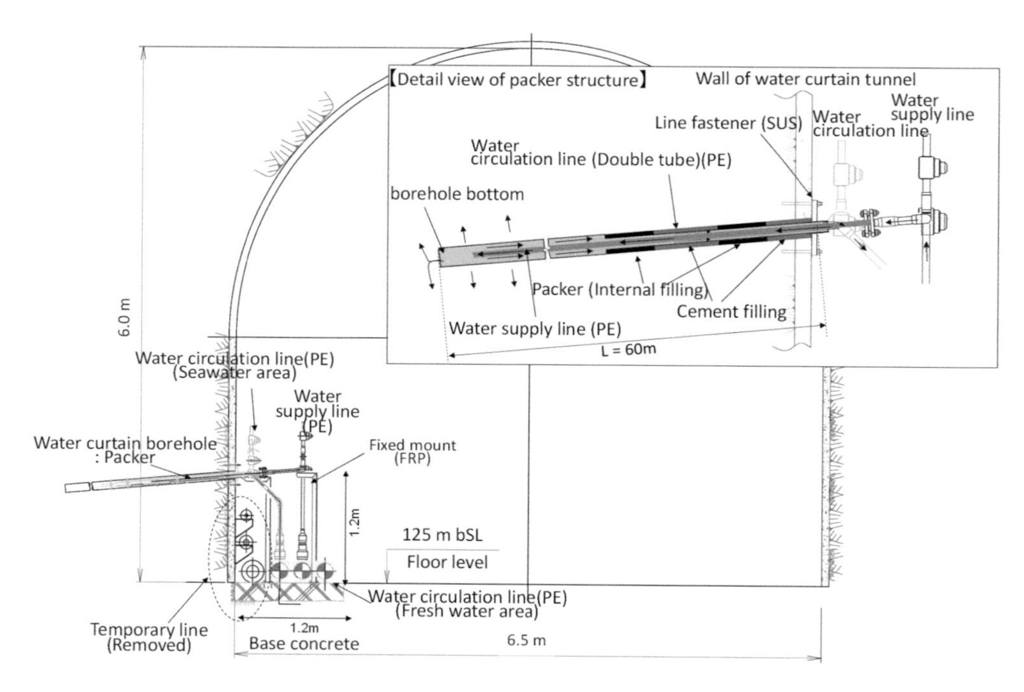

Figure 4.3 Structure of water supply, return pipes and circulation packer.

model and evaluating the water-tightness. Table 4.1 shows the test items used during and after drilling the water curtain boreholes. The geological structure was analysed from observation of the borehole walls by borehole television (BHTV) and investigation of cracks. Flowmeter logging was also used for continuous monitoring of the hydraulic conductivity and the distribution of the seepage rate. Thus, the continuity of the cracks causing seepage and their hydrological structure were analysed and the hydrogeological structure was estimated integrating the results of all these analyses.

The water passage was estimated from the response to drilling and Lugeon tests at adjacent boreholes. The "response to drilling" is a variation in the water pressure measured by a pressure sensor installed at the top of an existing borehole when a new borehole is drilled. Figure 4.4 shows an example of the data on the water pressure response to drilling. When new drilling penetrates a groundwater passage connected to an existing borehole, the internal pressure of the existing borehole drops sharply. The water passage can be estimated from this drop in pressure. These response measurements were taken sequentially with drilling and the

Table 4.1 Test items during and after water sealing boring

Test type	*Test item*
ROP logging	Drilling energy, speed and time
Seepage rate	Flow rate during drilling and at the permeability test
Pressure of seepage water	Seepage water pressure by segments and entire borehole
Borehole television	Geological division, crack, width, weathering, alteration, rock classification
Flowmeter logging	Characteristic of cracks in comparison with BHTV observation
Lugeon test	Segment seepage water pressure, volume and Lugeon value
Lugeon test response	Continuation of water passage
Water quality	Temperature, pH, electric conductivity, ORP, SS, TOC, Ca, Fe, etc.

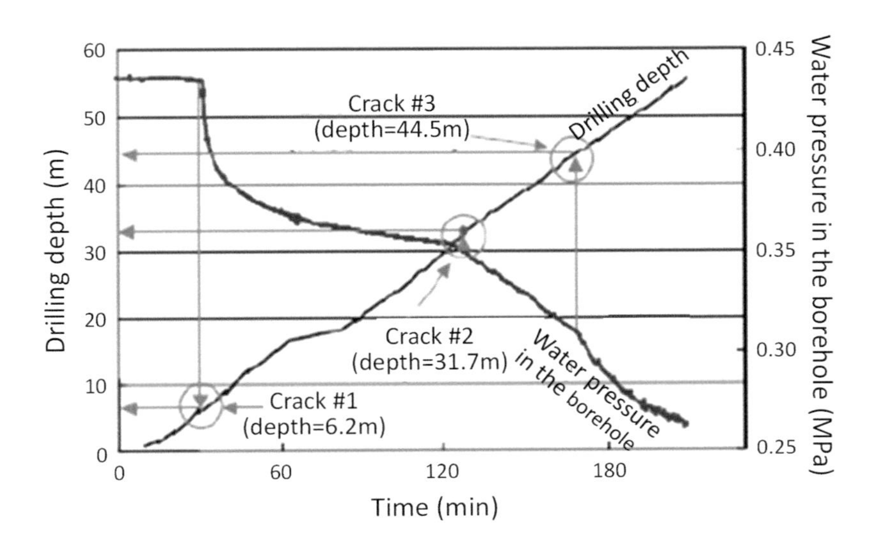

Figure 4.4 Example of water pressure response while drilling.

same measurements were taken at the time of the Lugeon tests. These results led to an estimation of the groundwater passage and the distribution of the hydraulic conductivity along the plane of boreholes (Figure 4.5).

4.1.2.2 Functionality Test of the Injection Boreholes

The test of the functionality of the injection boreholes evaluates the effect of grouting around the caverns and the continuity of the water curtain over all the caverns. It uses the water curtain boreholes systematically arranged on top and to the sides of the caverns.

As seen in Figure 4.6, the test measures the decrease in water pressure at the entry to the borehole after stopping the pressurised water injection through a water curtain borehole during cavern excavation. This decrease in water pressure is within a limited range if the hydraulic conductivity of the rocks between the water curtain borehole and the cavern is uniform or if the grouting around the cavern has successfully improved the pre-grouting zone to a low hydraulic conductivity. On the other hand, a large decrease in the water pressure is expected

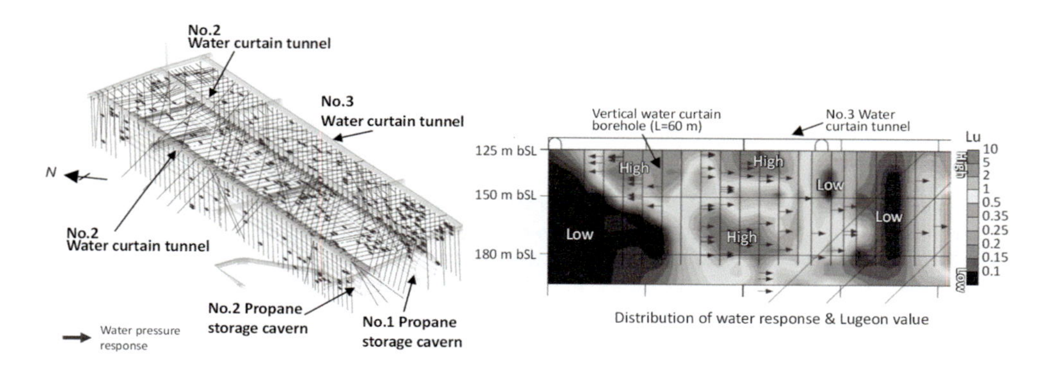

Figure 4.5 Distribution of water pressure response while drilling.

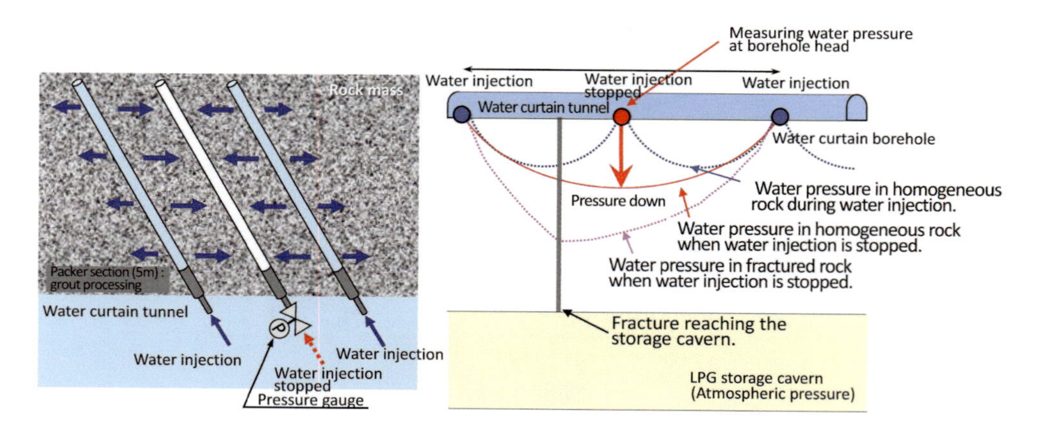

Figure 4.6 Conceptual diagram of the functionality test of the injection boreholes.

where there are highly permeable continuous cracks or heterogeneously permeable rocks dominate between the water curtain borehole and the cavern.

Judgement thresholds were set up by the prediction of the 3D hydrogeological model as the influence of structures such as the access tunnels between the caverns and shafts needs to be incorporated. Assuming that the hydraulic conductivity around the cavern was homogeneous, its standard value was set to the same as the target value of improvement of grouting. Figure 4.7 shows the analysis model for setting the standard judgement threshold values of the functionality test of the injection boreholes in the Kurashiki facility; Figure 4.8 is an

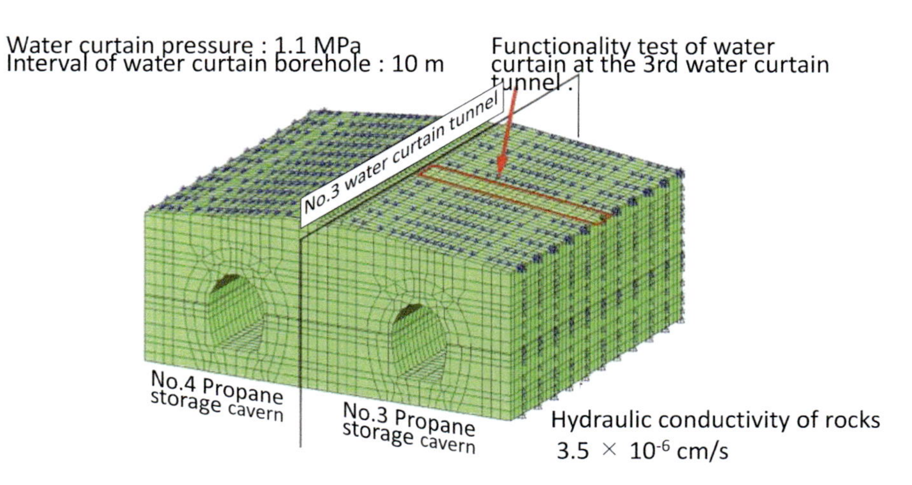

Figure 4.7 Analysis model for determining the judgement threshold of a functionality test on a water injection borehole in the Kurashiki facility.

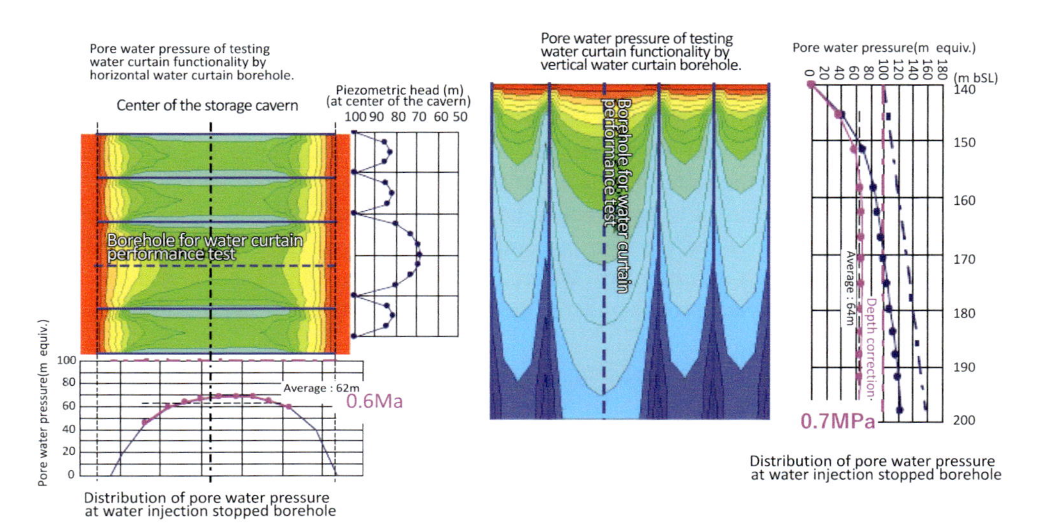

Figure 4.8 An example of a prediction for the judgement threshold of a functionality test on a water injection borehole in the Kurashiki facility. Water pressure is expressed in piezometric head value.

example of the distribution of the water pressure around the borehole when the injection was stopped; and Table 4.2 lists the values for judgement thresholds. When pressurised water injection through a single horizontal borehole was stopped, the average water pressure at the stopped water curtain borehole was expected to drop by about 0.6 MPa on average. In the corners, where the effect of the structure is great, the expected drop in pressure was around 0.3 MPa. From this expectation, the judgement thresholds for the functionality test of the horizontal injection boreholes were generally set to 0.6 MPa and 0.3 MPa for the corners, and if the water pressure dropped further, remedial measures were considered.

The functionality test of the vertical water curtain boreholes is influenced by the cavern space which increases with the progress of excavation, as cavern excavation takes place at multiple stages from arch to bench. Therefore, the greatest water pressure drop is when injection stops on completion of excavation of all four benches. Hence, the analysis predicted the result of the functionality test of the injection boreholes at the end of excavation of the fourth bench. The average water pressure at the stopped water curtain borehole was predicted to drop to about 0.7 MPa. A drop in the pressure to around 0.5 MPa was predicted between the caverns, where the effect of the structure is greater, and to around 0.3 MPa near the tunnels connecting the caverns, where the effect of the structure is greatest. These values were used as the judgement threshold, and additional water curtains were considered where the pressure dropped to lower than these values.

Unlike Kurashiki, the control threshold was not set by the standard in the Namikata facility, where the drop in the piezometric head was large and quick when the water injection was stopped, but it was decided by the rate of the water pressure drop ($\partial H/(\log_{10}\Delta t)$), where t is the elapsed time since the start of the test and ∂H is the drop in the piezometric head in meters in the borehole in which the pressurised water injection is stopped. The theoretical temporal variation in the piezometric head level was solved by transient analysis. Figure 4.9 shows the behaviour of the piezometric head dropping after stopping the pressurised injection predicted by the analysis against a time axis in logarithmic scale. The dropping behaviour of the piezometric head, as seen in Figure 4.6, has a linear relationship with the elapsed time in logarithmic scale after a certain elapsed time, regardless of cases where the hydraulic conductivity between the water curtain boreholes and the cavern was homogeneous or heterogeneous due to highly permeable cracks connecting the cavern. From this, the Namikata facility used the speed of the piezometric head drop (gradient of the graph) as the management standard for the functionality test of the water injection borehole.

Here, functionality tests of water injection boreholes were carried out on the entire cavern area before excavation of the cavern, on completion of cavern arch excavation and after excavation of each bench of the cavern. Where the speed of the drop in the piezometric head was higher than the management standard, in the hydrogeological structure, the quality of

Table 4.2 Judgement threshold values of the functionality test of water injection borehole in the Kurashiki facility

Borehole type	*Before excavation*	*After excavation (MPa)*
Horizontal borehole	0.8 MPa	General: 0.6 Corner: 0.3
Vertical borehole	–	Outermost: 0.7 Between cavern: 0.5 Around the connecting tunnel: 0.3

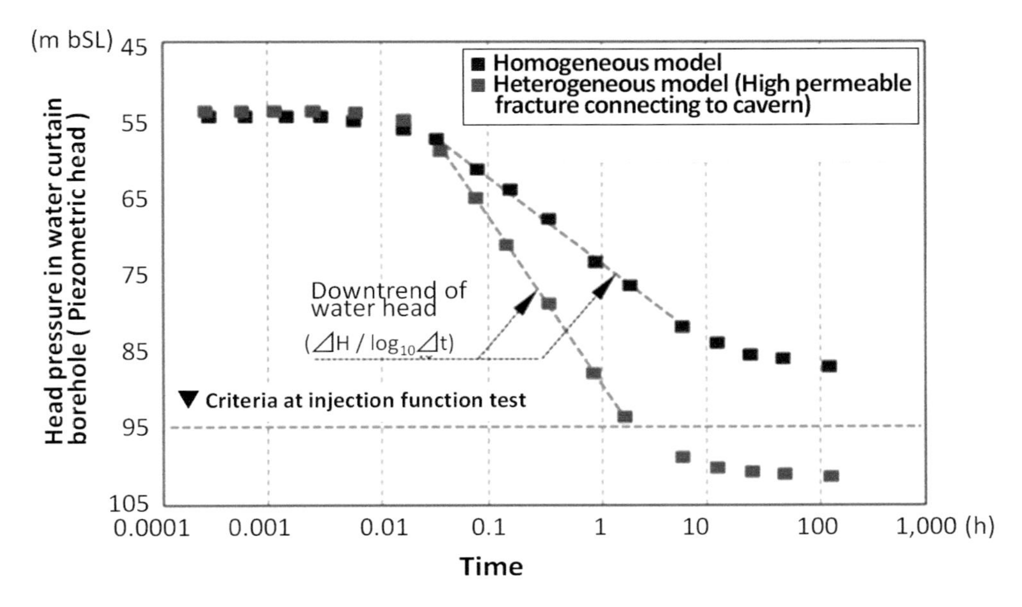

Figure 4.9 Curve of water pressure drop after stopping pressurised water injection by transient analysis in the Namikata facility.

the grout, the pore water pressure in the surrounding areas and the seepage rate in the cavern were reanalysed, and post-grouting and water curtain boreholes were added.

4.1.2.3 Evaluation of Functionality of the Water Curtain System

The hydrogeological characteristics were estimated by an integrated prediction using the 3D hydrogeological data collected prior to the excavation of the caverns and the actual measurements of the pore water pressure distribution, its temporal variation and seepage rate along the progress of excavation (Figure 4.10). In particular, the model estimated highly permeable zones, such as faults and crack zones around the caverns, to accurately understand the distribution of the water pressure. The actual measurement of the pore water pressure was compared with the values predicted by the 3D hydrogeological model to interpret the complex behaviouur of groundwater, and the comparison was used to assess the effectiveness of the water curtains and grouting constructed to control groundwater. The model was updated by data from additional water curtain boreholes and post-grouting constructed to improve the prediction of the pore water pressure. The result was used from completion of excavation to the air-tightness test, which is the final examination of the functionality of the storage caverns.

The variation in the groundwater during excavation is controlled based on the values predicted by the 3D hydrogeological model which reflect the complex heterogeneous hydrogeological structure. However, the values predicted by a homogeneous model were referred to in the early stage of excavation of the storage arch when the accumulation of the hydrogeological data was still poor.

Figure 4.11 shows an example of a comparison of the drop in water pressure predicted by the 3D hydrogeological model and by the homogeneous model with the actual measurement

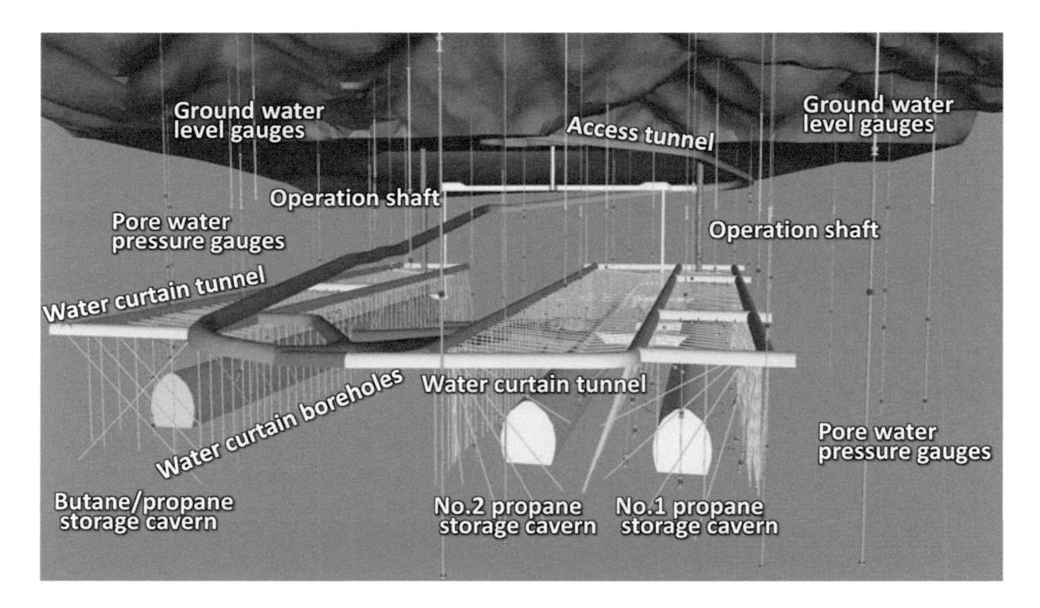

Figure 4.10 Perspective view of the groundwater monitoring system in the Namikata facility.

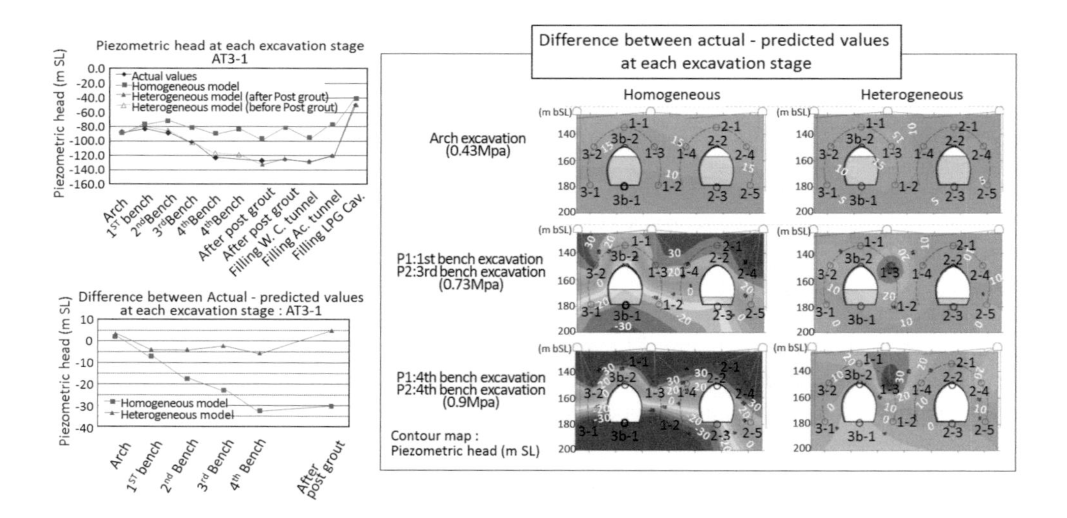

Figure 4.11 Example of a comparison of the predicted pore water pressure and the actual measurement from the excavation of a storage cavern and an air-tightness test in the Namikata facility.

along the progress of excavation. At the stage with the small cross section of excavation, the predicted values by the homogeneous model and the 3D hydrogeological model were similar with small differences from the actual measurement. This suggested that there was little problem in using the homogeneous model for evaluation at arch excavation which was an early stage of cavern construction as the influence of the heterogeneous rocks around the cavern was relatively small. As the bench excavation progressed and the cross section of the cavern excavation was extended, the prediction by the homogeneous model did not drop as much as the actual measurement and the difference gradually grew larger, while the prediction by the 3D hydrogeological model showed a similar drop to the actual measurement. With the accumulation of hydrogeological data, the 3D hydrogeological model was necessary in the area of a complex heterogeneous hydrogeological structure such as Kurashiki and Namikata. In particular, the groundwater behaviouur was compared with the prediction of the pore water pressure distribution, its temporal variation and seepage rate after the bench.

4.1.3 Confirmation of Effect of Water Curtain

4.1.3.1 Case of the Kurashiki Facility

Elements of the hydrogeological structure at the Kurashiki site include highly permeable zones around faults and microfractures; less permeable zones on the upper side of the faults; and water barriers of fault clays. The variation in the pore water pressure in response to excavation was measured for this complex heterogeneous hydrogeological characteristic and compared with the values predicted by the 3D hydrogeological model. Then, the water sealing functionality was assessed using the functionality test of the water injection boreholes. Based on this information, more water curtain boreholes were added to form effective water curtains. Figures 4.12 and 4.13 show the final arrangement of the water curtain boreholes. The management standard was satisfied at the functionality test of all the water injection boreholes of this arrangement (Figure 4.14). The pore water pressure measured was close to the values predicted by the 3D hydrogeological model (Figure 4.15). These confirmed that the water curtains were formed reliably and performed their water sealing functionality.

From the variation in the measured values of the pore water pressure during cavern excavation, the effectiveness of the water curtains was checked for two parts: above and below 140 m below sea level. Figure 4.16 shows the arrangement of the pore water pressure gauges

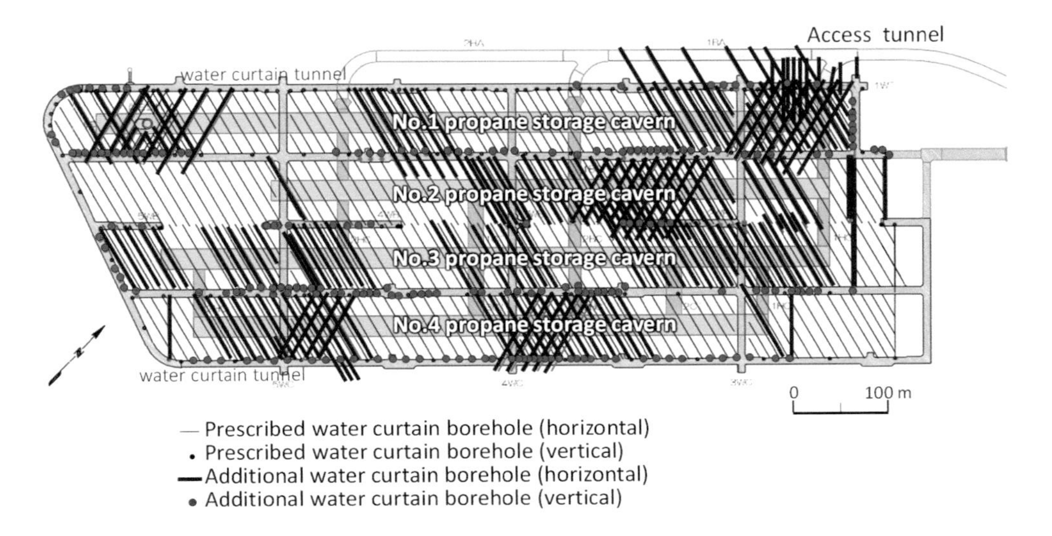

Figure 4.12 Plan view of the arrangement of the water curtain boreholes in the Kurashiki facility.

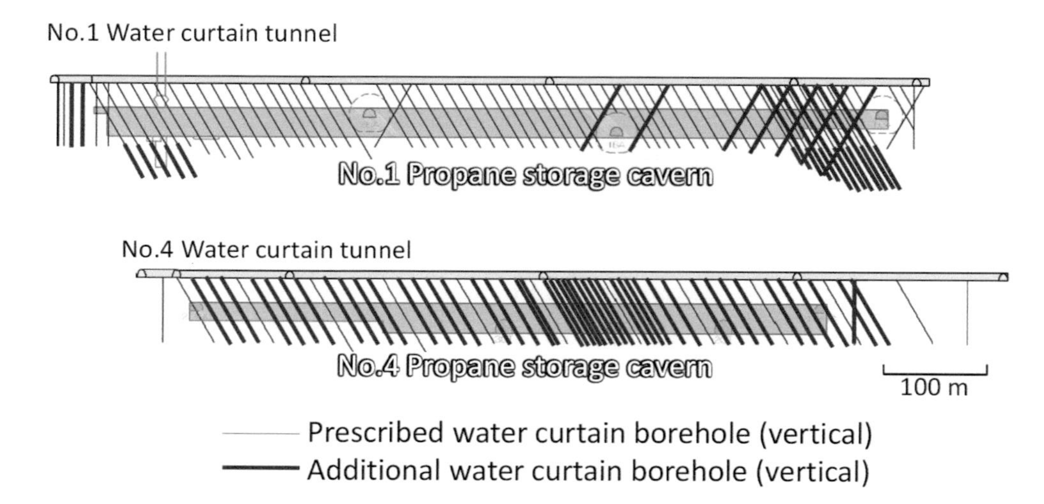

Figure 4.13 Cross section of the arrangement of the water curtain boreholes in the Kurashiki facility.

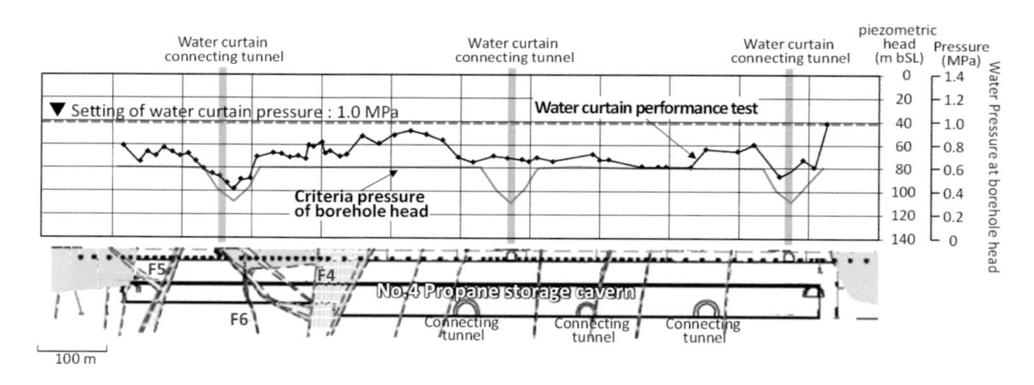

Figure 4.14 Comparison between the water pressure of a water curtain borehole on a functionality test of a water injection borehole after completion of excavation and the control threshold from a horizontal water curtain borehole on top of storage cavern No. 4 in the Kurashiki facility.

in the weathered granite zone at 100 m below sea level. The temporal variation in the pore water pressure measured above the water curtains during cavern excavation is shown in Figure 4.17. The decreasing tendency of the pore water pressure during excavation of the water curtain tunnels is shown. This tendency eases at the start of pressurised water injection through the water curtain boreholes. After the start of cavern excavation, the water pressure hardly dropped and the influence of the cavern excavation was not recognised by the measurement. These measurement data had the same trend as the values predicted by the 3D hydrogeological model (Figure 4.18), which confirmed that effective water curtains were formed.

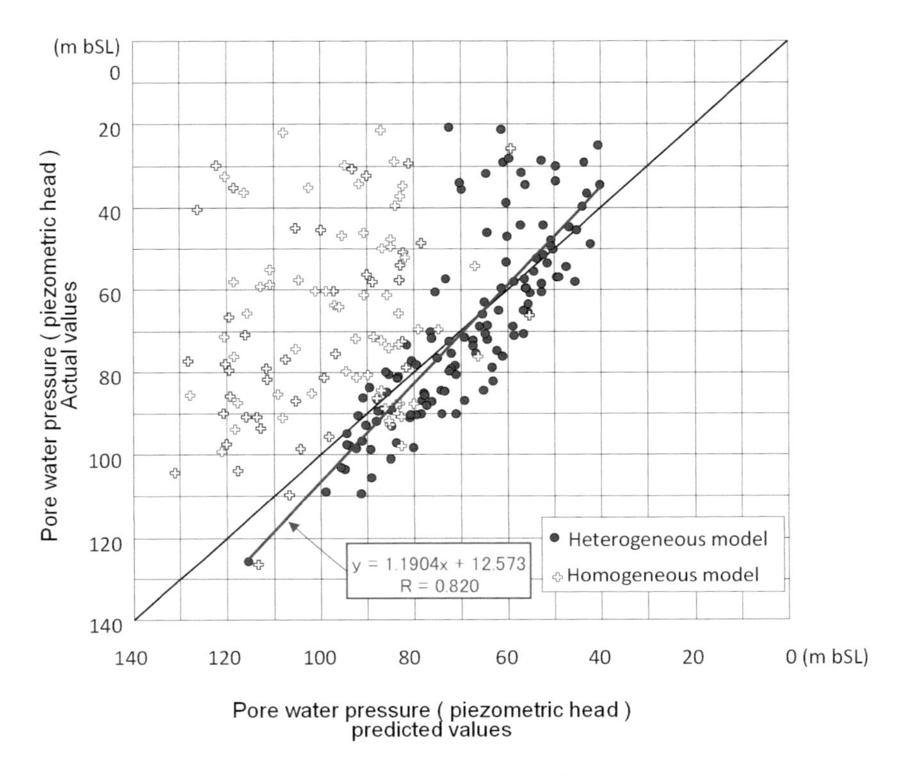

Figure 4.15 Comparison between the measured and predicted pore water pressure on completion of cavern excavation at the Kurashiki facility.

Figure 4.19 shows the temporal variation in the pore water pressure measured at the top of the caverns (148 m below sea level), which is lower than the water curtains. These measurements showed lowering tendencies as cavern excavation progressed. However, as seen in Figure 4.15, these measurements agree with the values predicted by the 3D hydrogeological model incorporating the effect of grouting and the water curtains. This confirmed the validity of the model.

In the Kurashiki facility, the water curtain boreholes were added in areas dominated by highly permeable cracks around the faults and microfracture zones. The increase in the pore water pressure by additional water curtain boreholes is shown in Figure 4.20 in contour lines. An increase in the pore water pressure up to 40 m was observed, demonstrating their effectiveness in water sealing.

The following sections describe the performance of additional water curtain boreholes in the three zones around faults and with microfractures (white polygons in Figure 4.21).

4.1.3.1.1 AROUND STORAGE CAVERN NO. 1 NEAR F-2 FAULT

Figure 4.22 shows the shut-in pressure and the measured pore water pressure of all the horizontal water curtain boreholes on the top of storage cavern No. 1 at the water injection functionality test and the measurements of the pore water pressure gauges at the top of the

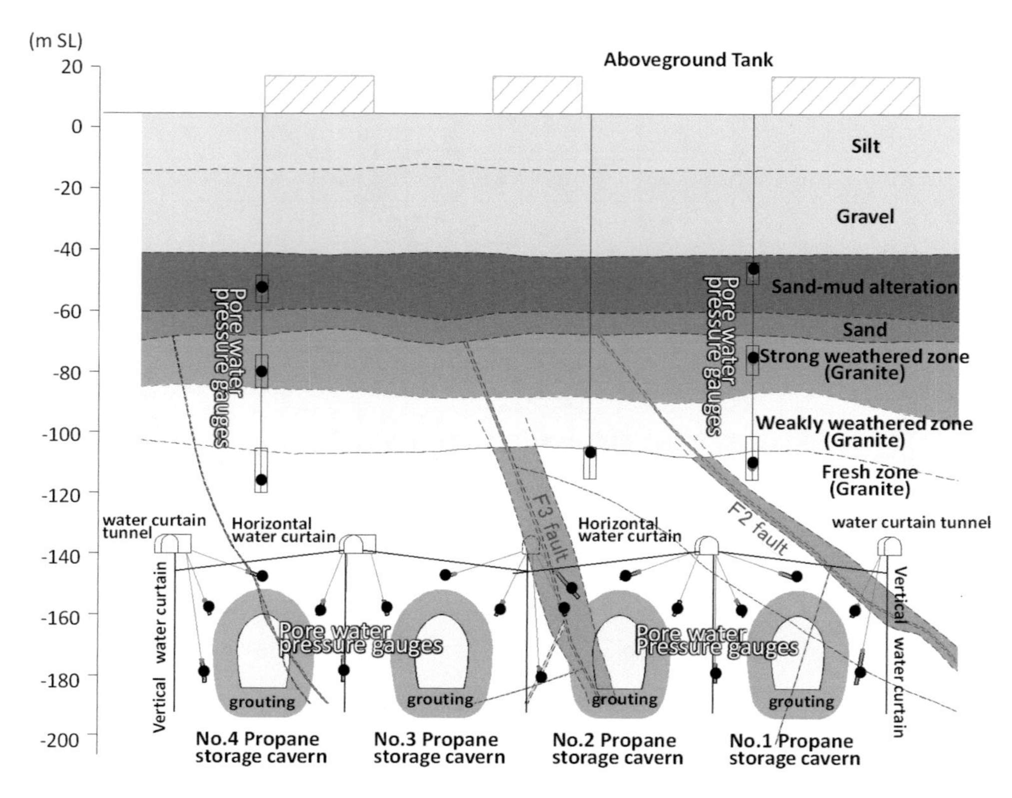

Figure 4.16 Arrangement of the pore water pressure gauge in the weathered granite zone in the Kurashiki facility.

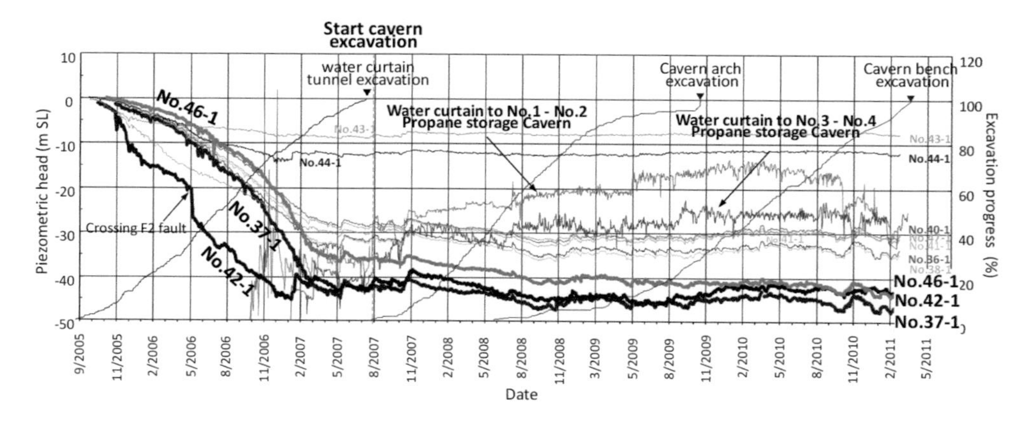

Figure 4.17 Pore water pressure expressed in piezometric head measured in the weathered granite zone (100 m bSL) from the excavation of a water curtain tunnel to cavern excavation in the Kurashiki facility.

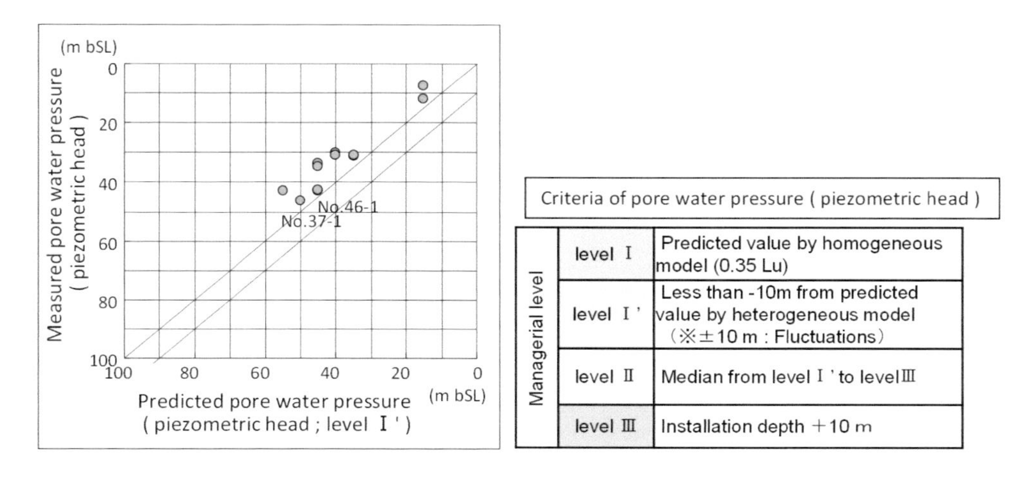

Figure 4.18 Comparison of the measured and predicted pore water pressure in the weathered granite zone in the Kurashiki facility.

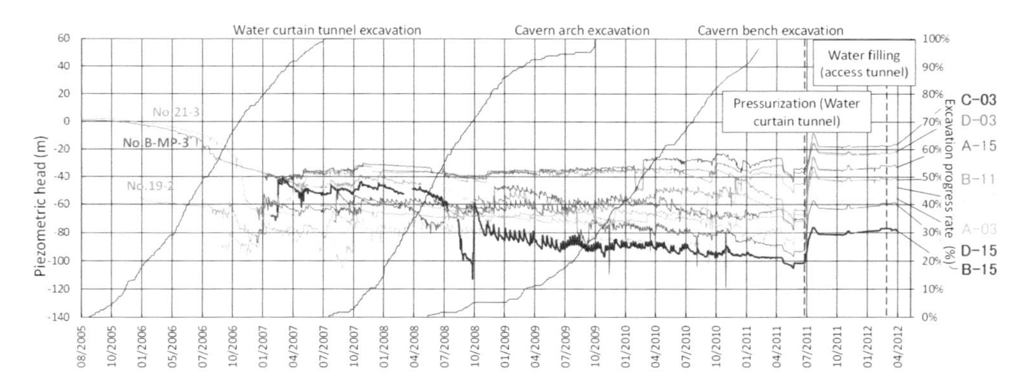

Figure 4.19 Pore water pressure expressed in piezometric head at the top of caverns (148 m bSL) measured during excavation at the Kurashiki facility.

storage caverns. The water injection functionality test in this figure was carried out when all the bench excavation was completed. The water curtain boreholes were drilled immediately after excavation of the storage caverns. Figure 4.22 includes the water pressure before and after the addition of water curtain boreholes and its assessment criterion. The pressure at the water injection borehole functionality test around the hanging wall of the F-2 fault ((1) in Figure 4.211) was lower than the management standard before drilling the additional water curtain boreholes, and its measured pore water pressure, A″-3 and A-3, also showed lower values.

Figure 4.23 shows the geology and distribution of hydraulic conductivity along section A (see Figure 4.21) near the area with low pressure at the water injection functionality test and low pore water pressure. This section is situated in the hanging wall of the highly permeable

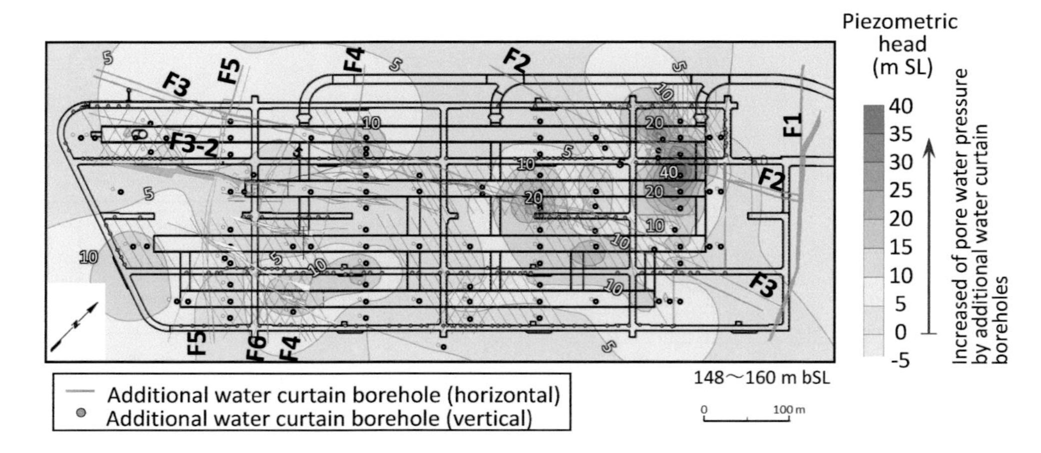

Figure 4.20 Contour map of the increase in pore water pressure expressed in piezometric head by additional water curtain boreholes in the Kurashiki facility.

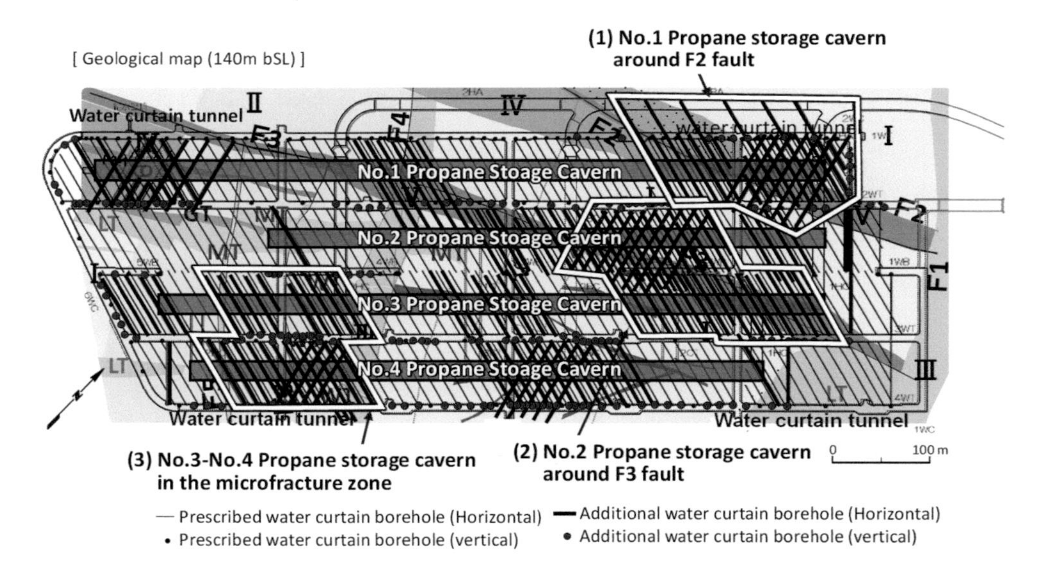

Figure 4.21 Locations of the additional water curtain boreholes in the cases mentioned in the Kurashiki facility.

F-2 fault. Along the section, hydraulic conductivity was low near the top of the cavern away from the F-2 fault. This was considered to be due to an insufficient water supply from the injection to the area near the F-2 fault. In order to enhance the supply of sealing water to the hanging wall of the F-2 fault, additional horizontal water curtain boreholes were drilled crossing the fault and the vertical water curtain boreholes were extended to cross the fault. As seen in Figure 4.24, the pore water pressure of A-1 increased by 26 m in the piezometric head by extending the vertical water curtain boreholes. The additional horizontal water curtain

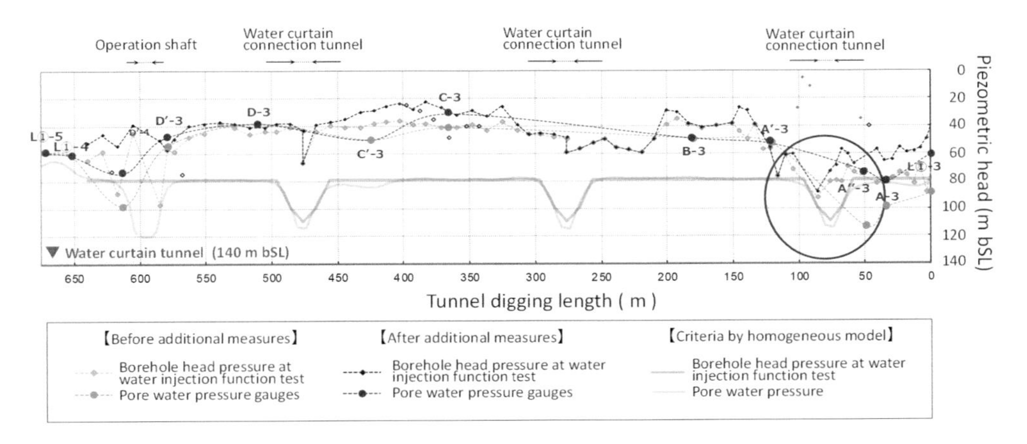

Figure 4.22 Borehole pressure on the water injection borehole functionality test in the horizontal water curtain borehole of storage cavern No. 1 of the Kurashiki facility (on completion of cavern excavation).

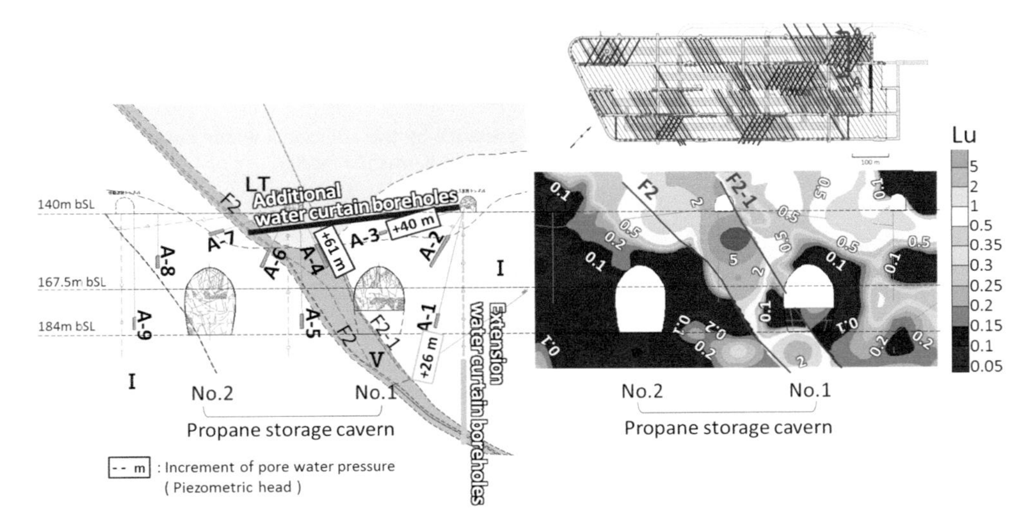

Figure 4.23 Geology and distribution of the Lugeon value along section A near storage cavern No. 1 of the Kurashiki facility.

boreholes increased the pore water pressure of A-4 by 61 m, A″-3 by 52 m and A-3 by 40 m in the piezometric head. This satisfied the management standard at the repeated water injection functionality test. Hence, effective water curtains were considered to be formed against the F-2 fault.

4.1.3.1.2 AROUND STORAGE CAVERN NO. 2 NEAR F-3 FAULT

The area around the F-3 fault near storage cavern No. 2 showed low water pressure at the water injection functionality test and had low pore water pressure. As steeply dipping cracks

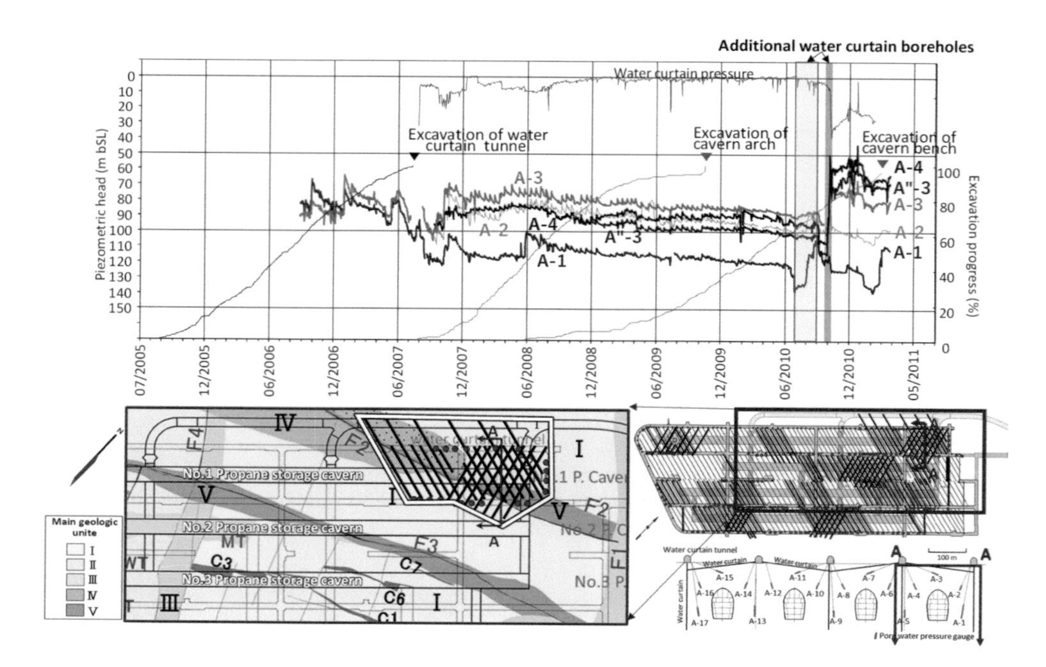

Figure 4.24 Temporal variation of the pore water pressure by the additional water curtain boreholes along section A of storage cavern No. 1 at the Kurashiki facility.

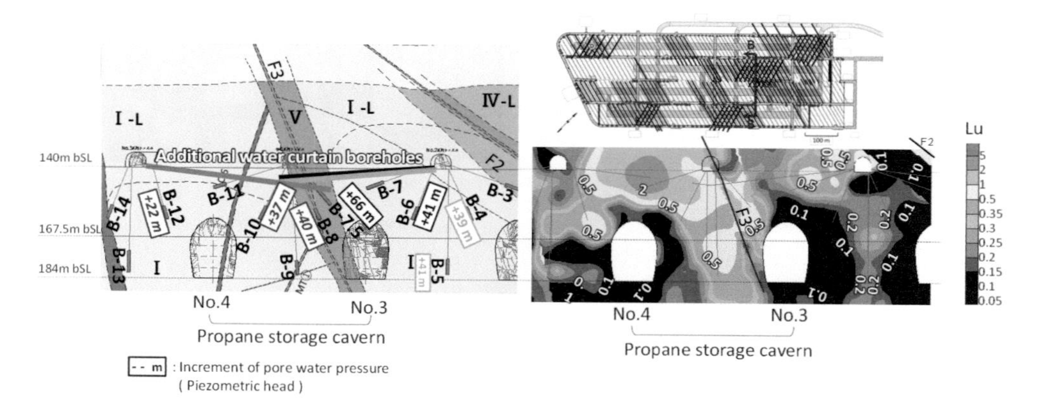

Figure 4.25 Geology and distribution of the Lugeon value along section B near storage cavern No. 2 of the Kurashiki facility.

were dominant around the F-3 fault, additional horizontal water curtain boreholes were drilled from water curtain tunnels No. 2 and 3 into both the foot wall and the hanging wall of the F-3 fault (Figure 4.25). With this additional drilling from water curtain tunnel No. 2 (Figure 4.26), the pore water pressure increased by 66 m in terms of the piezometric head at B-7.5 and 35 m at B-8. After this, the pore water pressure at B-7.5, B-8 and B-12 dropped during excavation of the cavern bench. Additional horizontal water curtain boreholes were

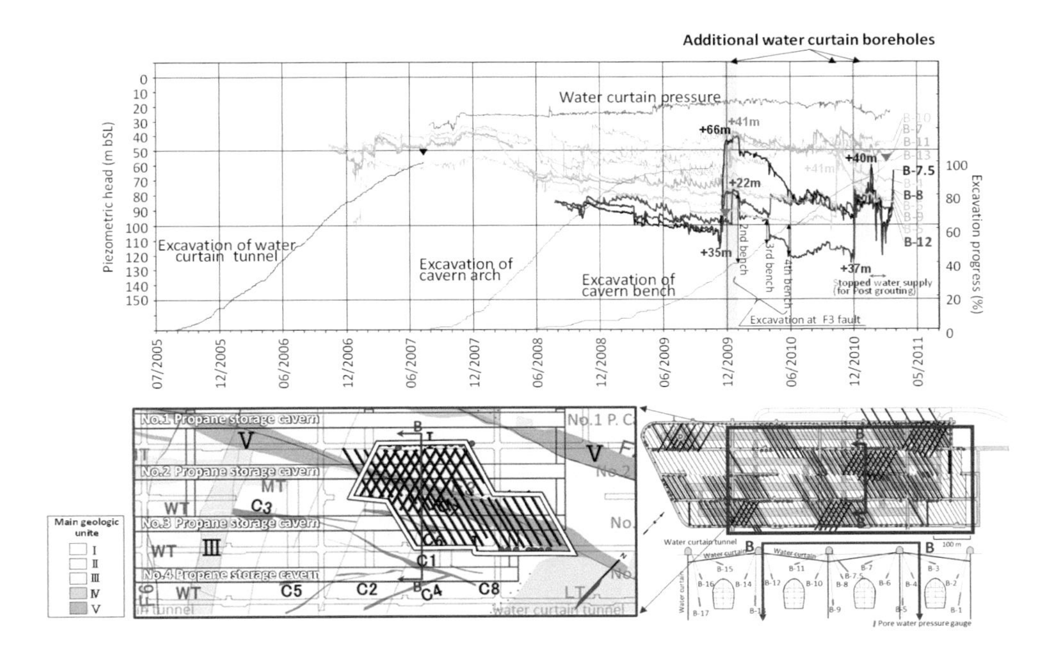

Figure 4.26 Temporal variation of the pore water pressure by the additional water curtain boreholes along section B of storage cavern No. 2 at the Kurashiki facility.

drilled from water curtain tunnel No. 3 to the foot wall of fault F-3. Then, the water pressure increased by 40 m at B-7.5, 37 m at B-8 and 22 m at B-12 in the piezometric head, satisfying the management standard of the water injection functionality test. Thus, effective water curtains were considered to be formed against the F-3 fault.

4.1.3.1.3 MICROFRACTURE ZONE NEAR STORAGE CAVERNS NO. 3 AND 4

The microfractures around storage caverns No. 3 and 4, shown as (3) in Figure 4.21, have a steep dip and cross the cavern axis at an acute angle. Here, the measured pore water pressure presented low values, and horizontal water curtain boreholes were added from water curtain tunnel No. 3 (Figure 4.27). This increased the piezometric head of storage cavern No. 3 by 40 m at C′-11 and 37 m at C″-11 and in storage cavern No. 4 by 20 m at C″-15 and 25 m at C-16. These increases are shown in Figure 4.28. As seen, the piezometric head (or pore water pressure) around the microfractures increased by adding water curtain boreholes, which satisfied the management standard of the water injection functionality test. Thus, it was concluded that effective water curtains were formed here.

4.1.3.2 Case of the Namikata Facility

The hydrogeological characteristics of the Namikata site include that hydraulic conductivity is generally low; there are geological subdivisions dominated by high-angle cracks; and there are continuous cracks providing flow passage to groundwater. Under this complex

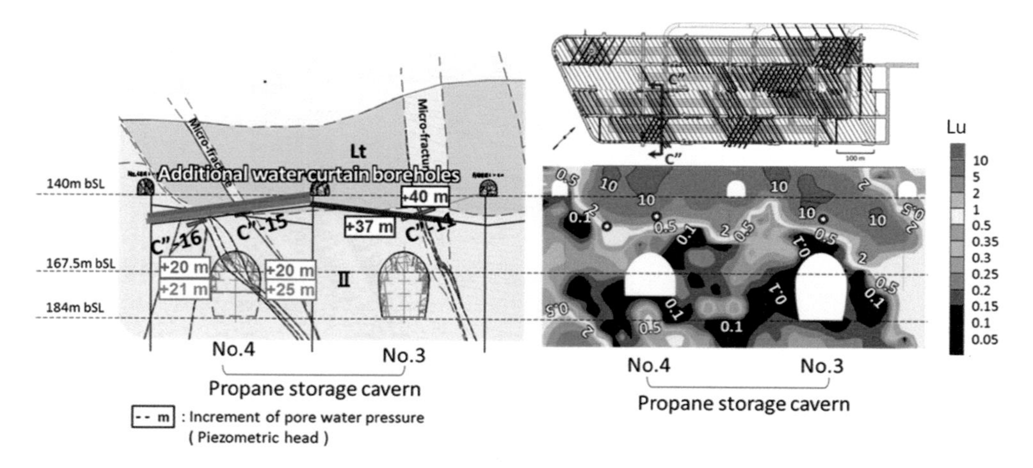

Figure 4.27 Geology and distribution of the Lugeon value along section C near storage cavern No. 3 and 4 of the Kurashiki facility.

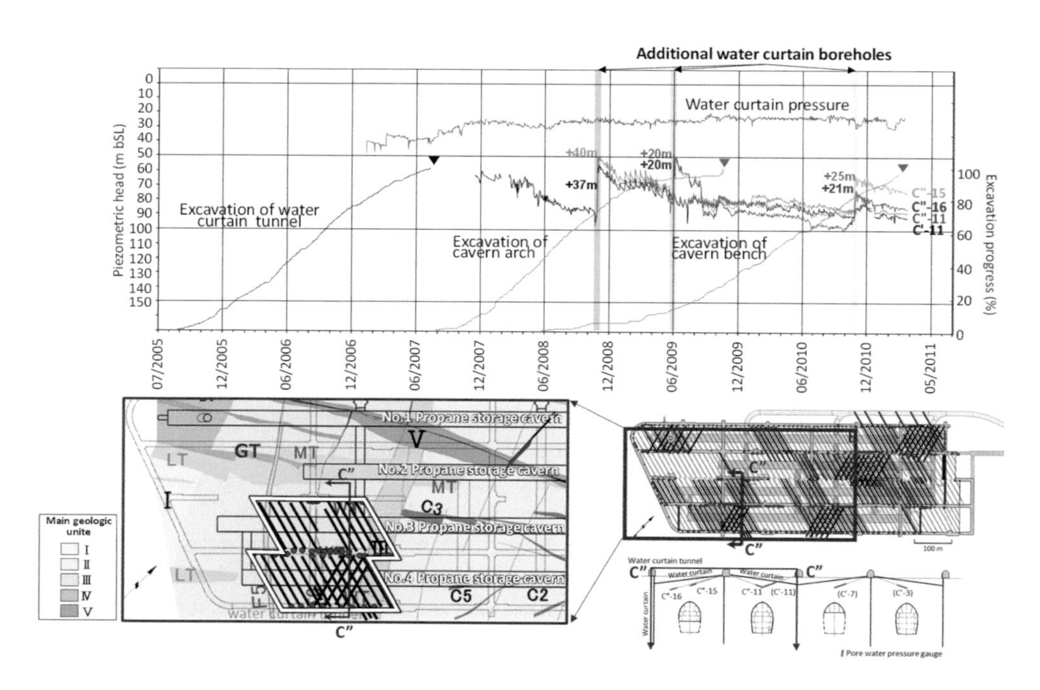

Figure 4.28 Temporal variation of the pore water pressure by the additional water curtain boreholes along section C of storage cavern No. 3 and 4 at the Kurashiki facility.

heterogeneous hydrogeological condition, water-tightness functionality was judged from the distribution and temporal variation in the measured pore water pressure during cavern excavation compared with the prediction by 3D hydrogeological model and the water injection functionality tests. According to the result, additional water curtain boreholes were drilled to construct water curtains. The final arrangement of the water curtain boreholes is shown in Figures 4.29 and 4.30. Figure 4.31 shows that all these boreholes satisfied the management standard at the water injection functionality tests after completion of the cavern excavation. The assessment of the pore water pressure, as seen in Figure 4.32, also presented values similar to or better than the prediction by the 3D hydrogeological model. These assessments confirmed the formation of effective water curtains and good water-tightness functionality.

The functionality of the water curtains was examined for two parts: above and below the water curtain situated 125 m below sea level. Figure 4.33 shows the temporal variation in the pore water pressure above the water curtains measured during the excavation of the storage cavern. The pore water pressure generally decreased during excavation of the access tunnels and water curtain tunnels. This tendency stopped as water injection commenced. When excavation of the storage caverns started, the pore water pressure hardly decreased, although it had minor fluctuations associated with the change in the water pressure of the water curtain boreholes. Therefore, the influence of the excavation of the storage caverns was hardly observed, and the water curtains were assessed effective.

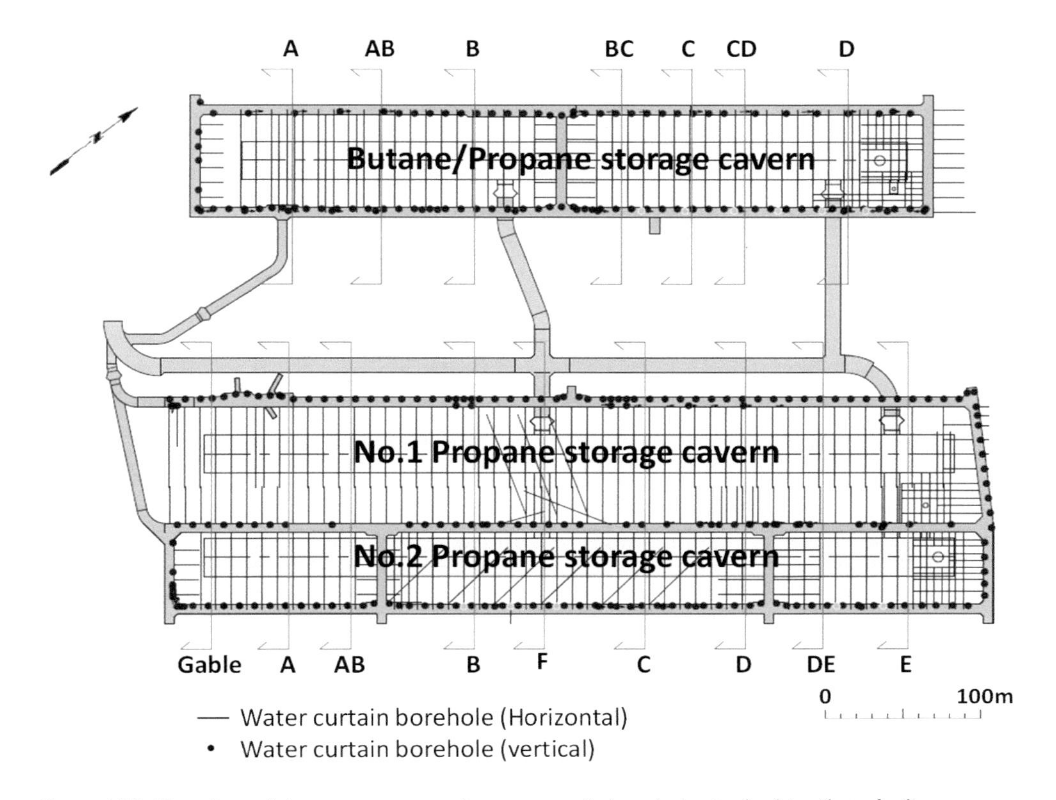

Figure 4.29 Plan view of the arrangement of water curtain boreholes in the Namikata facility.

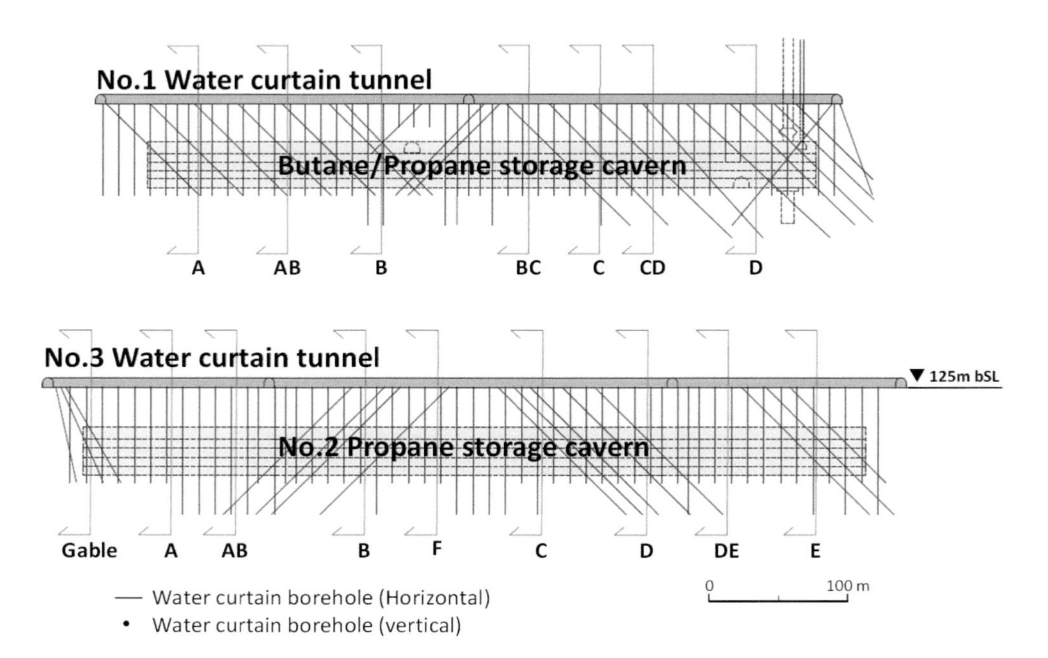

Figure 4.30 Cross section of the arrangement of water curtain boreholes in the Namikata facility.

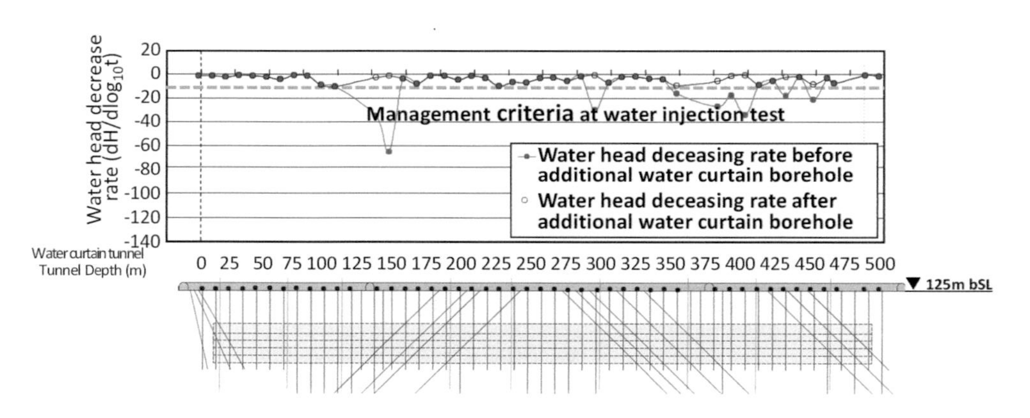

Figure 4.31 Comparison between the water pressure of a water curtain borehole at the water injection functionality test after completion of storage cavern excavation and judgement criteria value (on the horizontal water curtain borehole on top of propane storage No. 1) in the Namikata facility.

Figure 4.34 shows the temporal variation in the pore water pressure below the water curtains measured at the top of the storage cavern (135 m below sea level). The pressure had a decreasing tendency with the progress of the excavation of the storage caverns. This corresponds to the values predicted by the 3D hydrogeological model which incorporated the effects of grouting and the water curtains, as seen in Figure 4.32. This suggested the validity of the model.

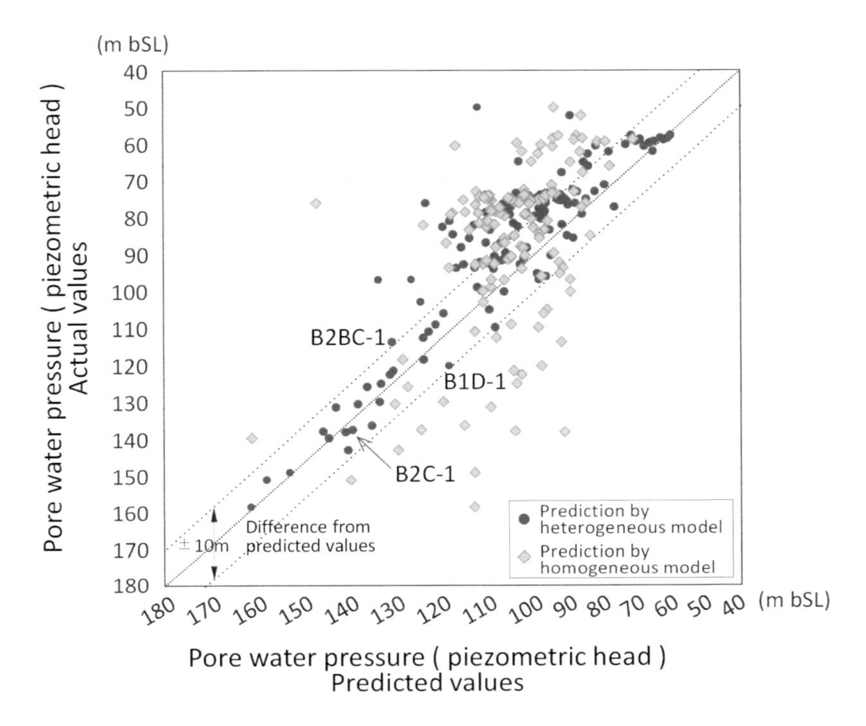

Figure 4.32 Comparison between the measured and predicted pore water pressure on completion of the excavation of the storage cavern in the Namikata facility.

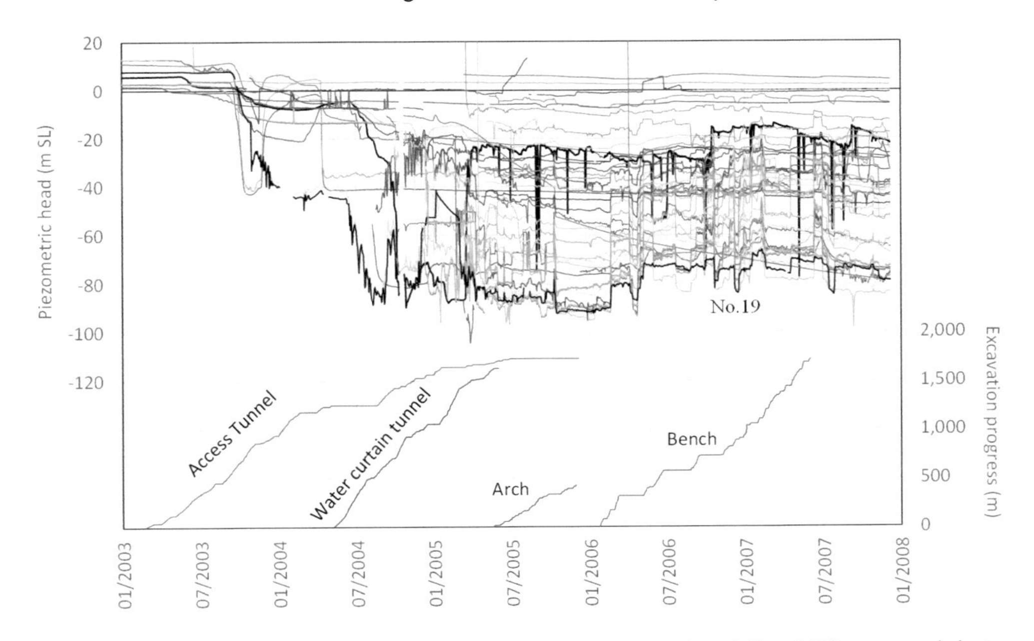

Figure 4.33 Pore water pressure at the top of caverns (shallower than 105 m bSL) measured during the excavation of the water curtain tunnel and storage caverns in the Namikata facility.

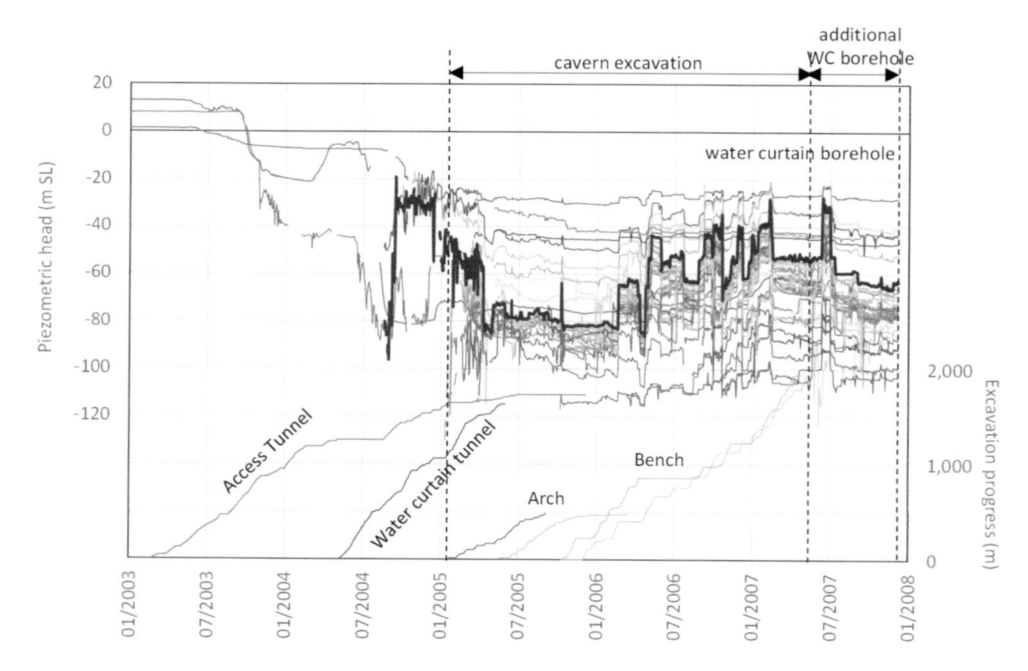

Figure 4.34 Pore water pressure at the top of a propane storage cavern (at 135 m bSL) measured during the excavation of the water curtain tunnel and storage cavern in the Namikata facility.

The geology of the Namikata site was classified into seven geological subdivisions (I–VII) by dominant orientation and the density of cracks found by investigating the water curtain boreholes and grouting holes. There were two dominant orientations of the cracks: N70E along the cavern and N60W across the cavern. The subdivisions II and VI were dominated by cracks of the N70E system. These cracks were open in parts and highly permeable. But the pressure in the boreholes at the water injection functionality test satisfied the management standard. On the other hand, the result of the shut-in test did not satisfy the judgement criterion in some zones in subdivision V, where the N60W fault system was dominant and the density of cracks was high. From this, it was apparent that the angle between the orientations of the dominant cracks and the water curtain boreholes influenced the water flow. Accordingly, water curtain boreholes were added in an oblique downward direction so that they would cross the cracks of the N60W system at a high angle to the cavern.

The following sections describe two examples of adding water curtain boreholes in subdivision V in the Namikata facility: for the propane storage cavern and the combined propane/butane storage cavern. Their locations are shown in Figure 4.35.

4.1.3.2.1 PROPANE STORAGE CAVERN

Figure 4.36 shows the result of the water injection functionality test of the vertical water curtain boreholes to the north of propane storage cavern No. 2. The area where the rate of change of the water pressure did not satisfy the management standard (Section 4.1.2.2)

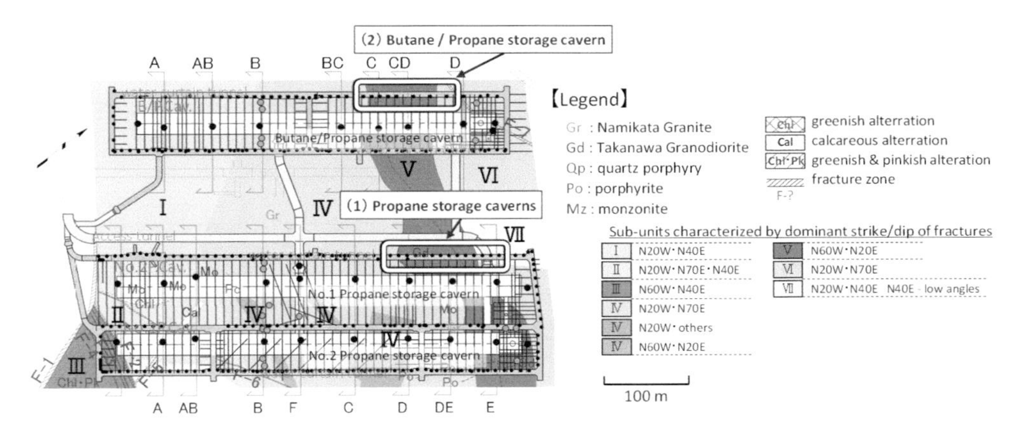

Figure 4.35 Locations of the case history of additional water curtain boreholes in the Namikata facility.

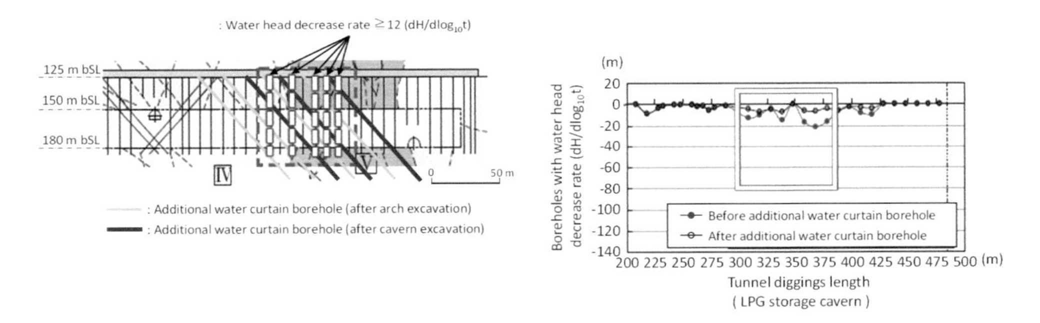

Figure 4.36 Water pressure gradient of a vertical water curtain borehole at the water injection functionality test of propane storage cavern No. 2 at the Namikata facility after excavation of a storage arch.

corresponds to geological subdivision V (Figure 4.35 (1)). Subdivision V is dominated by high-angle cracks across the orientation of the storage cavern. The additional water curtain boreholes were drilled obliquely downward crossing the cracks to enhance the supply of sealing water into the cracks (Figure 4.36). With this improvement, the water injection functionality tests satisfied the management standard, and the effectiveness of the water curtains was ascertained.

4.1.3.2.2 COMBINED PROPANE/BUTANE STORAGE CAVERN

As the geological structure of the combined propane/butane storage cavern was similar to the propane storage caverns, additional water curtain boreholes were drilled mainly in the subdivision V area. Where the rate of water pressure variation was lower than the management standard, additional water curtain boreholes were drilled obliquely downward to cross the cracks of subdivision V (Figure 4.37).

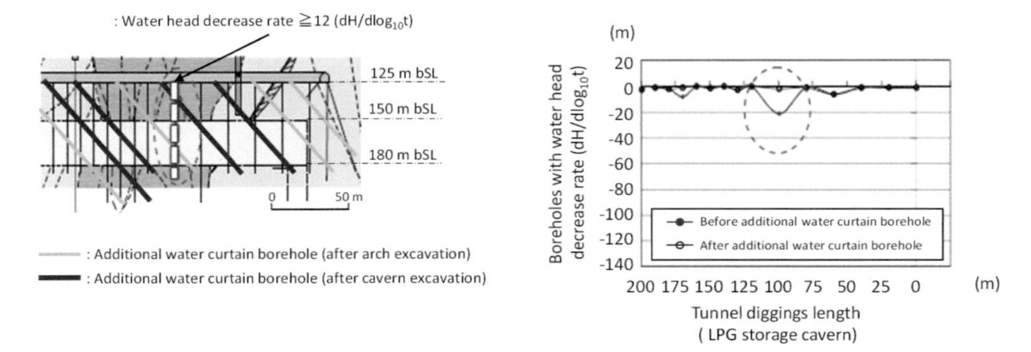

Figure 4.37 Water pressure gradient of a vertical water curtain borehole at the water injection functionality test of the combined propane/butane storage cavern at the Namikata facility after excavation of the storage cavern.

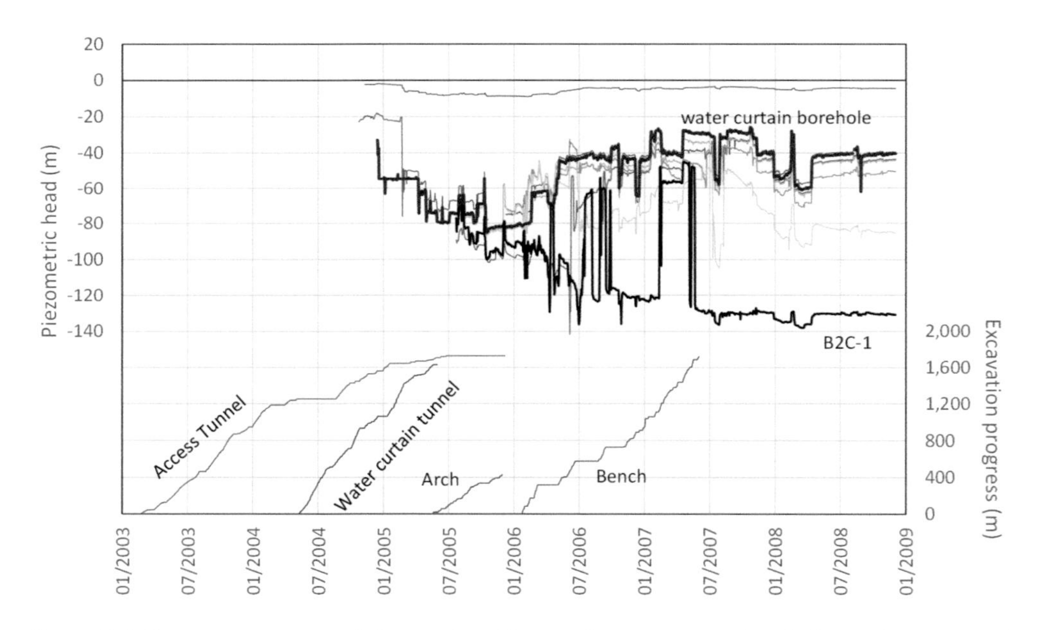

Figure 4.38 Pore water measured during the addition of water curtain boreholes at the combined butane/propane storage cavern of the Namikata facility.

Figure 4.38 shows the temporal variation in the measured pore water pressure along cross section C in association with the excavation of the combined butane/propane storage cavern situated in geological subdivision V. The locations of the measurement instruments are shown on a geological and Lugeon value map (Figure 4.39). This section was in geological subdivision V where high-angle cracks of the B20W and N60W systems cross the cavern at high density. Large drops of water pressure were observed here; in particular, the B2C-1

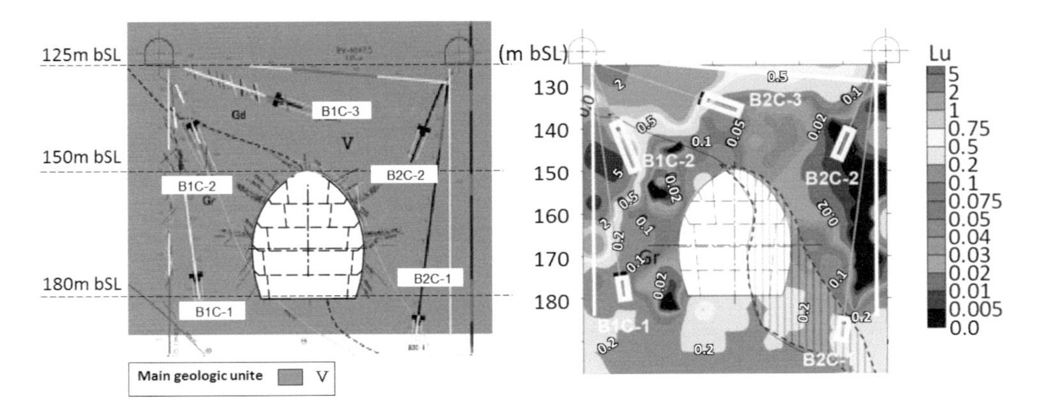

Figure 4.39 Geology and distribution of the Lugeon value around the combined butane/propane storage cavern of the Namikata facility.

instrument recorded a drop as large as 50 m at the time of excavation of the cavern. The water pressure increased by additional oblique water curtain boreholes as described above. It dropped again with the subsequent excavation, but the Lugeon value around the instruments was as low as less than 0.1 Lu, which is within the management standard of the water injection functionality test. In addition, the water pressure increased by adding water curtain boreholes and the behaviour of the pore water pressure agreed with the prediction by the 3D hydrogeological model. From these observations, the effectiveness of the water curtains formed was confirmed.

4.2 Construction of Groundwater Control System (Grouting)

Figure 4.40 shows the distribution of the hydraulic conductivity at the Kurashiki and Namikata facilities before and after improvement by grouting. As the Kurashiki facility has a complex heterogeneous hydrogeological structure, groundwater was injected through the water curtain boreholes on the top and sides of the storage cavern to pressurise it. In addition, grouting was necessary to improve the water-tightness around the cavern to control seepage and a drop in pore water pressure during excavation of the caverns. The design was based on the Lugeon value of the rocks, 0.35 Lu, obtained from the average of the Lugeon test of reconnaissance drilling during planning. However, a higher than average Lugeon value of the actual rocks, 0.53 Lu, was found during surveys and tests at the time of cavern excavation. The area of high hydraulic conductivity was widely distributed in the rocks: 15% of the rocks had a high Lugeon value over 1 Lu, while 10% had Lugeon values lower than 0.01 Lu; and 40% of the rocks had a Lugeon value higher than the initial design value of 0.35 Lu.

In the Namikata site, the design value of the rocks was set to 0.01 Lu based on the average value of the Lugeon test during the reconnaissance drilling survey. However, the surveys and tests at the time of cavern excavation presented Lugeon values higher than the design value: 0.31 Lu on average. The Lugeon values of the rocks in the Namikata site are largely low (less than 0.01 Lu) in 70% of the area but there are local highly permeable zones over 1 Lu. As

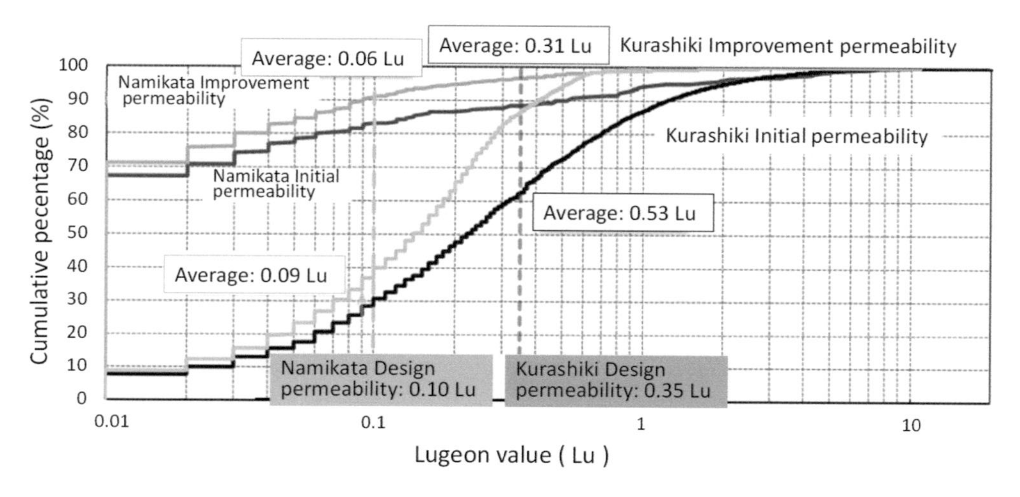

Figure 4.40 Distribution of the Lugeon values of the rock at the Kurashiki and Namikata facilities.

this is higher than the initial design value, grouting was applied around the storage cavern to reduce the average Lugeon value to lower than the design value. This was applied to all cracks with values over 1 Lu.

The functionality of the water-tightness of the rock cavern storage was ensured by the pressure of the water in the cracks around the storage cavern being higher than the cavern's internal pressure. Once unsaturation occurs around a cavern by lowering the pore water pressure caused by seepage, it is difficult to re-saturate it with groundwater. In order to maintain saturation of the rocks around the caverns throughout construction and operation, it is necessary to control seepage by grouting during excavation of the caverns. A pre-grouting system, which allows a high pressure injection, was applied to the areas surrounding the excavation to ensure improved zones and to satisfy the pre-determined target prior to excavation of the cavern. During pre-grouting, grouting boreholes were added until the Lugeon value of the improved zone reached the target value. Next, the storage cavern was excavated, and the quality of the zones improved by grouting around the storage cavern was monitored by measuring the seepage rate. If the seepage rate exceeded the management standard, post-grouting was added to the seep areas until the management standard was satisfied.

The grouting procedure is used to fill and seal the water passage by injecting grouting material, a mixture of cement and water, into the complex network of cracks. The cement particles invading the cracks will precipitate and be deposited in the cracks, reducing the width of the cracks. Eventually, the cracks are clogged, thus sealing is achieved. For an effective grout sealing, it is important to understand the spatial distribution of the cracks and the orientation, continuity, density, width and coarseness of the rocks. Then, a grouting injection is designed to suit the property of the rocks. Grout injection tests were carried out in the water curtain tunnels. Detailed surveys and analyses of the hydrogeological structure and the characteristics of the cracks were also carried out. Thus, the layout of the grouting holes and the injection specification were designed to suit the property of the cracks.

4.2.1 Design and Procedure of Grouting

4.2.1.1 Set Up of the Target Values and Extent of Grouting

In both the Kurashiki and Namikata facilities, during excavation it became clear that the permeability of the rocks around all the storage caverns was higher than the condition initially estimated. This necessitated an improvement in water-tightness by grouting. The target values of the improvement were established for the characteristics of the cracks and groutability from the results of grouting tests in the water curtain tunnels. The variation in the seepage rate and the groundwater pressure was analysed during excavation using the flow of the groundwater predicted by the 3D hydrogeological model. The model considered the target and the extent of grouting improvement on the complex heterogeneous hydrogeological structure of these sites.

The water-tightness was examined by the predicted value of the air-tightness test and the final test on completion of the cavern construction. These tests confirmed the applicability of the target value of grout improvement.

The extent of grouting must include the zones affected by excavation: relaxation by excavation, movement of opening of cracks and change in permeability caused by the release of stress. Throughout the excavation period, the extent of grouting improvement was established to include the outer perimeter of the zones directly affected by excavation to ensure the stability of the cavern against surrounding water pressure, to suppress the seepage into the cavern and to maintain the pore water pressure.

4.2.1.1.1 THE KURASHIKI FACILITY

4.2.1.1.1.1 Target Value for Improvement by Grouting Before discussing the target value and the extent of the grouting improvement, the hydrogeological characteristics of the Kurashiki facility area are outlined. Here, sediments of Tertiary Pleistocene to Quaternary Holocene are present down to around 70 m below sea level underlain by the Cretaceous granite. The top of the granite is well weathered and the fresh granites are mostly deeper than 120 m below sea level. The rock cavern storages were arranged between 160 and 184 m below sea level. As the fresh granite cover in this area is thin, around 40 m, the rocks around the cavern developed cracks, and highly permeable faults and cracks were also present, providing groundwater passage from the weathered layer.

Figure 4.41 shows the distribution of the Lugeon values of the bedrocks before grouting on the geological map at the level of the base of the storage arch. In the Kurashiki facility area, the hydrogeological features were classified into five subdivisions (I–V) and a microfracture zone in respect of the dominant orientation and the density of cracks. Subdivision I, the most common subdivision, is dominated by cracks in the NNW-SSE orientation, the density is about 0.5/m and its greatest Lugeon value is 4.3 Lu (average 0.3 Lu). Subdivisions II and III also have cracks predominantly in the NNW-SSE orientation, the density is about 1–2/m and its greatest Lugeon value is 8.4 Lu (average 0.5 Lu). Subdivision IV has cracks predominantly in the ENE-WSW orientation, the density is about 1/m and its greatest Lugeon value is 2.1 Lu (average 0.3 Lu). Subdivision V has cracks along the F-2 and F-3 fault systems, the density is about 2/m and its greatest Lugeon value is 13 Lu (average 0.9 Lu). The microfracture zone is an area with densely distributed fine thin cracks with very small openings, and its greatest Lugeon value is 3.3 Lu (average 0.6 Lu).

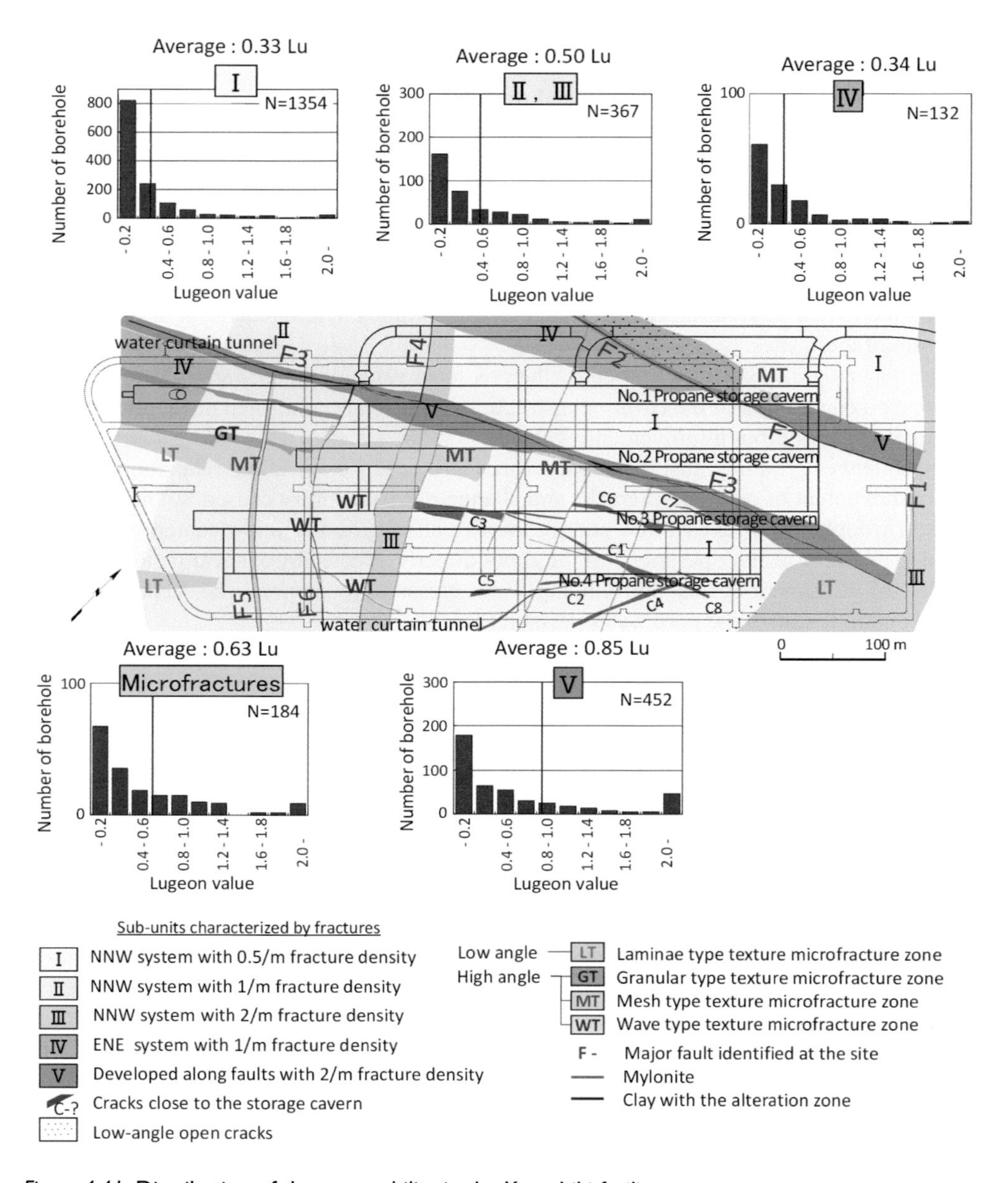

Figure 4.41 Distribution of the permeability in the Kurashiki facility area.

For such subdivisions based on the characteristics of the cracks and permeability, the grouting was designed to improve the Lugeon values to less than 0.35 Lu on average, the specification. Grouting was applied to all areas with high permeability with values higher than 1 Lu, which may provide a passage to the groundwater. Grouting tests were carried out on each subdivision in the water curtain tunnels prior to excavation of the storage cavern. The grouting procedures were designed according to the tested groutability which varied depending on the characteristics of the cracks of the bedrock. On the analysis of groutability from the grouting tests in the water curtain tunnels, the bedrock around the cavern was classified into four zones: F-1 and F-2 crack system zones; the mylonite zone around faults; and the microfractured zone (Figure 4.42). In the F-1 and F-2 crack system zones, which correspond to geological subdivisions I–IV, the permeability dropped as additional grouting boreholes were drilled. The groutability of these zones was regarded as good. The mylonite zone along the F2 and F3 faults, which correspond to the geological subdivision V, needed a relatively large number of grouting boreholes, while the permeability dropped when grouting boreholes were added. On the other hand, the microfractured zone showed little improvement with additional grouting boreholes, necessitating a large number of additional grouting boreholes.

Target values for grout improvement at the Kurashiki facility were set as shown in Figure 4.43 from the characteristics of the cracks and groutability. On the side walls in the general bedrock, the target values were 0.3 Lu for outside and 0.15 Lu for inside; on the side walls in the faulted bedrock, the target values were 0.65 Lu for outside and 0.25 Lu for inside; and on the side walls in the microfracture zone, the target values were 0.65 Lu for outside and

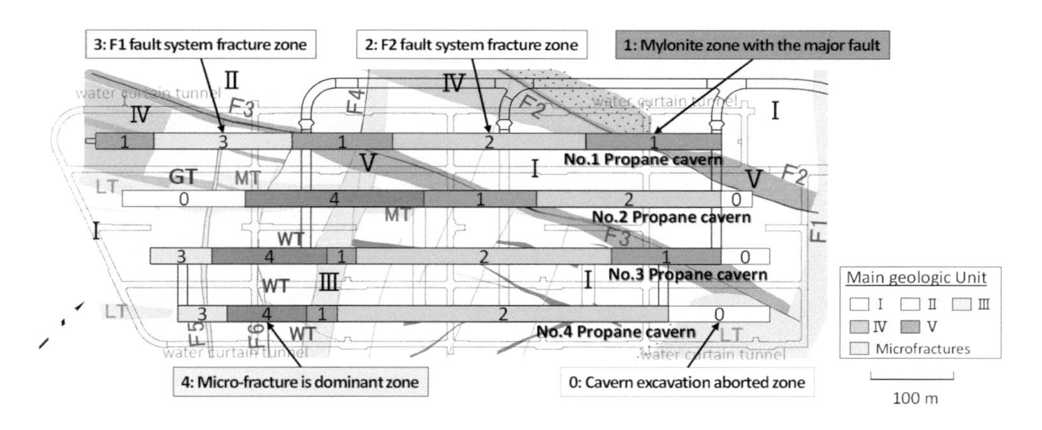

Figure 4.42 Groutability zones of the Kurashiki facility.

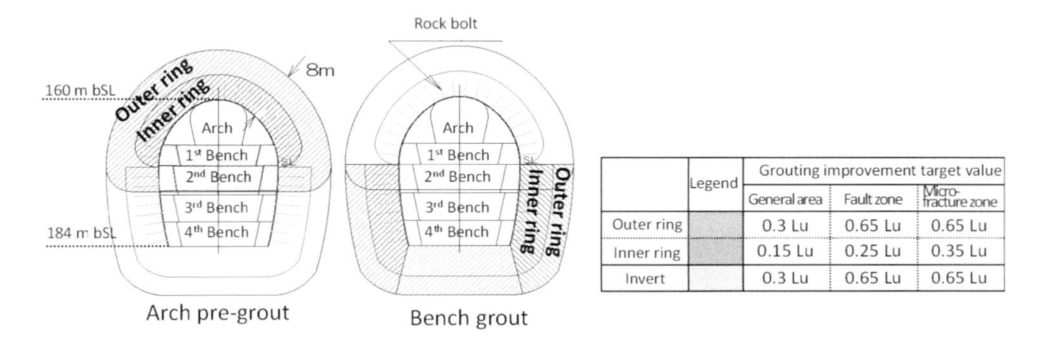

	Legend	Grouting improvement target value		
		General area	Fault zone	Micro-fracture zone
Outer ring		0.3 Lu	0.65 Lu	0.65 Lu
Inner ring		0.15 Lu	0.25 Lu	0.35 Lu
Invert		0.3 Lu	0.65 Lu	0.65 Lu

Figure 4.43 Target Lugeon values by grouting improvement at the Kurashiki facility.

0.35 Lu for inside. While the floors of benches 1 and 3 were the site for subsequent excavation, improvement by grouting was carried out to prevent excessive seepage, a drop in the pore water pressure and the occurrence of unsaturated zones. The target Lugeon values of these floors were similar to the outside of walls: 0.3 Lu for the general bedrock and 0.65 Lu for the fault and microfracture zones.

For the fault and microfracture zones with low groutability, the applicability of the target of improvement by grouting was confirmed in terms of air-tightness by air flow tests in the boreholes.

To ensure that improved zones were formed, the grouting improvement was applied to two rings: inner and outer (Figure 4.43). The grouting boreholes were first arranged in a fan shape to form the outer improved ring in the direction of the excavation. The other fan-shaped layout of grouting boreholes was added overlapping the outer ring to form the inner ring. This procedure allowed high pressure injection to the inner ring as the previously improved outer ring prevented leakage. This protection zone is called a "cover rock". The outer ring acted as a "cover rock" resulting in an improvement in efficiency.

Figure 4.44 shows the result of grouting the water curtain tunnels by correlating the Lugeon values inside and outside of the tunnels. The Lugeon value of the common zone of geological subdivision I was 0.3 Lu outside and 0.15 Lu inside. In the faulted subdivisions IV and V, the Lugeon value was 0.65 Lu outside and 0.25 Lu inside. From this result,

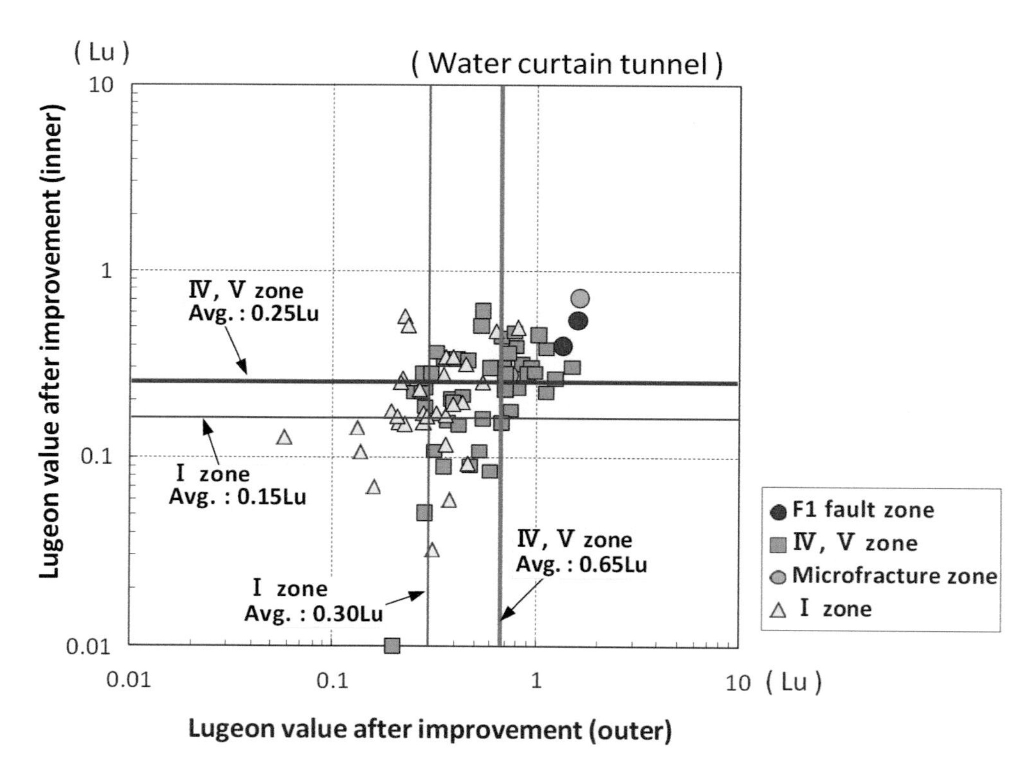

Figure 4.44 Result of an improvement by a two-ring grouting system in the water curtain tunnels in the Kurashiki facility.

the improvement using the two-ring grouting system was clear and it was reflected in the target of improved values.

4.2.1.1.1.2 Validation of the Target Value of Grouting Improvement The validity of the target values of grouting improvement was examined by the air-flow test of the boreholes in order to confirm the air-tightness after grouting in the zones of faults and microfractures because these are the zones of low groutability in the Kurashiki facility. The borehole gas permeability test analyses the distribution of the pore water pressure around the air chambers near the critical air-tightness while pressurising by injecting compressed air. The air chamber is a section of a borehole isolated by packers and it simulates a cavern. Three observation boreholes were drilled 1.5 m from the surrounding borehole with air chambers to measure the pore water pressure (Figure 4.45).

Figure 4.46 shows the hydrogeological data from borehole gas permeability tests of the zones affected by faults. From the observation of the core and BTV data at the test hole, a two-meter-long part with a concentration of fractures was identified. This part used an air chamber to apply pressure in the pressure test. The response to this pressure was observed in three observation holes during the test. In these observation holes, a pressure response test was run first to select highly responsive segment. The variation of pore water pressure responding to the pressure applied in the test hole was measured. From this measurement, the effectiveness of grouting was estimated. The Lugeon value at the borehole gas permeability test after grouting was 0.25 Lu in the fault-affected zone, which was close to the target values of improvement.

Figure 4.47 shows a schematic diagram of the borehole gas permeability test. This test measures the change in the internal pressure, ΔP, after stopping the stepwise air injection while the water level is kept constant. From this change, it evaluates air-tightness by finding

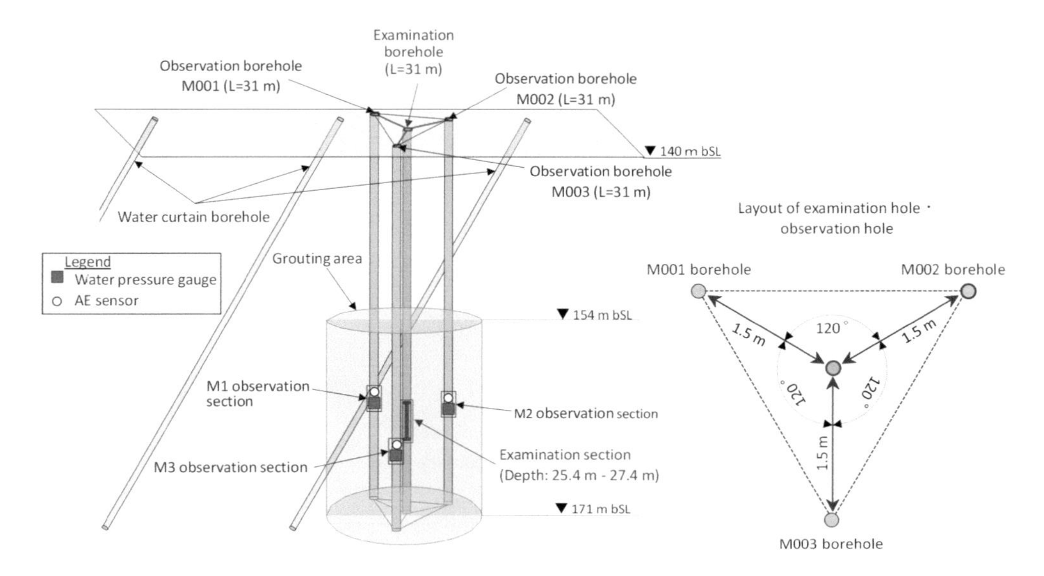

Figure 4.45 Layout of an air chamber and observation boreholes for a borehole gas permeability test in the fault-affected zone.

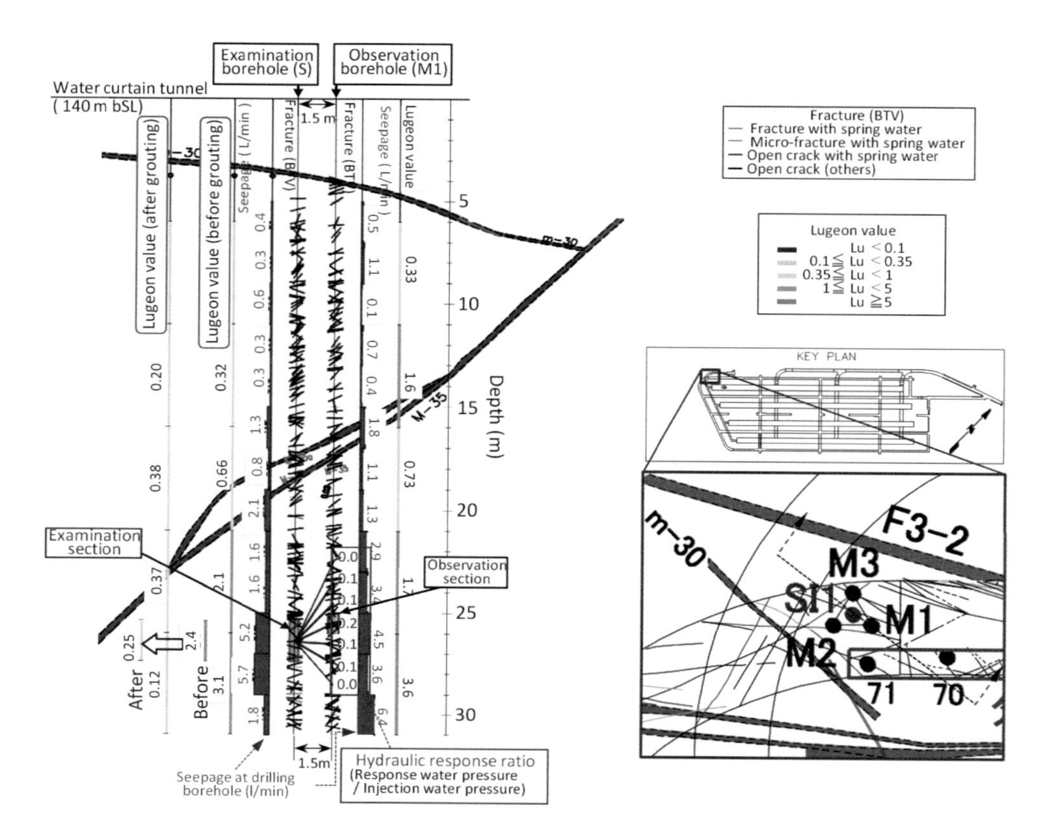

Figure 4.46 Hydrogeological data of the test section and observation sections at the borehole gas permeability test in the fault-affected zone.

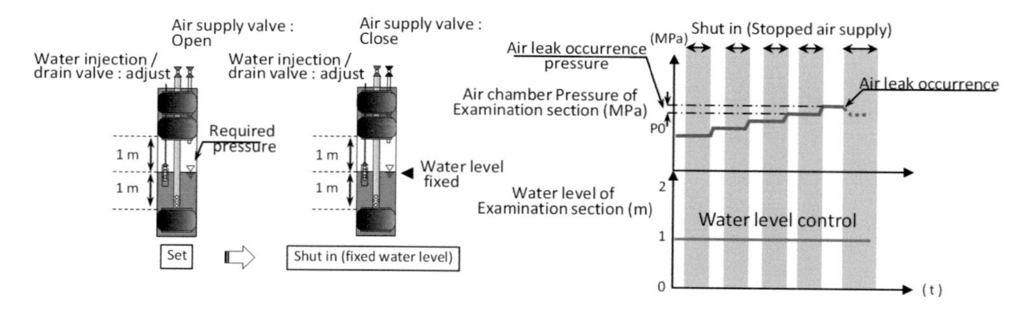

Figure 4.47 Schematic diagram of a borehole gas permeability test.

an air leak. This method of evaluation is based on the same principle as the air-tightness test of the actual storage cavern described in Chapter 6. The presence of air leakage was judged by the change in the internal pressure, ΔP, as corrected for temperature, flow rate and dissolved air in the drained water.

Figure 4.48 shows the internal pressure of the air chamber, the pore water pressure around the air chamber and the differential pressure between them at the test in the microfracture zone. The graphs at the top are the internal pressure measured and corrected for temperature, flow rate and dissolved air during the water draining process. The figure shows the results of the tests at the four stages of internal pressure of the chamber. All the tests showed a decreasing trend in the measured internal pressure after stopping air injection. On the other hand, ΔP did not decrease in Tests 1–3, but did in Test 4. From this behaviour of ΔP, the limit of the air-tightness of the microfractured zone was considered somewhere between Test 3 (internal pressure 1.263 MPa) and Test 4 (internal pressure 1.266 MPa; equivalent of initial water pressure). The corresponding differences from the surrounding pore water pressure were 0.0–1.9 m in Test 3 and 0.6–1.3 m in Test 4.

From the above results in the test interval of the microfractured zone where Lugeon values had been improved to 0.78 Lu, air-tightness was assured if the difference from the

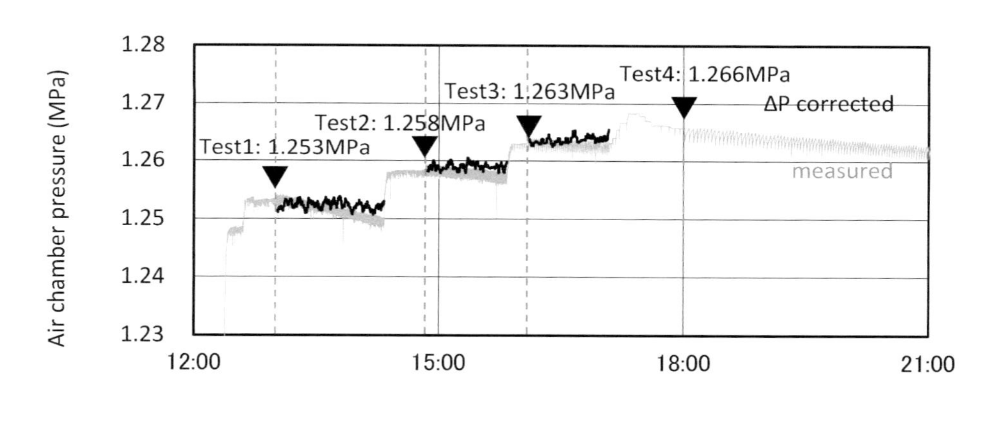

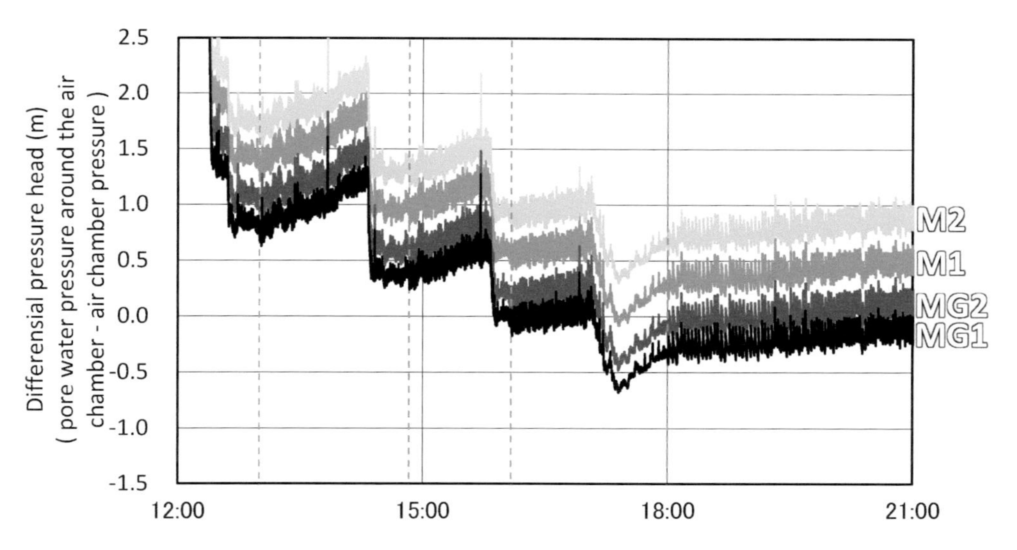

Figure 4.48 Internal pressure of the chamber, the pore water pressure and the water head. The difference between the values at borehole permeability test in the microfractured zone.

surrounding pore water pressure was 0 m or the internal pressure of the chamber was less than the surrounding pore water pressure. Similarly, in the test interval of the fault zone where the Lugeon value was improved to 0.25 Lu, air-tightness was assured if the difference from the surrounding pore water pressure was 0.6 m or the internal pressure of the chamber was 0.6 m less than the surrounding pore water pressure.

In both the microfractured zones and the fault-affected zones, the surrounding pore water pressure was higher than the internal pressure of the air chamber, thus their air-tightness was assured. In practical operation, the pressure of the storage cavern is maintained lower than the pore water pressure around the cavern. Although groutability in the microfractured zone is low because of the presence of many fine cracks, the target of improvement, 0.65 Lu, proved to be sufficient to maintain air-tightness. Thus, the target values of improvement of the low-groutability fault and microfractured zones were proved to be appropriate in terms of air-tightness.

4.2.1.1.1.3 Extent of Improved Area by Grouting To design the extent of grouting around the cavern, the influence of cavern excavation was considered including relaxation by the displacement of the opening of cracks due to the release of stress and the change in hydraulic conductivity. Figure 4.49 shows the prediction of the extent of change in permeability by analysis on the opening of cracks by excavation of the cavern. The excavation is expected to influence the areas where the stress level of the cracks due to relaxation by excavation

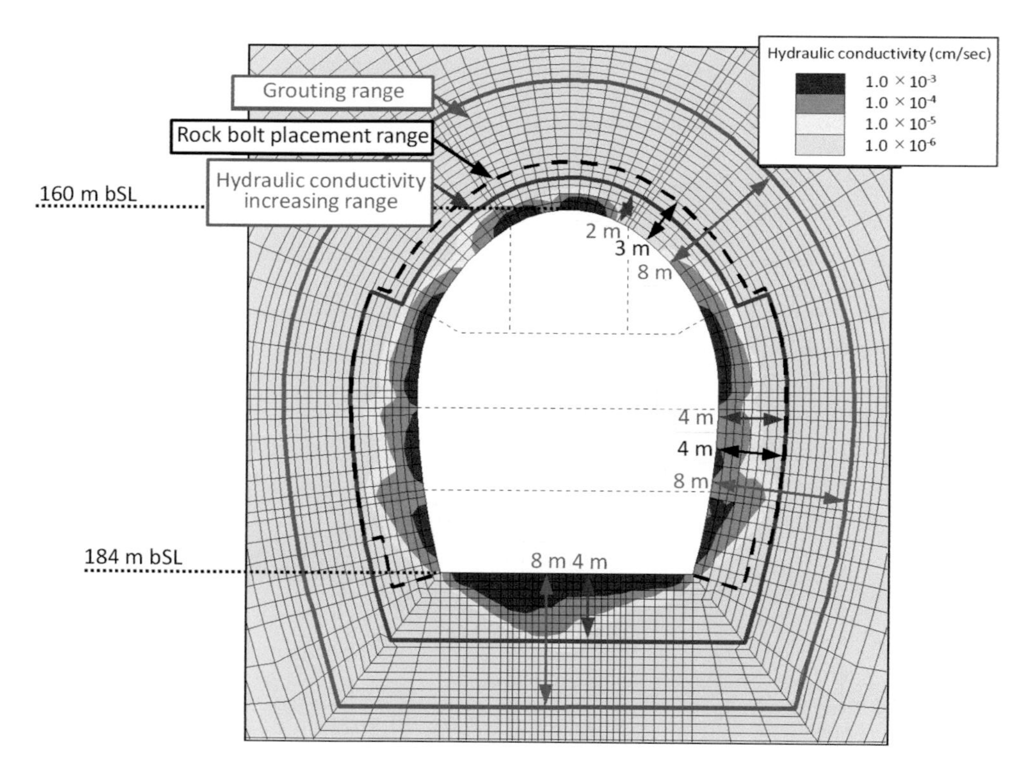

Figure 4.49 Areas where hydraulic conductivity changes due to the opening of cracks during excavation in the Kurashiki facility (predicted value).

reaches the shear strength and the areas where the cracks open. The extent of the former was about 2 m on the arch, 3 m on the side walls and 4 m on the floor, and the extent of the latter was about 2 m on the arch, 4 m on the side walls and 4 m on the floor. The hydraulic conductivity changed with the opening of cracks due to excavation, and its extent was estimated to be about 2 m on the arch, 4 m on the side walls and 4 m on the floor.

From the above analysis, it was considered that the area within 4 m radius of the cavern was subject to relaxation and the hydraulic conductivity changed within this region. Rock bolts 3–5 m in length were considered necessary to ensure the stability of the storage cavern. It was decided to apply grouting improvement to the area where kinetic and hydrological characteristics changed in response to the excavation. This grouting improvement was further applied to the outer perimeter of the area affected by excavation to mitigate seepage into the cavern. It was decided that the area on which to apply grouting was 8 m from the cavern walls.

4.2.1.1.2 THE NAMIKATA FACILITY

4.2.1.1.2.1 Target Value of Improvement by Grouting Figure 4.50 shows the distribution of the permeability of the bedrocks before the grouting procedure in Lugeon values on the

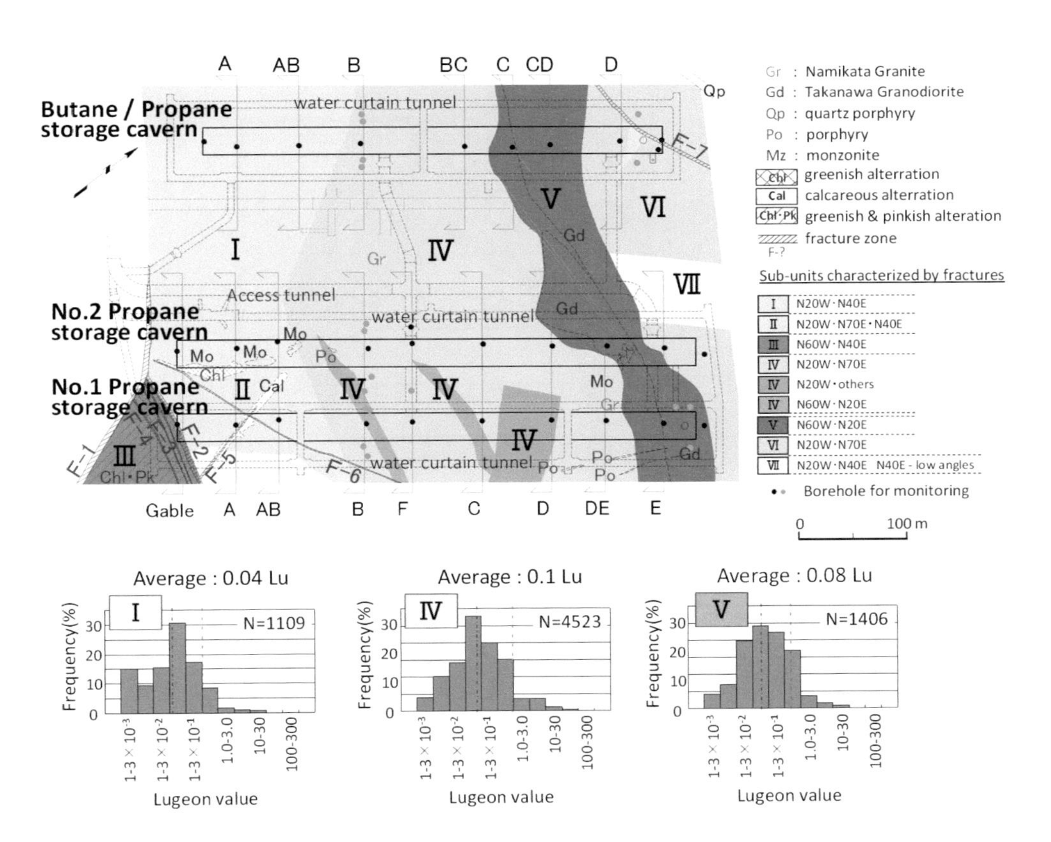

Figure 4.50 Distribution of permeability in the Namikata facility area by bedrock subdivision.

geological plan view at the level of the base of the storage arch. The average Lugeon values were around 0.1 Lu and lower than Kurashiki for the subdivisions I, IV and V classified by the crack density. However, occurrences of a Lugeon value exceeding 1 Lu varied among the zone subdivisions: about 4% for Zone I, 10% for Zone IV and 6% in Zone V. In some cases, the surrounding pore water pressure suddenly dropped when the grouting borehole encountered highly permeable cracks. This suggested that cracks with high continuity locally provided clear passage to groundwater flow. Accordingly, the cracks with a Lugeon value higher than 1 Lu were improved by grouting to less than 0.1 Lu, making the average Lugeon value of the entire area lower than 0.1 Lu. The target value of the grouting improvement was thus set to 0.1 Lu.

4.2.1.1.2.2 Extent of Improvement by Grouting The extent of grouting improvement in the Namikata facility was set based on an examination similar to the Kurashiki facility. Figure 4.51 shows the area of hydraulic conductivity increase where the cracks open by excavation of the cavern as analysed. The extent of the area where the cracks open was about 3 m from the arch, 5 m from the side walls and 6 m from the floor. These were considered to be the extent of the influence of the excavation. The extent of the area where the stress of the cracks due to relaxation reached the level of shear strength was about 3 m from the arch, 4 m from the side walls and 6 m from the floor. The hydraulic conductivity changed with the

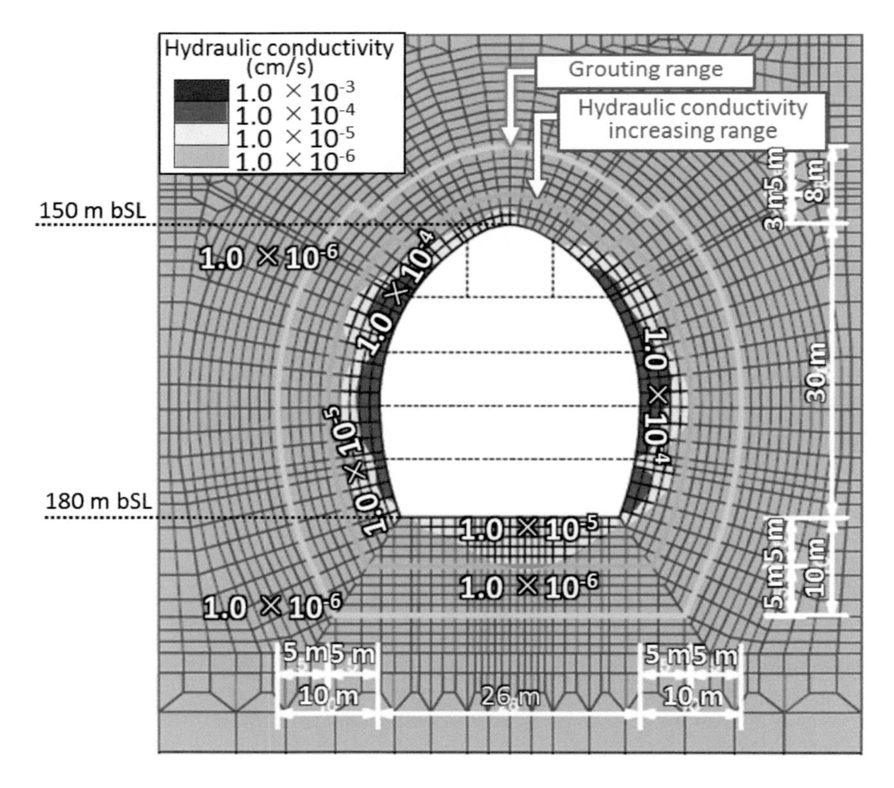

Figure 4.51 Areas where the hydraulic conductivity changes due to opening of the cracks during excavation in the Namikata facility (predicted value).

opening of cracks due to excavation, and its extent was estimated to be about 3 m from the arch, 5 m from the side walls and 5 m from the floor.

From the above analysis, the area within 5 m radius of the cavern was considered to be subject to relaxation and the hydraulic conductivity changes within this region. Therefore, rock bolts 4–6 m in length were considered necessary to ensure the stability of the storage cavern. It was decided to grout into the areas where kinetic and hydrological characteristics change in response to excavation. This grouting was further applied to the outer perimeter of the area affected by excavation to mitigate seepage into the cavern. Eventually, it was decided that the area to grout was 10 m from the cavern walls.

4.2.1.2 Grouting Pattern and Specification of Injection

4.2.1.2.1 GROUTING PROCEDURE

In order to keep the bedrock around the cavern saturated with groundwater throughout the construction period, water curtains with high water curtain pressure (0.5–1.2 MPa) were kept active before the start of excavation of the cavern. Therefore, the grouting operation needed to take place under high water pressure in the bedrock around the cavern. The "pre-grouting method" was used before cavern excavation. This method enables high pressure injection into rocks surrounding the excavation area.

Figure 4.52 shows the excavation–grouting cycle. First, the cavern arch and bench were concurrently pre-grouted during arch excavation. This improved the regions around the arch

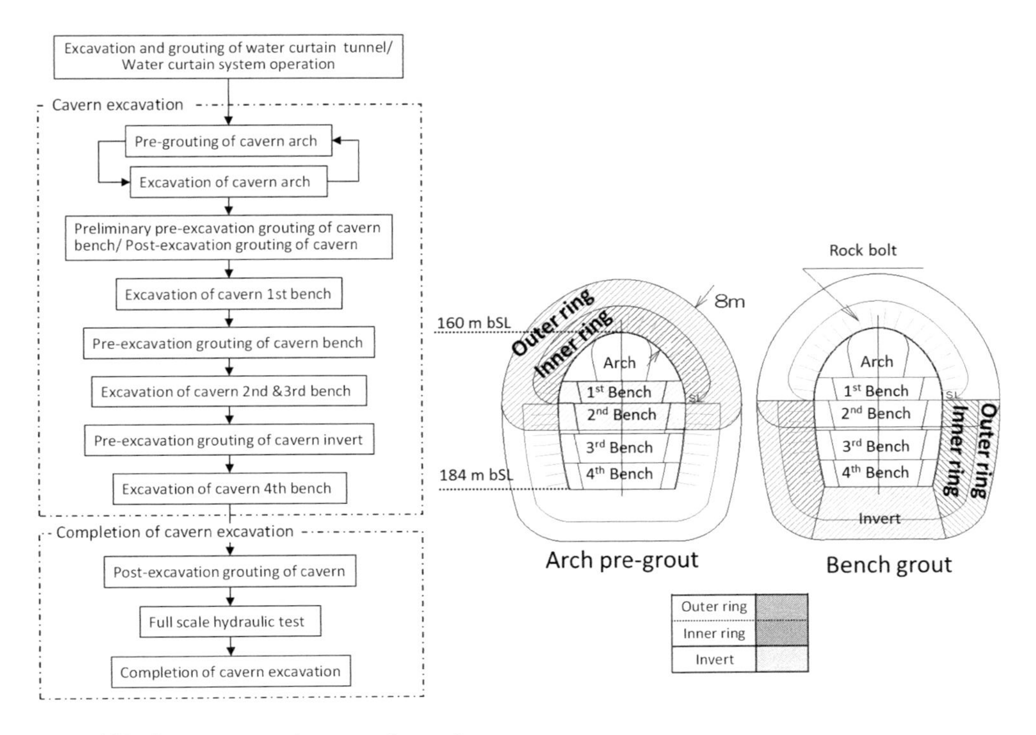

Figure 4.52 Cavern excavation–grouting cycle.

and the floor of benches No. 1 and 2. Then, bench No. 1 was excavated. Next, the side walls and the floor were improved by grouting and benches No. 2 and 3 were excavated. Then, bench No. 4 was excavated after grouting improvement of the floor of bench No. 4. Thus, grouting and excavation took place alternately, and excavation was limited to the areas improved by grouting in order to mitigate seepage and a drop in the pore water pressure.

Figure 4.53 shows the workflow of the construction of a cavern arch which includes concurrent pre-grouting and excavation. First, 35-m long boreholes were drilled in a fan shape from the cutting front toward the direction of the excavation. The first 3 m was left unexcavated as "cover rock" to prevent failure of the drilling surface and leakage of grout material. Next, an improvement ring was formed by injecting grout into the next 30 m interval. Excavating the next 15 m, the pre-grouting procedure was repeated. In the pre-grouting process at arches and benches, more grouting boreholes were added in between until the permeability of the boreholes before grouting reached the standard criterion to stop adding boreholes. When the grouted interval was excavated, the seepage rate and the pore water pressure surrounding the cavern were checked and assured to be lower than the standard management criteria. If they exceeded the criteria, post-grouting was applied to the interval to form a secure improvement zone.

4.2.1.2.2 GROUTING PATTERN AND SPECIFICATION OF INJECTION AT THE KURASHIKI FACILITY

4.2.1.2.2.1 Characteristics of Cracks and Grout Injection Grout injection tests were carried out in the water curtain tunnels to analyse the characteristics of cracks and grout injection. In the following sections, the characteristics of the cracks and the performance of the grout injection are described comparing the mylonite zones around faults where groutability is low and the microfractured zones where groutability is extremely low.

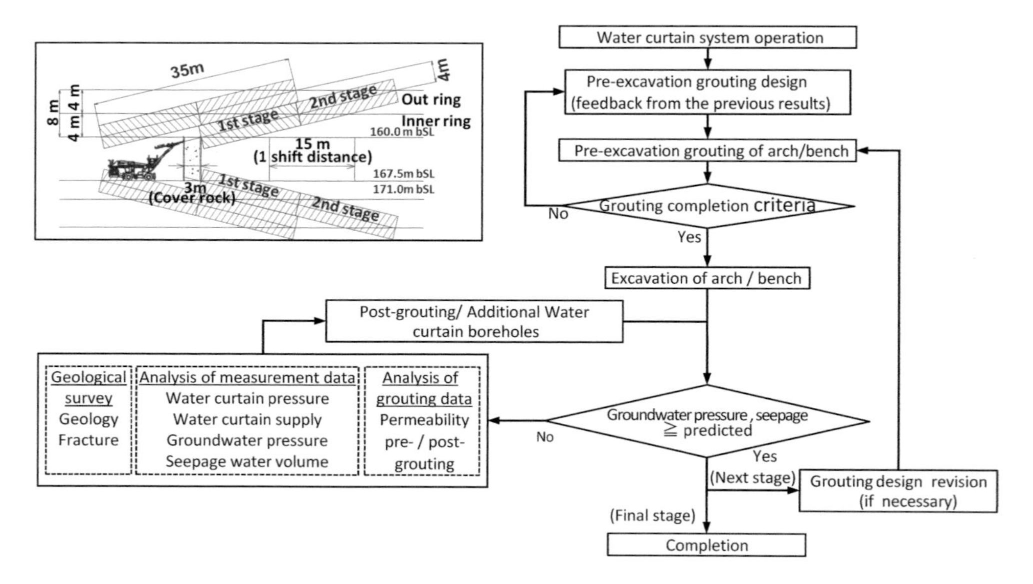

Figure 4.53 Workflow of grouting and excavation of a cavern arch.

4.2.1.2.2.1.1 The Mylonite Zones around Faults The mylonite zones around faults were classified as subdivision V with a crack density of about two per meter along faults F-2 and F-3. The maximum Lugeon value in these zones was 13 Lu (average 0.9 Lu). The mylonite had high continuity and high permeability.

Figure 4.54 shows an example of the injection pattern, the temporal variation in the pressure and flow rate of grouting, the number of injections and the variation in the Lugeon values at drilling of grouting drillholes. A good injection curve was observed in the relationship between the injection pressure and the injected flow rate: the injection flow rate did not decrease until the maximum pressure was reached and the injection flow rate gradually decreased after the maximum pressure was reached. The graph of the number of stages of grouting and improvement of the Lugeon values shows that supplementing with the first (12 m interval) and second (6 m interval) stages of additional grouting drillholes was not sufficient. In this case, the addition of up to a fifth stage (0.75 m interval) of grouting was necessary. While the permeability of the grouted zone decreased with the addition of grouting bores, this area needed a relatively large number of additional grouting holes, which showed that groutability was low.

Figure 4.55 shows polarising microscope images of mineral fill of the crack in mylonite after grout injection. These photographs show the good grout injection characteristic of cement fill in the smooth opening of a crack about 20 μm wide.

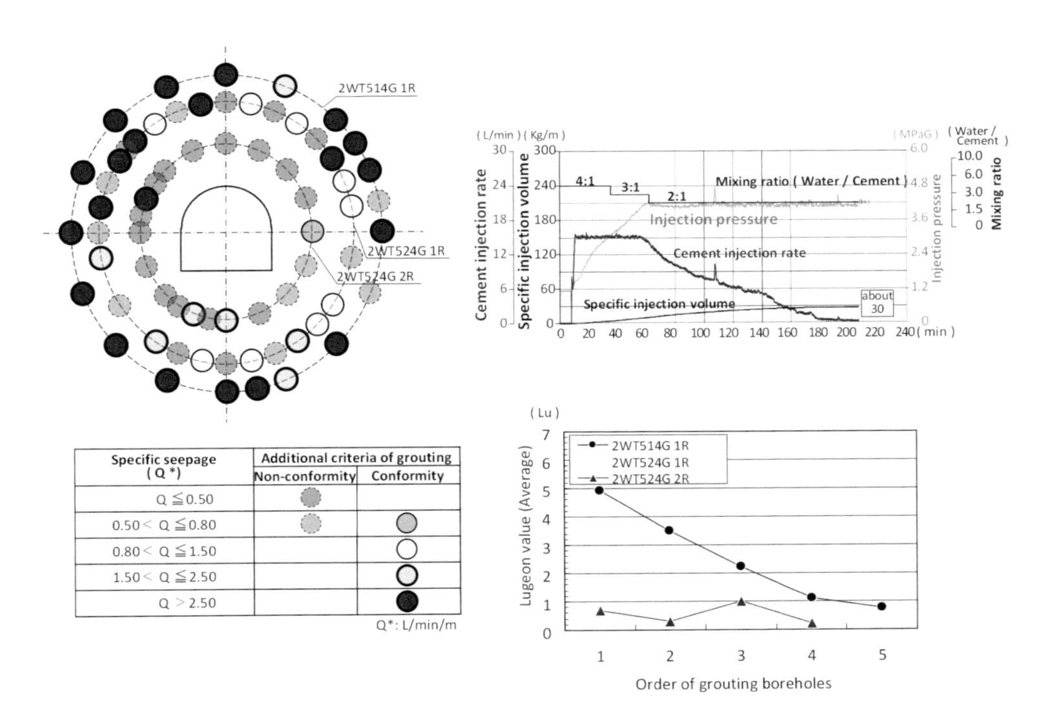

Figure 4.54 Example of the injection pattern, temporal variation of pressure and flow rate of grouting, the number of injections and the variation in Lugeon values following drilling of grouting drillholes in the mylonite zone of the Kurashiki facility.

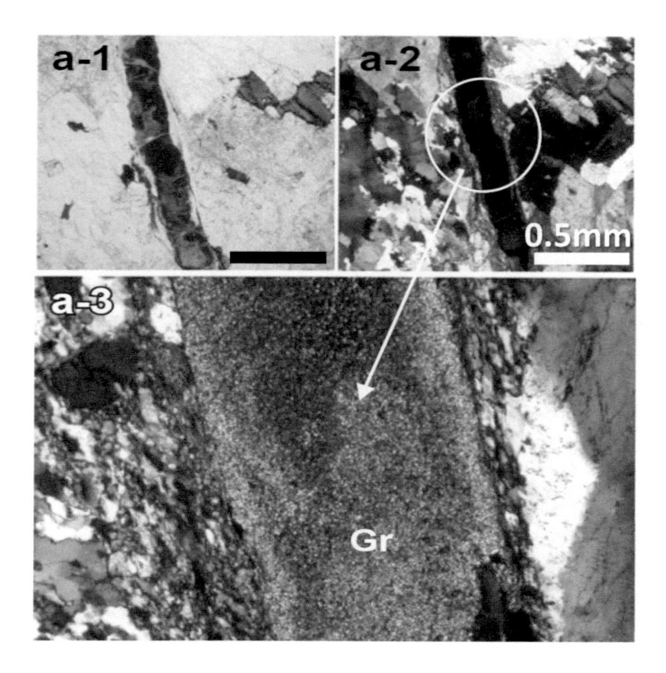

Figure 4.55 Polarising microscope images of the mineral fill of the crack in mylonite after grout injection in the Kurashiki facility.

4.2.1.2.2.1.2 Microfractured Zone The microfractured zone had fine cracks densely populated with extremely small openings, and its maximum Lugeon value was 3.3 Lu (0.6 Lu on average). Figure 4.56 shows the injection pattern, the temporal variation in the pressure and flow rate of grouting, the number of injections and an example of a variation of the Lugeon value at drilling of the grouting boreholes. The injected flow rate sharply increased as the injection pressure increased; the flow rate decreased before the maximum pressure was reached, which suggested that the cement milk had clogged the cracks without penetrating into the cracks. A graph of the number of stages of grouting and improvement of the Lugeon values shows that the seven stages of additional grouting boreholes were necessary and three or four rings were needed, compared with the planned two, to achieve the target of improvement. Many additional grouting boreholes were considered necessary due to the small improvement area achieved by each grouting borehole as the penetration range of the cement milk was small in this zone. As seen, the microfractured zone had extremely low groutability. The permeability test before grout injection also recorded a sudden decrease in the water flow while the water flow increased as the water injection pressure increased after the start of the test. This decrease of the water flow was also observed at the time of grout injection. It is a typical characteristic of dense microfractures with extremely small openings.

The texture and structure of the microfractures were investigated under the polarising microscope for the cause of low groutability. Figure 4.57 shows photographs of microfractures at outcrop and under the microscope. Microsheetings densely developed in zones a few centimeters wide. Microfractures b-1 and b-2 had cracks along the edges of mineral grains forming a network of fractures a few micrometers wide by expansion and contraction.

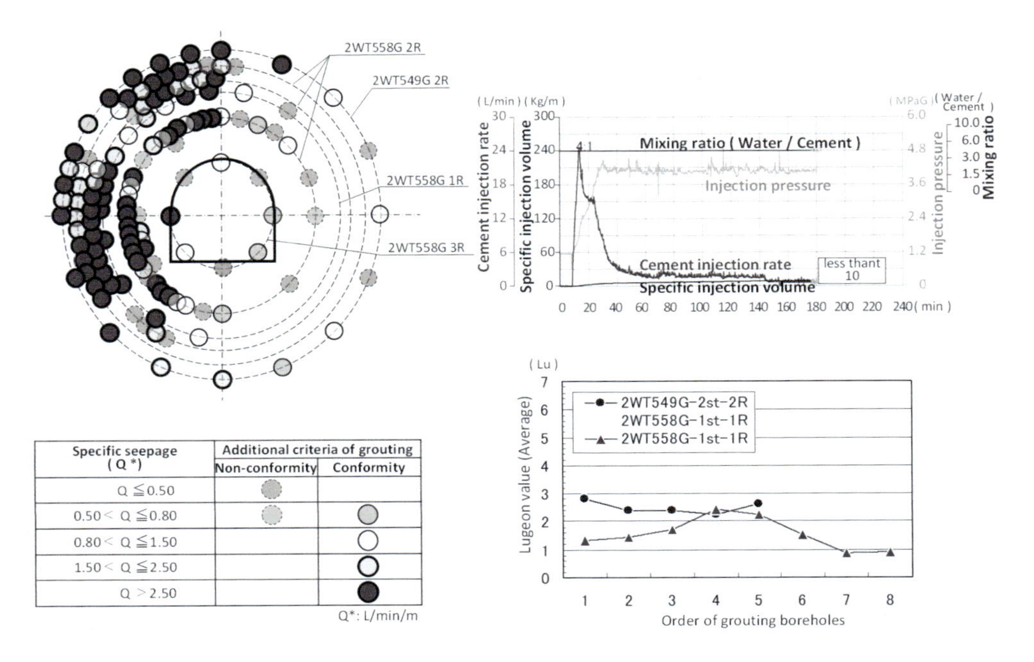

Figure 4.56 Example of the injection pattern, temporal variation in pressure and flow rate of grouting, the number of injections and the variation in Lugeon values following drilling of grouting drillholes in the microfractured zone of the Kurashiki facility.

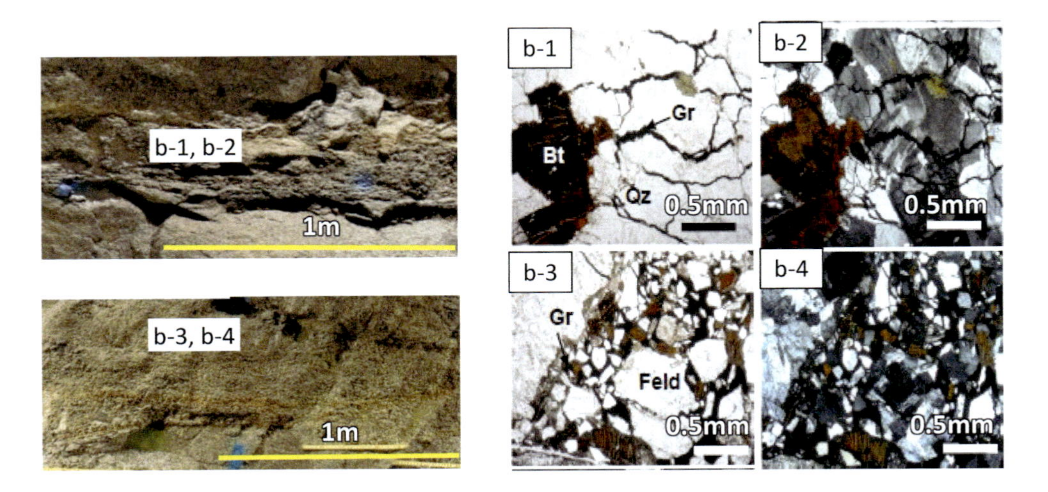

Figure 4.57 Photographs of microsheetings at outcrop and under the microscope (Kurashiki).

In microfractures b-3 and b-4, the mineral grains were crushed and separated, and the voids between scattered angular grains were filled by cementing. Such a network of fractures was considered instantly closed by the cementing material of grout, and the penetration area of the grout became small as a result.

Observation under the microscope recognised that the structure of the angular minerals in the cracks was due to extension stress rather than shear. In other words, microsheetings were considered a result of the stress relief associated with uplifting and the intrusion of granite. Accordingly, the presence of two kinds of microsheeting was considered: low-angle microsheeting parallel to the ground surface which acted as a stress release plane; and high-angle microsheeting along the faults in which the stress was released.

From these observations, geological surveys of the tunnel face and BTV of boreholes were conducted to investigate the distribution and continuity of microfractures. Figures 4.58 and 4.59 show the results of surveys of the microfractured zone encountered in cavern No. 3. The zones where microfractures densely developed continuously meandered for a few meters to tens of meters, changing their widths between a few centimeters to a few meters. These continuous microfractured bands were cut by faults in the mylonite. This network of mylonites and microfractures was considered to provide the groundwater passage. Therefore, it was necessary to identify the distribution of the microfractures and to arrange grouting boreholes suitable for the zones, to improve grouting in the microfractured zones.

4.2.1.2.2.2 Pattern of Grouting Drillholes The layout of the grouting boreholes at the cavern arch is a two-circle system of initial outer grouting boreholes and subsequent inner grouting boreholes, as seen in Figure 4.60. The 35 m grouting boreholes are arranged in a spiral shape with a 15 m overlap. The injection boreholes are the first and second stage boreholes and their intervals are 6 m for the outer circle and 4 m for the inner circle. Where the permeability at drilling exceeded the standard, intermediate grouting boreholes were added. Adding

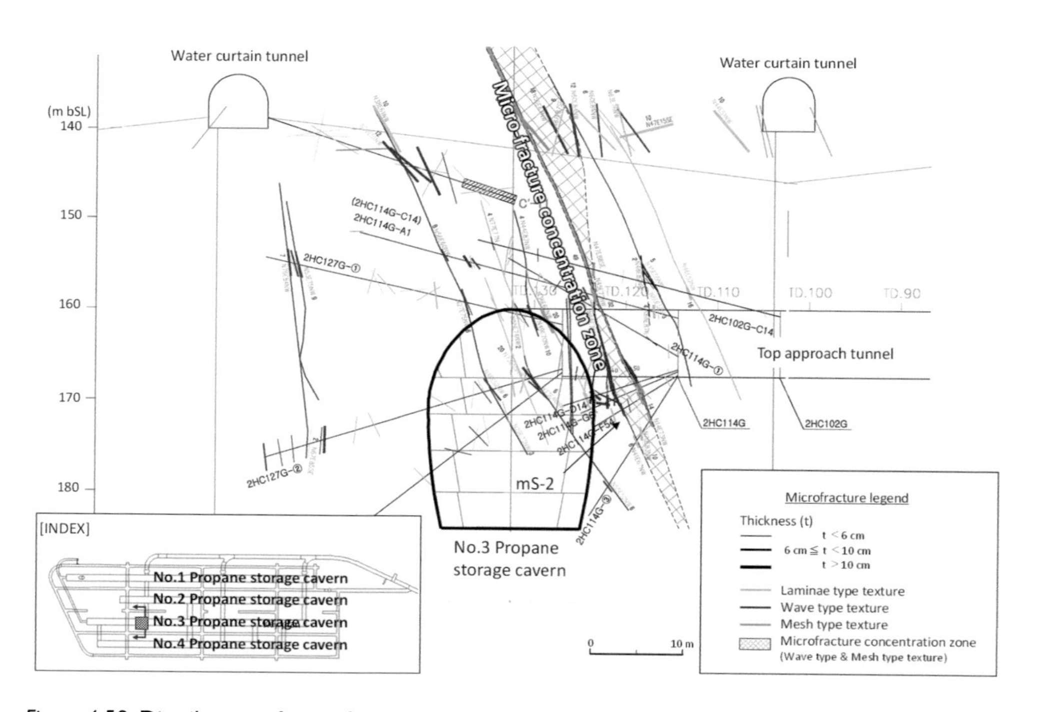

Figure 4.58 Distribution of microfractures around cavern No. 3: cross section at the Kurashiki facility.

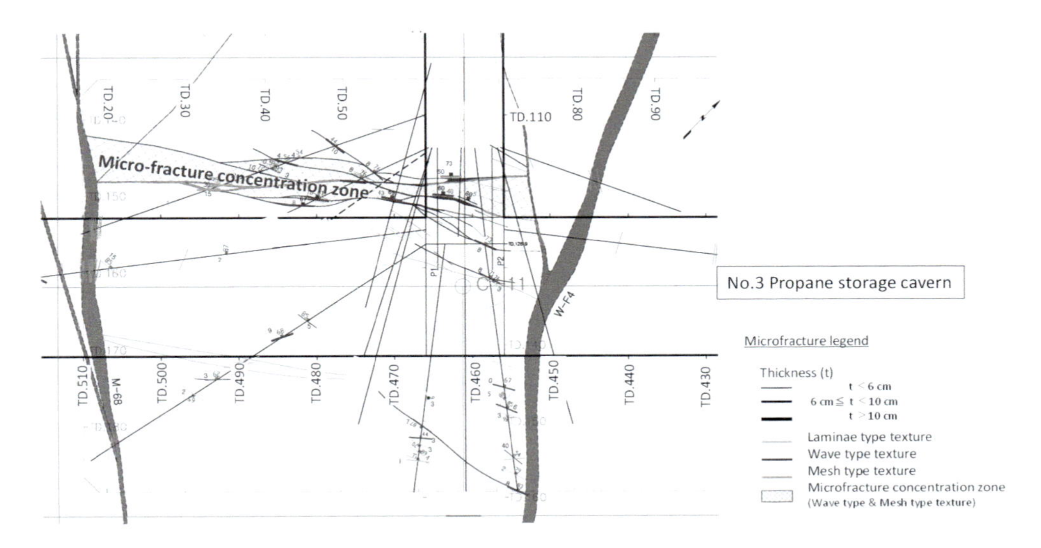

Figure 4.59 Distribution of microfractures around cavern No. 3: plan view at the Kurashiki facility.

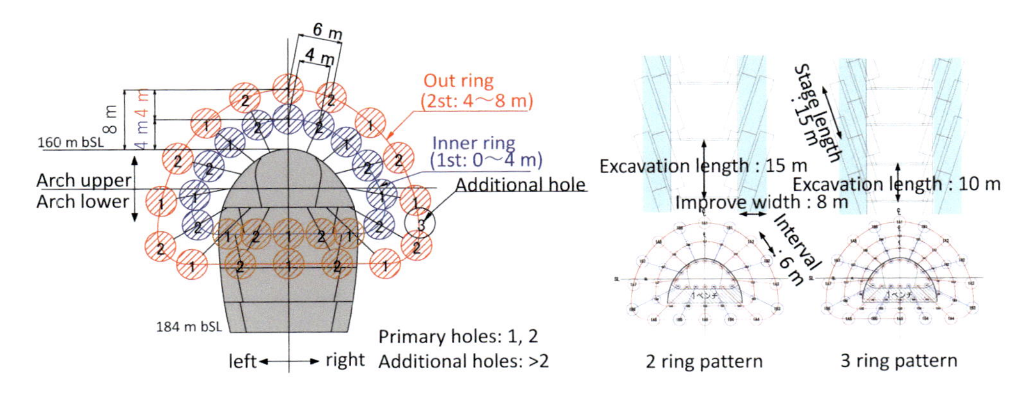

Figure 4.60 Injection pattern of a cavern arch at the Kurashiki facility.

these intermediate boreholes continued until the standard criterion was satisfied, halving the interval. The boreholes were arranged so that the intervals between the inner and outer circles were 4 m, in respect of the improvement zone of 8 m from the cavern wall.

In the fault and microfracture zones where groutability was low, another circle was added between the inner and outer circles. This resulted in three rings. If the standard was still not reached, a fourth stage of grouting was not feasible, as the drilling interval became too small at 1.5 m. Instead, the borehole interval of the third ring was reduced to 10 m at the cavern arch and the overlap of 35-m long boreholes was increased three-fold.

Figure 4.61 shows the layout of the grouting boreholes on the side walls and floor of the bench. The grouting boreholes were arranged at 4 m intervals, namely Columns A–E and K–O from bench floor 1 into the side walls. Within each column, the injection boreholes were

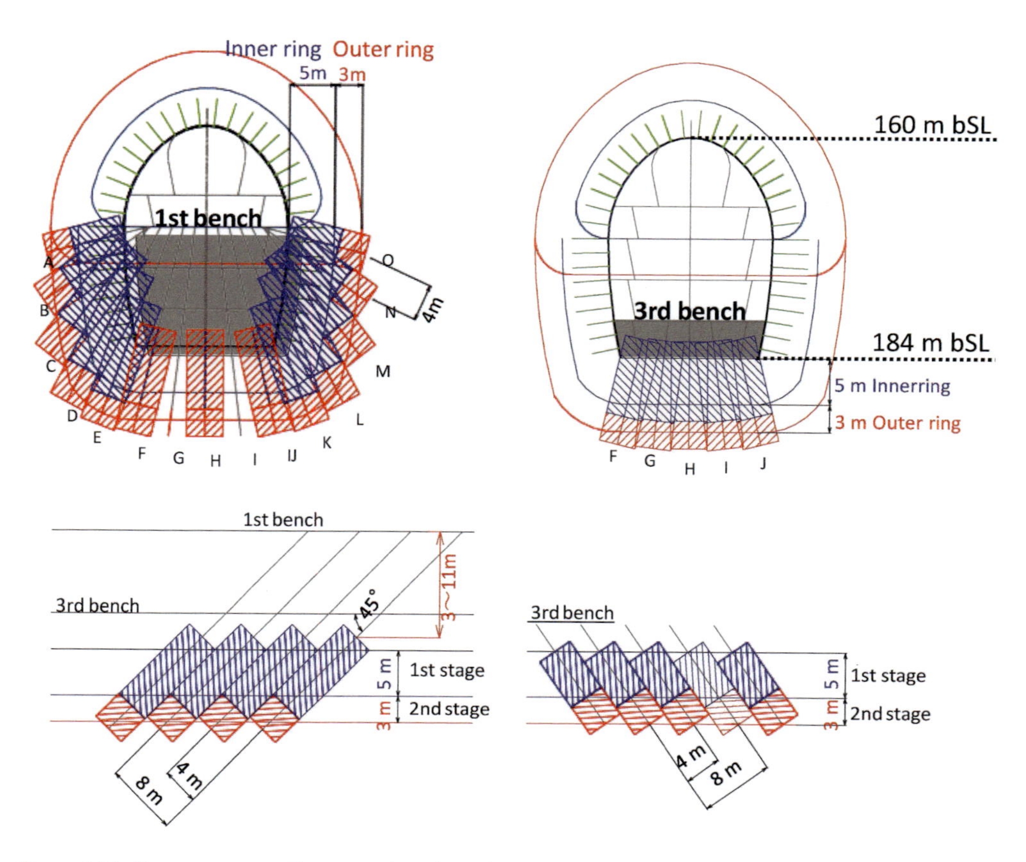

Figure 4.61 Grout pattern of a cavern bench at the Kurashiki facility.

drilled at 8 m intervals. If the permeability measured during drilling exceeded the standard criterion, additional boreholes were drilled in between so that the standard criterion was satisfied for both inner and outer circles. In the fault and microfracture zones, where groutability was low, the interval between injection boreholes was made to 4 m according to the record of the grouting in the arch. The improvement zone under the floor (Columns F–J) was formed by adding grouting boreholes across the cavern axis. The floor grouting from bench No. 1 was regarded complete when the outer circle reached the target Lugeon value, and the floor grouting from bench No. 3 was regarded complete when the inner circle reached the target Lugeon value. While the width of the improvement zone from the cavern walls was 8 m, the inner and outer circles of the bench grouting were set to 5 and 3 m, respectively. The grouting injection interval was set to cover this improvement zone.

The layout of the injection boreholes was designed to cross the NNW-SSE strike of the dominant cracks to ensure that injections crossed these cracks. On the other hand, the orientation of the cracks was so complex that a "cross layout" was used in the areas where ENE-WSW cracks were dominant in addition to the NNW-SSE strike. The "cross layout" is designed for the boreholes to cross each other (Figure 4.62). When the dip of the cracks is steep, it is difficult for grouting boreholes with a basic layout to cross the cracks. In this

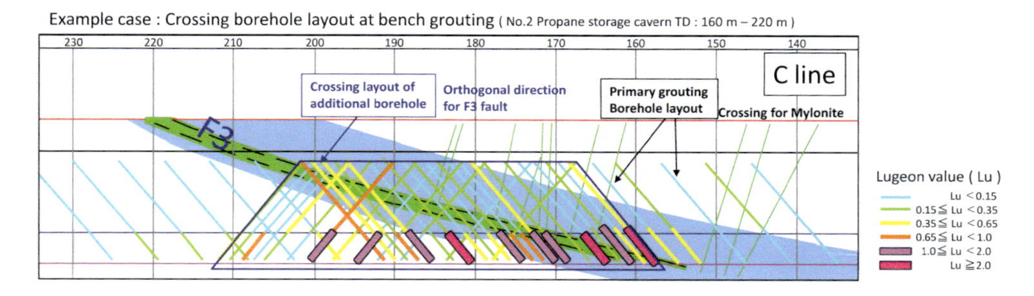

Figure 4.62 Example of the cross layout of bench grouting drillholes at the Kurashiki facility (plan view).

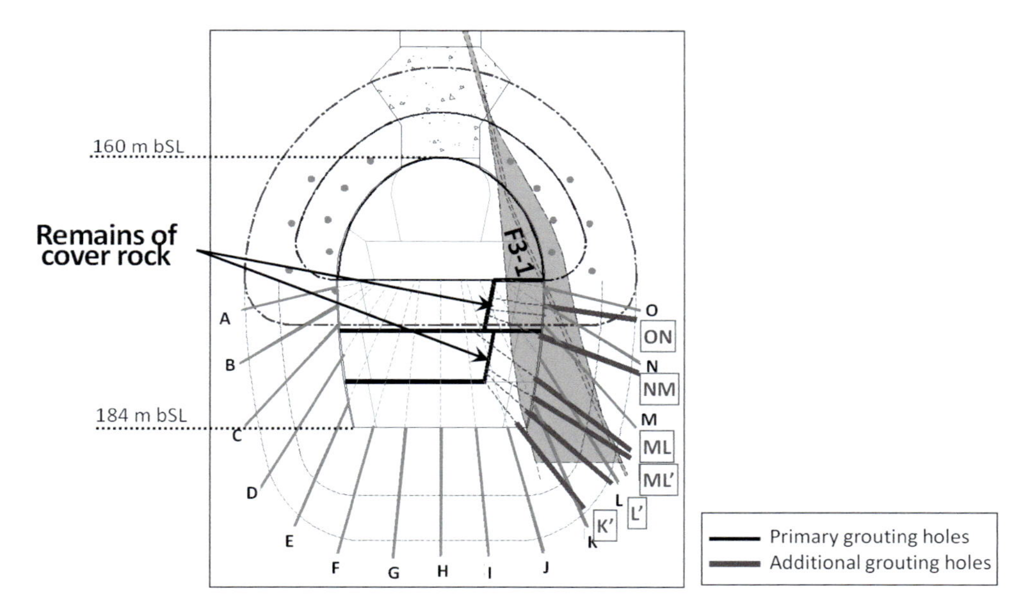

Figure 4.63 Example of the layout of bench grouting drillholes with unexcavated floor at the Kurashiki facility.

case, the improvement effect was enhanced by two means: an additional intermediate layout to cross the cracks; and leaving a part of the floor unexcavated so that high pressure injection was possible (Figure 4.63).

4.2.1.2.2.3 Specification of Injection The permeability was high in the fault and microfracture zones, where groutability was low. In these zones, fine cracks with openings from 10 to 30 μm were densely developed as seen in the photograph on the left in Figure 4.64. The cement particles of the grouting material must penetrate into these fine cracks to be effective in improving these rocks. Accordingly, ultrafine cement material with a maximum size of 10 μm was chosen for the grouting (right of Figure 4.64).

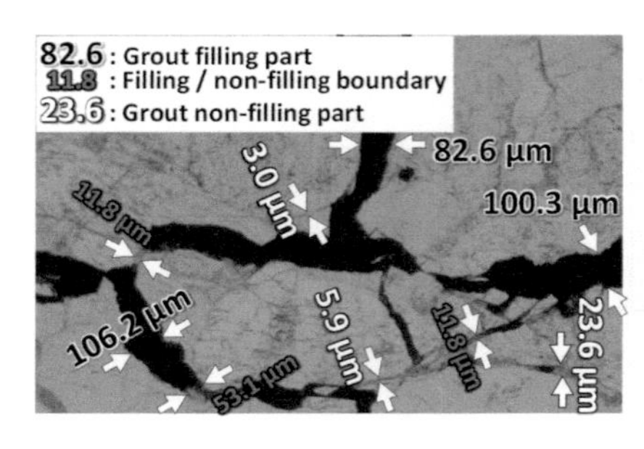

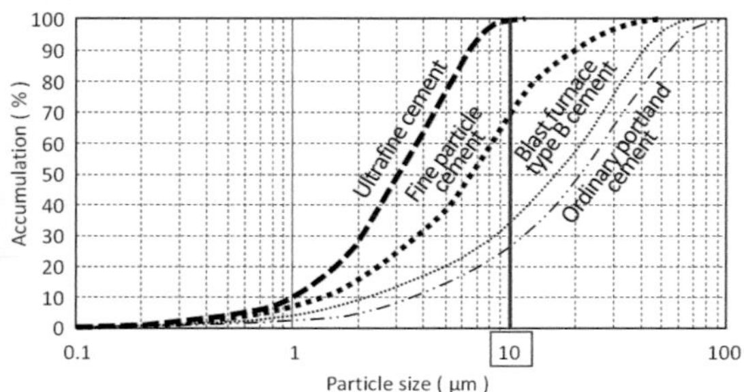

Figure 4.64 Microscopic photograph of cracks in the Kurashiki facility and the distribution of the cement particle size.

When excessive precipitation and agglutination occur during cement injection into the opening of cracks the size of the maximum cement particle, penetration of the cement into the cracks is limited, resulting in ineffective grouting improvement. The dynamic injection method was introduced to improve the dispersion and penetration of cement particles.

The dynamic injection adds vibrating pressure to static pressure during grout injection, and is expected to solve local clogging and to improve the flow of grout. This has been proven by laboratory and in-situ tests. Its superiority in improving rocks with low permeability over the conventional static pressure injection has been demonstrated by an increase in the flow rate of the injected grout. The injection pressure was centred at 4 MPa, being the sum of the seepage pressure 1 MPa and an additional 3 MPa, and the vibration pressure was ±0.6 MPa (15%) at 6 Hz. The rate of injection was up to 10 L/min. This pressure was maintained for 10 min after the rate dropped to 1 L/min to conclude injection. A high-speed shear mixer with a high rotation was used to disperse the cement particles. The rotation speed used was 1500 rpm, nearly ten times faster than the conventional 170 rpm. The water to cement ratio (W/C) at the start was 400% when there was a stage with Lugeon values over 1 Lu, and 600% when it was under 1 Lu. The switching point of the composition was set to 1000–2000 L, and the upper limit of each injection was set to 8000 L.

As ultrafine cement has high liquidity, it takes more than 8 h to solidify. This caused the issue of increasing construction days. Some coagulant was added to the ultrafine cement. A test compared the ultrafine cements with and without coagulant at locations with a similar Lugeon value (0.4 Lu). Figure 4.65 shows the temporal variation in the injection flow rate. The sample with coagulant showed a rapid flow rate drop at 3 h after start of injection. This confirmed a reduction in the duration of the injection, i.e., waiting time.

The standard criteria of adding extra grouting boreholes were decided as in Table 4.3, considering analyses of the record of the initial pre-grouting of the arches, the seepage rate and the pore water pressure.

In grouting, when the Lugeon value at drilling and the seepage rate at excavation were less than the standard criteria, this stage of grouting boreholes was regarded as final. Otherwise, an additional grouting hole was drilled at the mid-point to the adjacent holes.

When all the grouting boreholes satisfied the standard criteria, grouting of that segment was complete. This is one cycle in the operation of drilling a grouting bore.

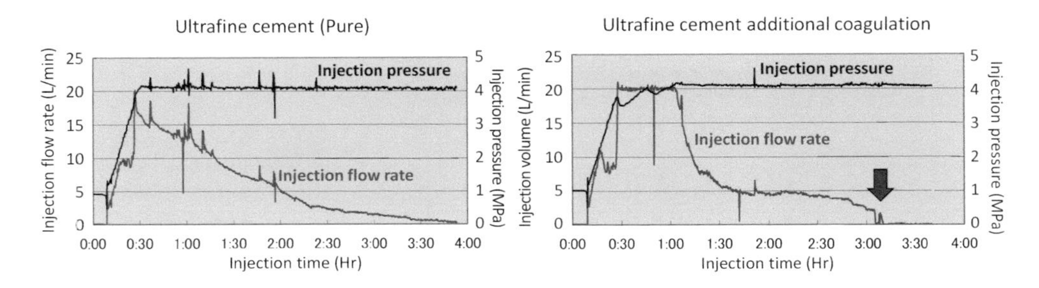

Figure 4.65 Comparison of the injection flow rate with and without coagulant at a water curtain tunnel.

Table 4.3 Standard criteria of adding extra grouting drillholes at the Kurashiki facility

(a) Arch			*(b) Bench*			
Zone	*I*	*II–V*	*Zone*	*I, IV*	*II, III, V*	*Microfractured*
Inner ring	0.35 Lu	0.35 Lu	Inner ring	0.25 Lu	0.25 Lu	0.35 Lu
Outer ring	0.65 Lu	1.0 Lu	Outer ring	0.65 Lu	0.65 Lu	0.65 Lu

4.2.1.3 Grouting Pattern and Specification of Injection at the Namikata Facility

4.2.1.3.1 PATTERN OF GROUTING DRILLHOLES

The average Lugeon value of the rocks of the caverns in the Namikata facility was 0.3 Lu, and 70% of the rocks had lower than 0.01 Lu. On the other hand, 10% of the rocks locally had values over 1 Lu. From this, continuous highly permeable cracks were considered to be present, forming passages for groundwater to flow. Accordingly, grouting at the Namikata facility required the accurate detection of local highly permeable cracks within the low permeability rocks to cross by the grouting boreholes.

Unlike the Kurashiki facility, the distribution of permeability in the Namikata facility could be characterised by geological subdivision. Therefore, the injection boreholes, doubling with the reconnaissance holes, were laid out according to the permeability of the rocks. As seen in Figure 4.66, the areas for arch grouting were classified into three sections: highly permeable sections with Lu $\geqq$ 1; medium sections with 0.5 $\leqq$ Lu < 1.0; and low sections with Lu < 0.5. The layout of the grouting boreholes in the arch area is shown in Figure 4.67. In the high permeability sections, 21 grout injection boreholes were drilled in the A, B and C circles for improvement width 10 m; eight grout injection boreholes in the medium permeability sections for improvement width 5 m; and only three boreholes for confirmation in the precursor drilling zones (Z circle) in the low permeability sections. After pre-grouting, the caverns were excavated for 21 m. This pre-grouting–excavation cycle was repeated.

Figure 4.68 shows the layout of the pre-grouting boreholes at the bench. From the result of the grouting at the arches, a 12 m (24 m for low permeability sections) interval between boreholes was decided with a 10 m improvement width. As done for the pre-grouting of the arches, the area was classed into sections of high, medium and low permeability and the injection design and standard criteria for additional drilling were designed.

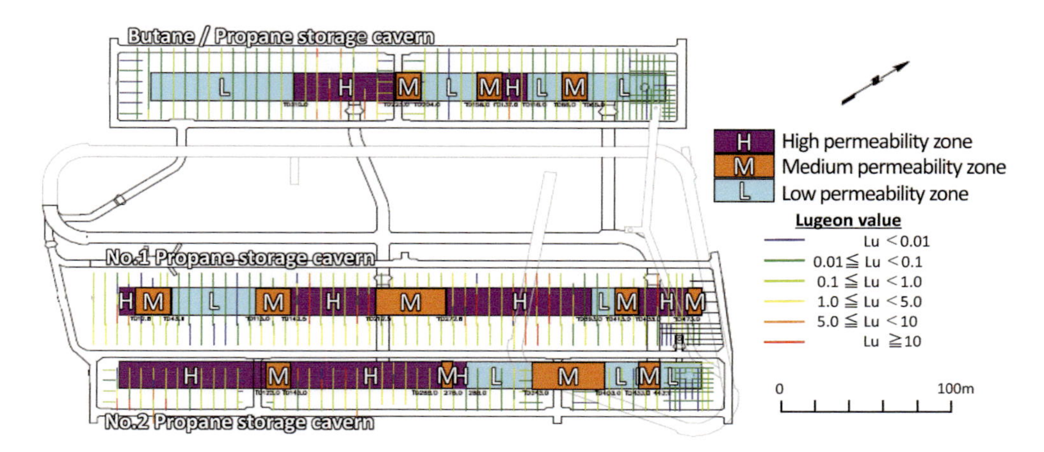

Figure 4.66 Classification by Lugeon value of the pre-grouting pattern at an arch in the Namikata facility.

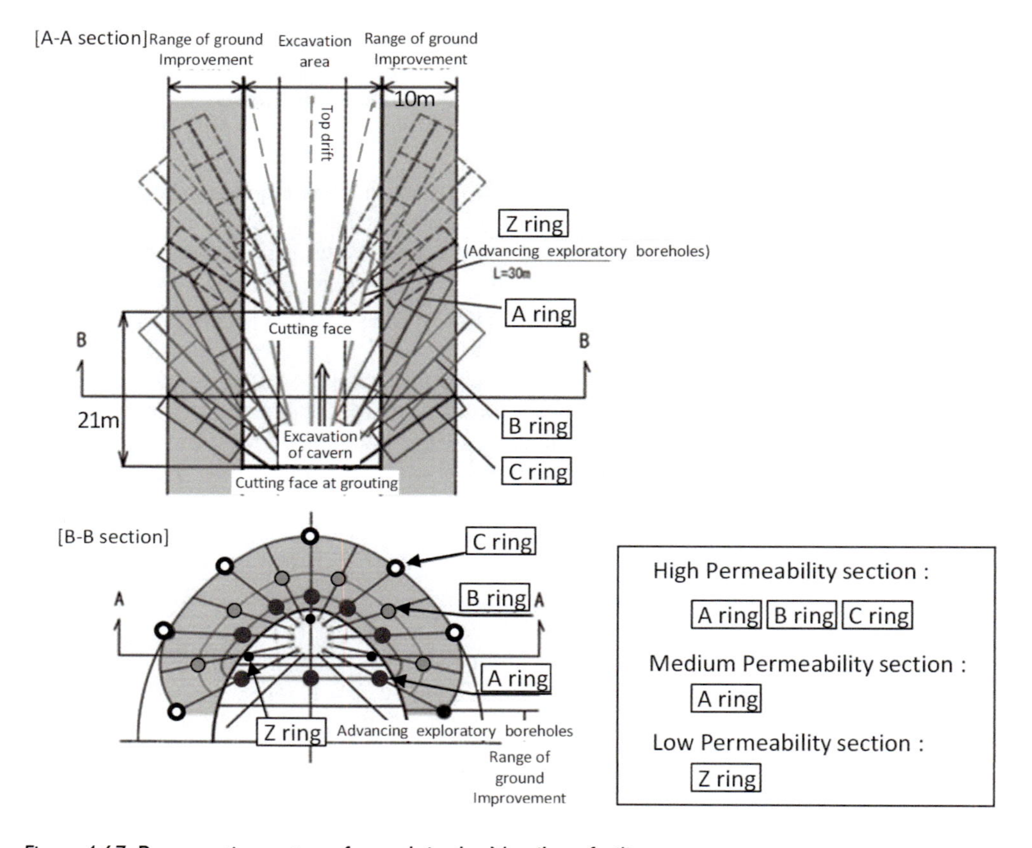

Figure 4.67 Pre-grouting pattern for arch in the Namikata facility.

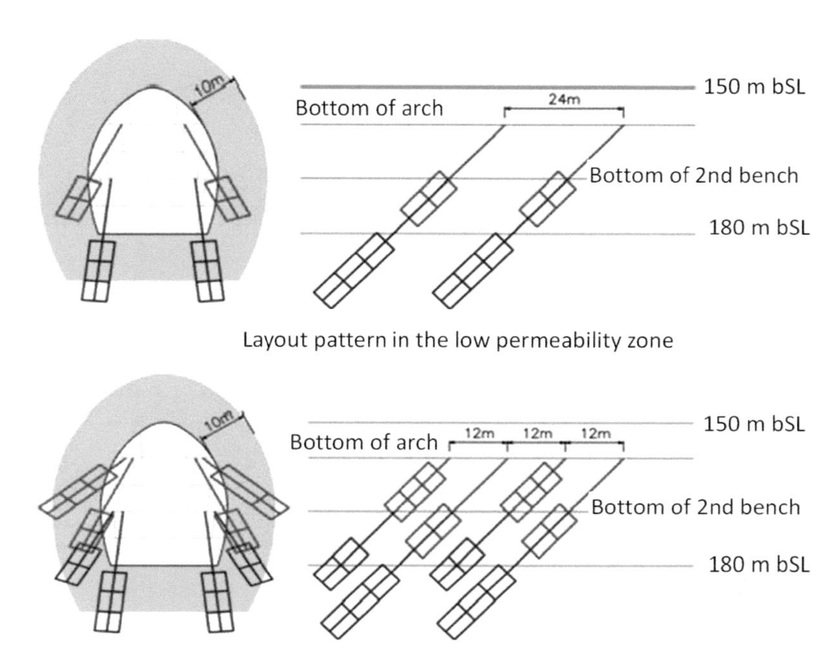

Figure 4.68 Injection pattern of pre-grouting of benches in the Namikata facility.

Figure 4.69 shows the workflow of the grouting. Grouting boreholes were set out at an initial interval of 12 m. When the Lugeon value and the ratio between the seepage rate and the injection exceeded the standard criteria, additional grouting boreholes were drilled at the centre points between the grouting boreholes. This procedure was repeated until these values satisfied the standard criteria. These intermediate grouting boreholes were added when one of the boreholes in the low permeability sections showed seepage of more than 2.5 L/min or permeability over 1.0 Lu in a 6 m segment, and when one of the boreholes in the medium to high permeability sections recorded a cement injection flow rate of more than 10 kg/m as well as the above two conditions. When all the grouting boreholes satisfied the standard criteria, pre-grouting of that section was regarded complete and excavation started. After excavating the pre-grouted section, the behaviour of the groundwater including the seepage rate and the pore water pressure was monitored. If the management standard criteria were satisfied, then the post-grouting procedure was taken to complete all the improvement of the section.

4.2.1.3.2 SPECIFICATION OF INJECTION

As in the Kurashiki facility, the dynamic injection method was introduced to ensure better sealing of local highly permeable cracks. Considering the surrounding high water pressure, the central injection pressure was set to 2.5 MPa with a vibration pressure of ±0.5 MPa at 5–7 Hz. The injection was terminated after maintaining the pressure for 30 min, when the injection pressure reached the standard pressure and the injection rate dropped to less than 0.2 L/min/m. Material for the grouting was chosen by preceding field tests, considering

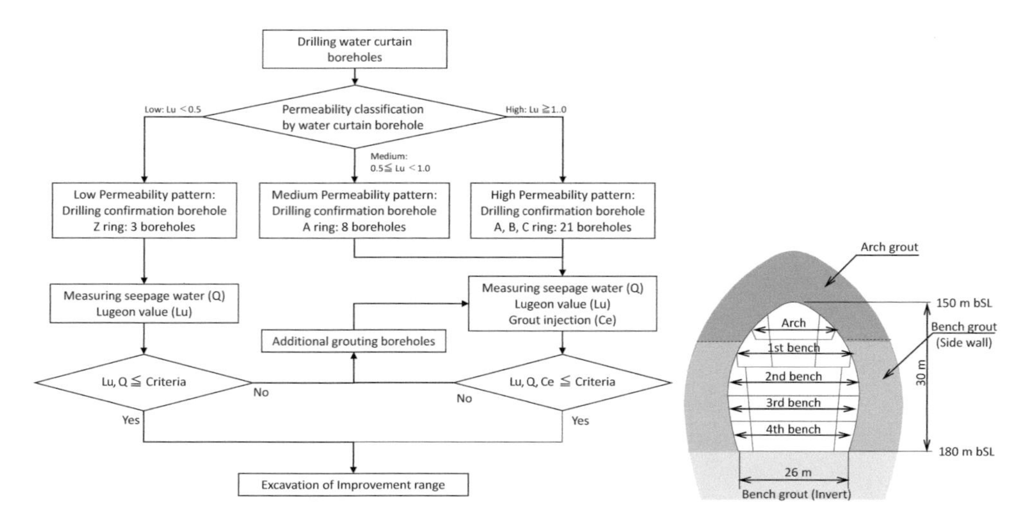

Figure 4.69 Workflow of the pre-grouting of arches at the Namikata facility.

long-term durability, chemical stability against stored LPG and workability. High-Early-Strength Portland Cement was used as a base, Portland Blast-Furnace Slag Cement for the areas where saline water invaded and ultrafine cement for low permeability zones.

4.2.2 Performance of Grouting and Improvement Effect

4.2.2.1 Performance in the Kurashiki Facility

4.2.2.1.1 GROUTING IMPROVEMENT OF THE ENTIRE STORAGE CAVERN

Figures 4.70 and 4.71 show the improvement in pre-grouting around the caverns of the Kurashiki facility at arch and bench, respectively, for each rock subdivision and circle. The horizontal axis is the Lugeon value surrounding the cavern and the vertical axis is the ratio of the grouting boreholes that reached the improvement target. The light line presents the initial state of the distribution and the dark line is the final distribution. In the arch grouting, all the boreholes on the inner circle in Zone I satisfied the grouting target: the maximum Lugeon value after the improvement was under 0.35 Lu and the average was 0.11 Lu, which satisfied the target of 0.15 Lu. The maximum Lugeon value of the boreholes on the outer circle in Zone I after the improvement was under 0.65 Lu and the average was 0.22 Lu, which satisfied the target of 0.30 Lu. Among the boreholes on the inner circle in areas other than Zone I, the maximum Lugeon value after the improvement was under 0.35 Lu and the average was 0.13 Lu, which satisfied the target of 0.15 Lu. On the outer circle in areas other than Zone I, the maximum Lugeon value after the improvement was under 1 Lu and the average was 0.29 Lu, which satisfied the target of 0.65 Lu. Thus, the improvement target was achieved in all the rock subdivision zones.

In the bench grouting, all the boreholes on the inner circle in Zone I satisfied the grouting target. The maximum Lugeon value after the improvement was under 0.25 Lu and the average was 0.08 Lu, which satisfied the target of 0.15 Lu. The maximum Lugeon value of the boreholes on the outer circle in Zone I after the improvement was under 0.65 Lu and the

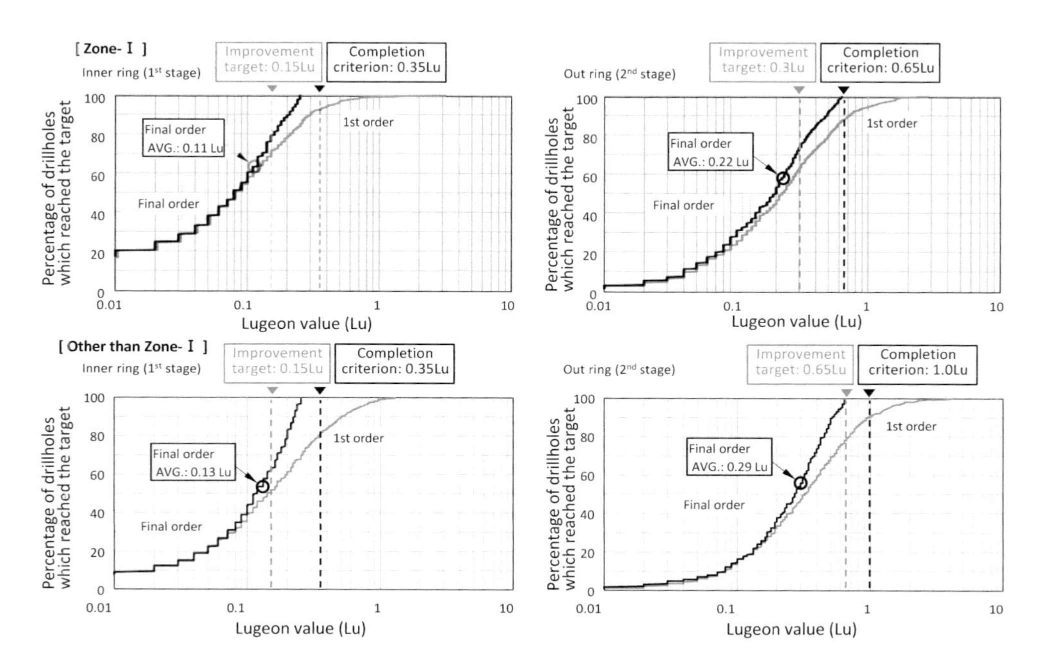

Figure 4.70 Distribution of the permeability before and after pre-grouting of the arches in the Kurashiki facility.

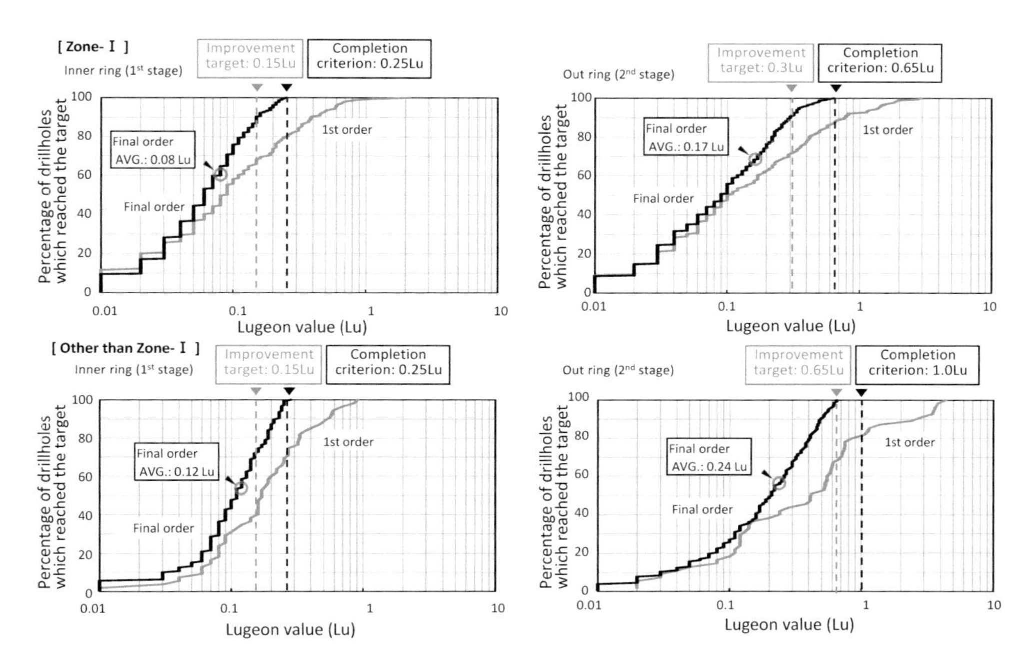

Figure 4.71 Distribution of the permeability before and after pre-grouting of the benches in the Kurashiki facility.

average was 0.17 Lu, which satisfied the target of 0.30 Lu. Among the boreholes on the inner circle of Zone V, the maximum Lugeon value after the improvement was under 0.25 Lu and the average was 0.12 Lu, which satisfied the target of 0.15 Lu. On the outer circle of Zone V, the maximum Lugeon value after the improvement was under 0.65 Lu and the average was 0.24 Lu, which satisfied the target of 0.65 Lu. Thus, the improvement target was achieved in all the rock subdivision zones.

After completing one section of pre-grouting of the arch and benches, that section was excavated. Then, the seepage rate into the cavern and the pore water pressure of the surrounding rocks were compared with the specified standard for management. The seepage rate is an important measurement in judging the overall effect of pre-grouting. The performance of grouting improvement was incorporated into the 3D hydrogeological model to predict the seepage rate at subsequent arch and bench excavation. This prediction was compared with the actual measurement and used in the management of excavation.

Figure 4.72 shows the comparison between the predicted and measured flow rate by excavation steps of each cavern and the whole site. The flow rate at caverns No. 1–4 greatly increased when the arch was excavated, while the increasing trend converged at the excavation of the benches. All the measured values were generally within the predicted range under 1.0 MPa of the sealing pressure. This confirmed that the improvement in the grouting zones reached the target permeability. In cavern No. 1, the seepage rate dropped from the stage of excavation of bench 1 to bench 2. This was due to the additional grouting on the floor of the benches and the post-grouting at the F-3 fault zone. It reflects both measured and predicted values. This post-grouting will be discussed in Sections 4.2.2.1.2 and 4.2.2.1.3.

Figure 4.73 shows a comparison between the measured and predicted values of the seepage rate at each stage of excavation for each cavern. In most cases, the measured water flow was within the predicted value. However, the measured value exceeded the predicted value in some parts of cavern No. 2, albeit small. As also seen in Figure 4.73, the actual seepage rate in cavern No. 2 slightly exceeded the prediction after excavation of the third bench. This section was a zone of microfractures with high permeability and low groutability. The performance of the grouting in this section will be discussed in Section 4.2.2.1.4. On completion of excavation, the water curtain pressure was 1.0 MPa and the rate of seepage into the caverns

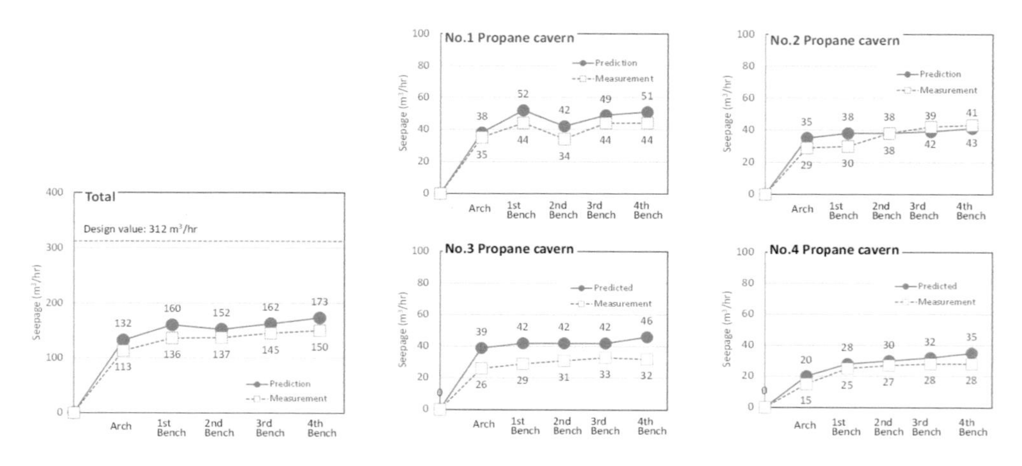

Figure 4.72 Comparison between the measured and predicted seepage rate by excavation steps in the Kurashiki facility.

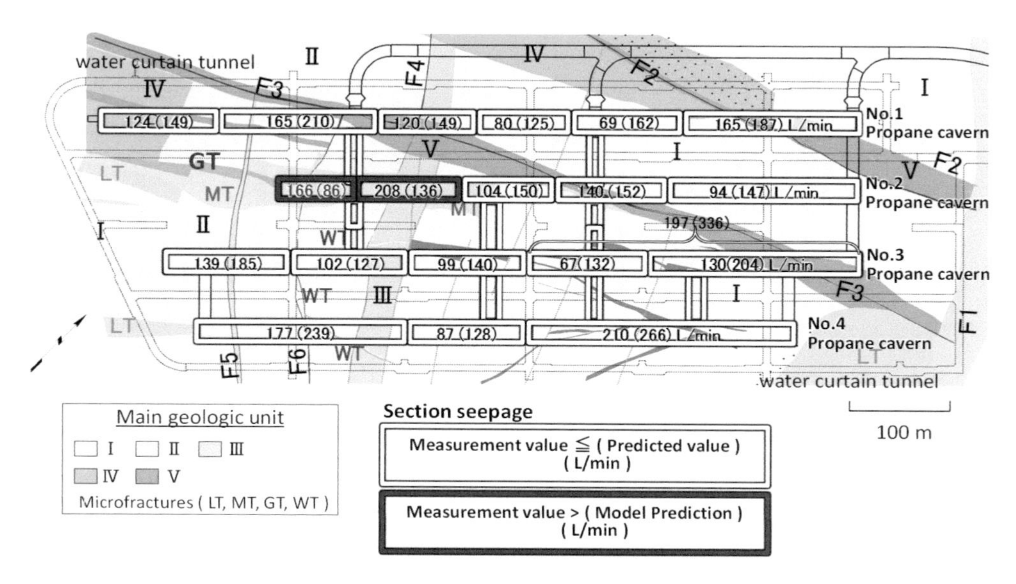

Figure 4.73 Comparison between predicted and actual seepage rate on completion of excavation by section in the Kurashiki facility. Actual measurement in front of brackets and predicted value in brackets (unit: liter/minute).

was 150 m^3/h, which is lower than the predicted 173 m^3/h. The seepage rate at each step in each section of the caverns was generally within the prediction. From this, the effectiveness of the grouting over the entire area was confirmed.

4.2.2.1.2 GROUTING THE CAVERN ARCHES

The seepage in several parts of cavern No. 1 locally exceeded the standard criteria, while the grouting target value was satisfied. This cavern was the first to be pre-grouted and excavated. The seepage into the cavern was controlled by post-grouting. This section describes its case study.

The causes of excessive local and sectional water flow were analysed. Figure 4.74 shows the geological section of the areas where excessive local and sectional water flow occurred at excavation of the arch and provides an example of the relationship between the stages of pre-grouting of the arch and the Lugeon values. As seen in the cross section on the left, the F3-1 faults cross the cavern axis at an acute angle. The horizontal axis of the graph on the right is the stage and the corresponding grouting borehole interval and the vertical axis is the average Lugeon values at the outer circle of the faulted parts alone and of the floor alone. The average Lugeon value over the entire section presented a decreasing trend with the progress of pre-grouting, reaching 0.54 Lu by the fourth stage, which satisfied the target value of 0.65 Lu. On the other hand, the average Lugeon values of the faulted parts and the floor did not show an appreciable decrease with progress at the grouting stages: it did not reach the target even after the fourth stage. The criterion to add grouting boreholes to the outer circle, 1 Lu, which was based on the assumption that the average Lugeon value of the circles would be under the target of 0.65 Lu, was not adequate in areas with low groutability such as fault

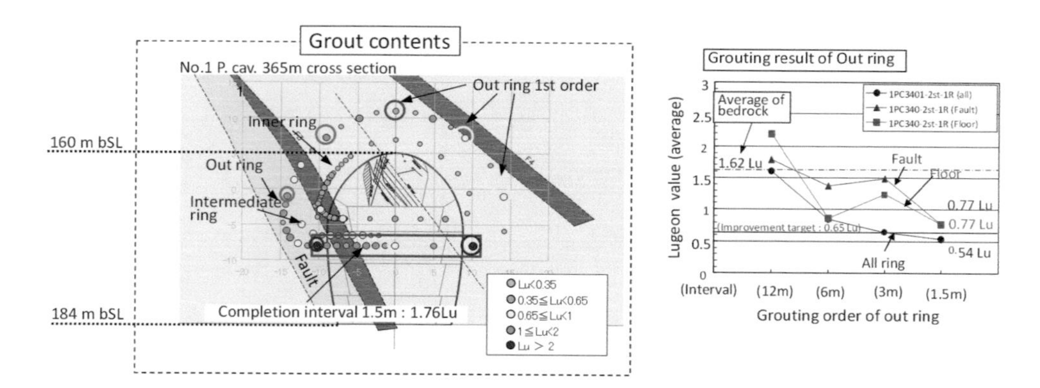

Figure 4.74 An example of arch pre-grouting of cavern No. 1 of the Kurashiki facility (outer ring).

zones. Accordingly, the criterion to add grouting boreholes to the outer circle was reduced from 1 to 0.65 Lu, thus the seepage rate was reduced. From this result, the criterion to add grouting boreholes to the outer circle was set to 0.65 Lu for all arch pre-grouting and bench pre-grouting of subsequent caverns.

4.2.2.1.3 GROUTING BENCHES OF CAVERNS

At cavern No. 1, after pre-grouting cavern No. 2 above, bench No. 1 was excavated and post-grouting was applied. At post-grouting, the seepage rate exceeded the predicted value in the section near high-angle F-3-1 faults while all the grouting boreholes reached the standard criteria values for additional boreholes. An investigation into the permeability of the F-3-1 faults at a test borehole revealed that the permeability of the grouting improvement zones did not satisfy the target improvement in the F-3-1 fault zone around benches 1 and 2: 0.15 and 0.30 Lu for the inner and outer circle, respectively. A detailed investigation using borehole television in the test borehole revealed that the cracks were densely distributed in the F-3-1 fault zone and crossed the cavern at a high angle. Then, the orientations of the post-grouting boreholes became coincident with the cracks reducing the frequency of intersecting each other, and some cracks were left ungrouted.

From the above observation, as seen in Figure 4.75, two columns of post-grouting, O and N, were added; these had a relatively good chance of crossing the F-3-1 faults. As a result, the inner and outer circles of Column O and the inner circle of Column N satisfied the target value, while the outer part of Column N failed. This was thought due to Column O crossing the cracks of the fault that Column N did not adequately intersect. For this, instead of drilling additional grouting boreholes from the floor of bench No. 1, bench No. 2 was excavated leaving the lower side wall of the F-3-1 fault and pre-grouting was applied from bench No. 2 utilising the remaining lower side wall as "cover rock". Thus, the grouting improvement of the zone managed to appropriately cross the cracks of the fault zones, and the improvement target was achieved over the fault zones where improvement from bench No. 1 had been inadequate.

As the cracks were not reached from bench No. 2, the same procedure was repeated utilising the "cover rocks" of the lower side wall of bench 3 for excavation of benches No. 3 and 4.

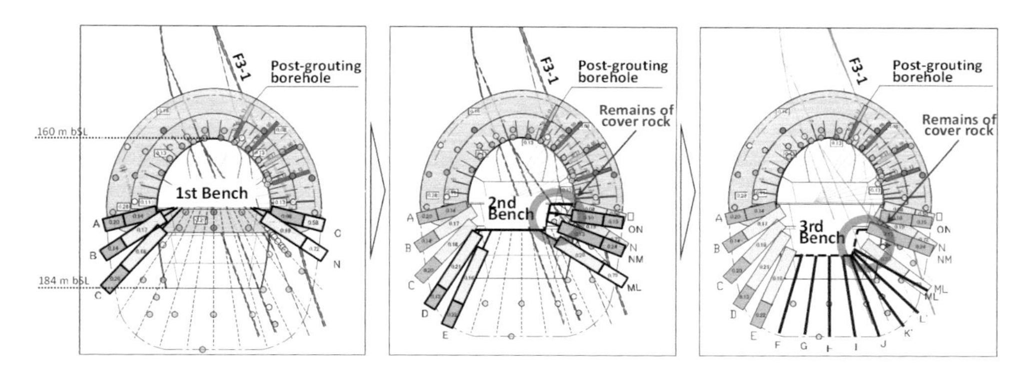

Figure 4.75 Revised pre- and post-grouting plan for the fault zone during bench excavation.

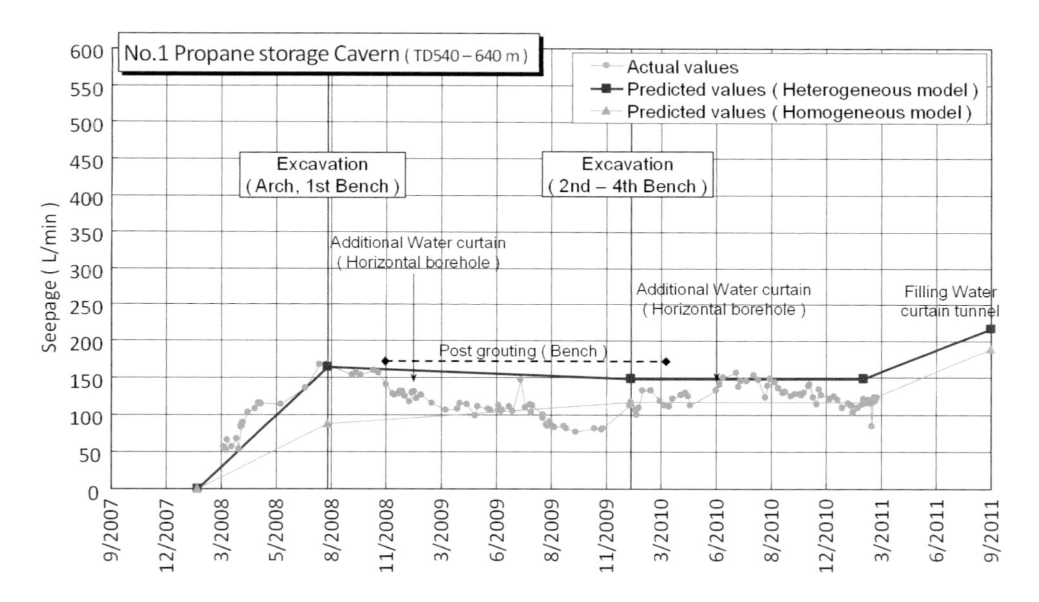

Figure 4.76 Comparison between predicted and measured seepage rate in the cavern during excavation in the zone where F-3-1 faults cross at a high angle in the Kurashiki facility.

Figure 4.76 shows the temporal variation in the predicted and measured seepage rate in the cavern at excavation in the zone where faults cross at a high angle. As a result of the revised grouting procedure, the seepage rate in the cavern decreased rather than increased during bench excavation. This led to the belief that an appropriate grout improvement zone was formed from the side wall to the floor in the fault zone.

4.2.2.1.4 GROUTING THE MICROFRACTURE ZONE

The dense microfracture zones containing groups of microfractures with openings of a few micrometers, meander changing their width from a few centimeters to a few meters. Their lengths range from a few meters to tens of meters. The microfracture zones require grouting as they have high permeability up to 3.3 Lu (0.6 Lu on average). Their groutability was extremely low because the grouting cement particles could only penetrate into a small area as the openings of the cracks were very small. This required many additional grouting boreholes.

The microfracture zones required five stages (interval to 75 cm) of grouting along the standard inner and outer circles. In addition, an intermediate circle between the inner and outer circles was arranged according to the investigation of the distribution and characteristics of the microfractures. Further, even more grouting boreholes were added parallel to the strike of the fractures if this intermediate circle did not achieve the target value of improvement (Figure 4.77). In order for effective grouting, the 3D distribution of the microfractures was investigated by borehole television and detailed observation of the cutting edges.

Figure 4.78 shows the cumulative percentage of grouting holes against the threshold Lugeon values by the stages of grouting in the microfractured zones. After the above additional grouting, the permeability of the microfracture zones satisfied the target value of improvement.

4.2.2.2 Performance in the Namikata Facility

4.2.2.2.1 GROUTING IMPROVEMENT OF THE ENTIRE STORAGE CAVERN

The performance of grouting in the propane storage cavern is described first. Figure 4.79 shows the cumulative percentage of grouting holes against the threshold Lugeon values before and after arch grouting of the propane storage cavern. In the high permeability zone, the maximum Lugeon value dropped from 74 to 1 Lu by grouting improvement, satisfying the completion criterion. The average Lugeon value also dropped from 0.61 to 0.1 Lu by grouting improvement, which satisfied the completion criterion. In the medium to low permeability zone, the maximum Lugeon value dropped from 22 to 1.0 Lu by grouting improvement, satisfying the completion criterion. The average Lugeon value also dropped from 0.34 to 0.08 Lu, which satisfied the completion criterion of 0.1 Lu.

Figure 4.80 shows the cumulative percentage of grouting holes against the threshold Lugeon values before and after bench grouting of the propane storage cavern. In the high

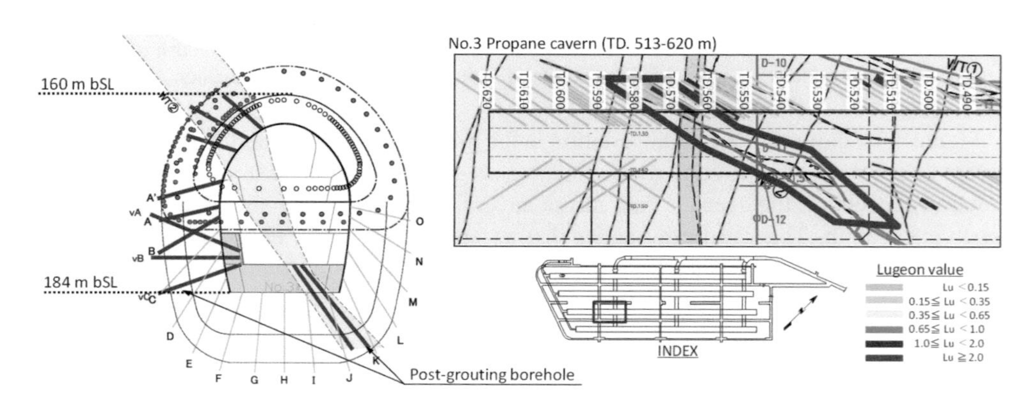

Figure 4.77 Post-grouting of the microfracture zone in cavern No. 3 of the Kurashiki facility.

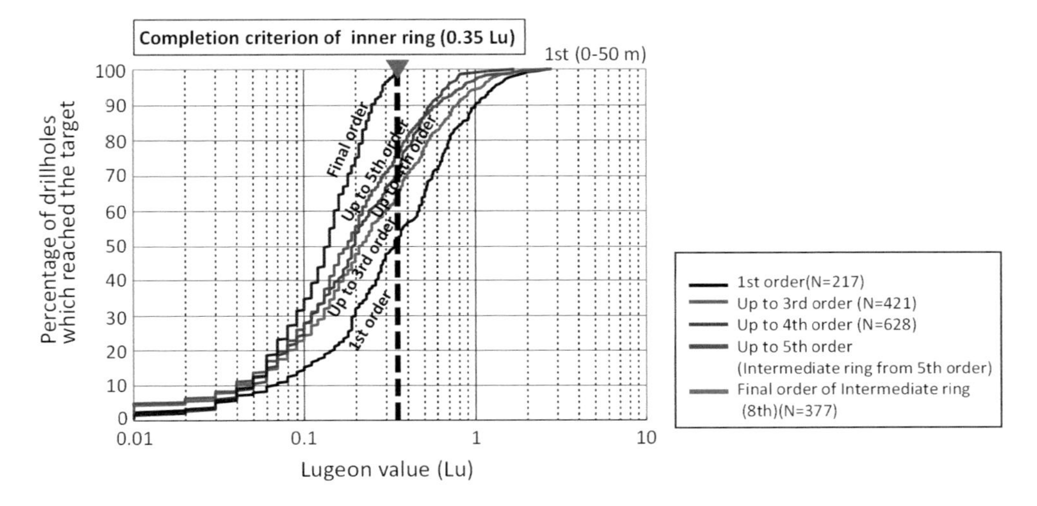

Figure 4.78 Percentage of grouting holes that reached Lugeon values lower than the values on the horizontal axis for each grouting stage in the Kurashiki facility.

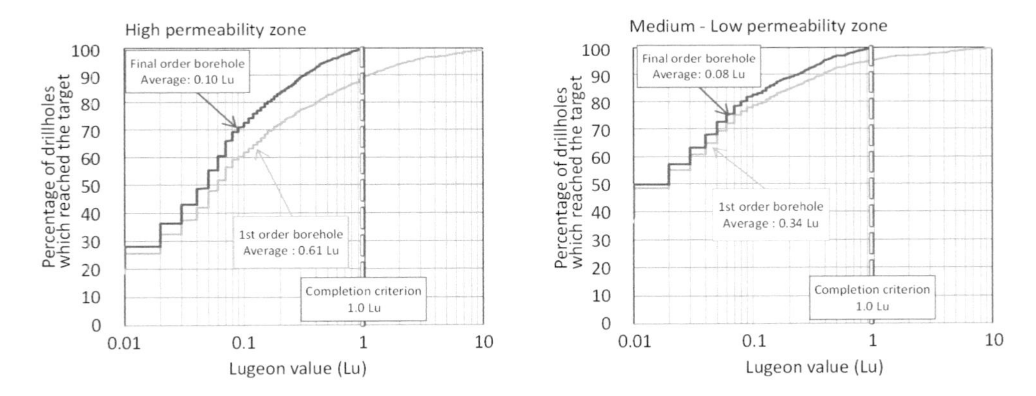

Figure 4.79 Percentage of grouting holes that reached Lugeon values lower than the values on the horizontal axis before and after arch grouting of the propane storage cavern in the Namikata facility.

permeability zone, the maximum Lugeon value was 16 Lu, but the permeability dropped to 1 Lu by grouting improvement and satisfied the standard criterion. The average Lugeon value also dropped from 0.39 to 0.09 Lu by grouting improvement, which satisfied the completion criterion. In the medium to low permeability zone, the maximum Lugeon value dropped from 7 Lu to below 1.0 Lu by grouting improvement, satisfying the completion criterion. The average Lugeon value also dropped from 0.22 to 0.10 Lu, which satisfied the completion criterion of 0.1 Lu.

Figure 4.81 shows the cumulative percentage of grouting holes against the threshold Lugeon values before and after arch grouting of the combined butane/propane storage cavern. The maximum Lugeon value was 8 Lu, but the permeability dropped to 1.0 Lu and concluded

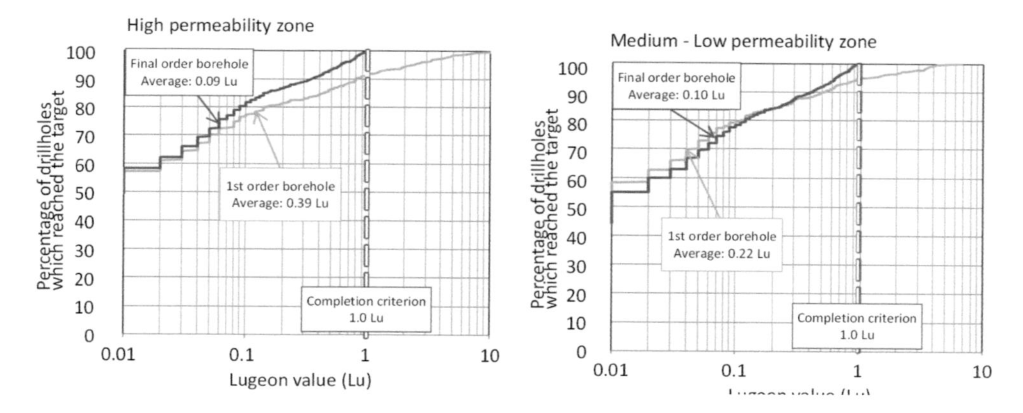

Figure 4.80 Percentage of grouting holes that reached Lugeon values lower than the values on the horizontal axis before and after bench grouting of the propane storage cavern in the Namikata facility.

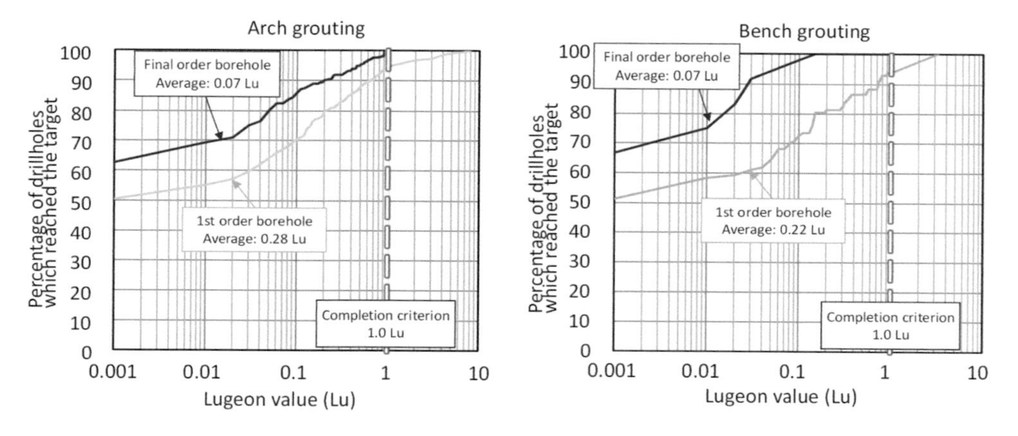

Figure 4.81 Percentage of grouting holes that reached Lugeon values lower than the values on the horizontal axis before and after arch and bench grouting of the combined butane/propane storage cavern in the Namikata facility.

at 0.07 Lu. In the bench grouting, the maximum Lugeon value dropped by grouting from 3.2 Lu to below 1.0 Lu and concluded at 0.02 Lu. At the final stage of arch and bench grouting of the combined butane/propane storage cavern, the target values of permeability, maximum 1.0 Lu and average 0.1 Lu, were achieved.

Figure 4.82 shows the comparison between the predicted and measured flow rate by excavation steps of each cavern. Both caverns presented measured flow rates lower than prediction at all stages of excavation, which were the result of grouting improvement around the caverns. The flow rate of the groundwater on completion of excavation was 5 and 20 m^3/h for the combined butane/propane storage cavern and propane storage cavern, respectively, which were generally consistent with the values predicted by the 3D hydrogeological model.

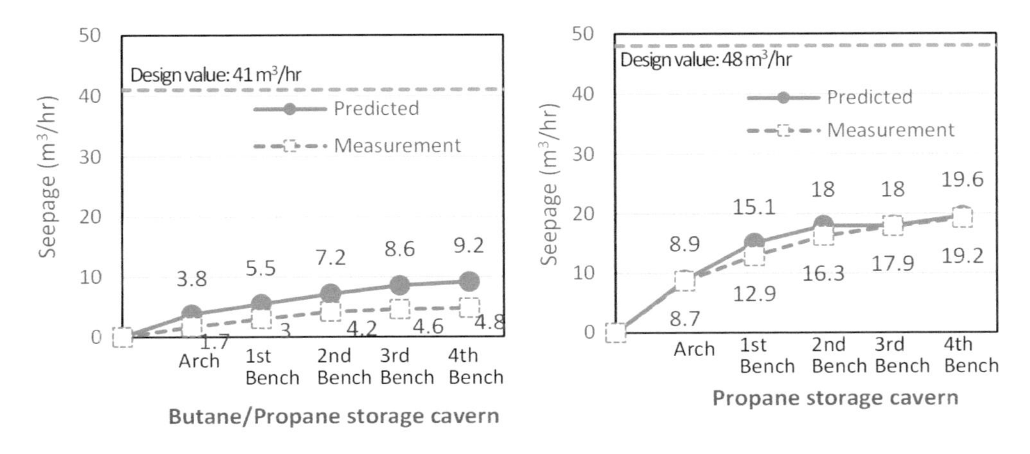

Figure 4.82 Comparison between the measured and predicted seepage rate by excavation steps in the Namikata facility.

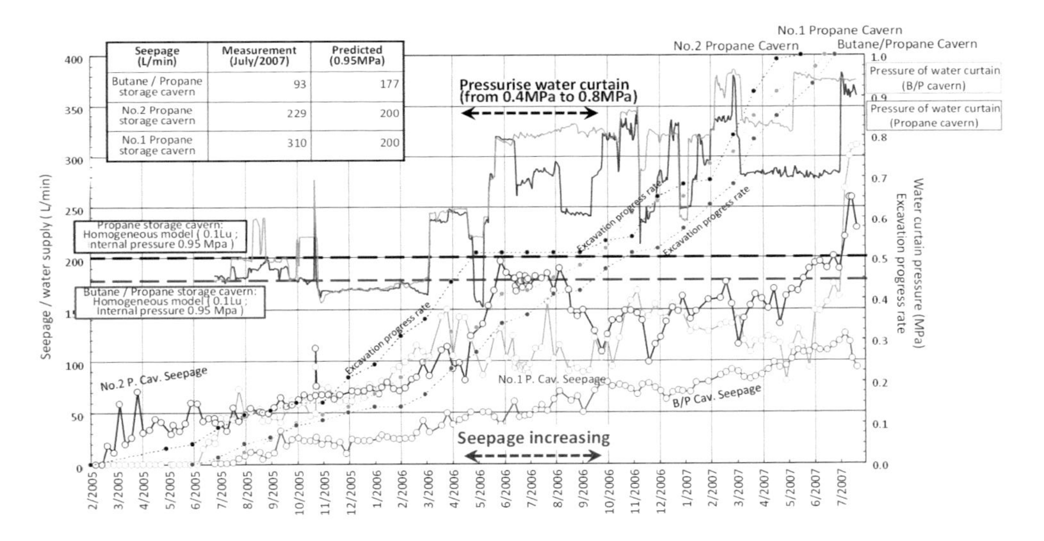

Figure 4.83 Measurement of the seepage rate and confining water pressure during cavern excavation in the Namikata facility.

4.2.2.2.2 POST-GROUTING

This section briefly describes the post-grouting of No. 2 propane storage cavern in the Namikata facility as an example. Figure 4.83 shows the measurement of the seepage rate and the confining water pressure during the cavern excavation. When the confining water pressure was increased from 0.4 to 0.8 MPa during excavation of bench No. 1 in No. 2 propane storage cavern, the seepage rate increased. For this, post-grouting was applied.

Figure 4.84 shows a map of No. 2 propane storage cavern with Lugeon values before post-grouting and the localities of seeps on the ceiling of the arch when the sealing water pressure

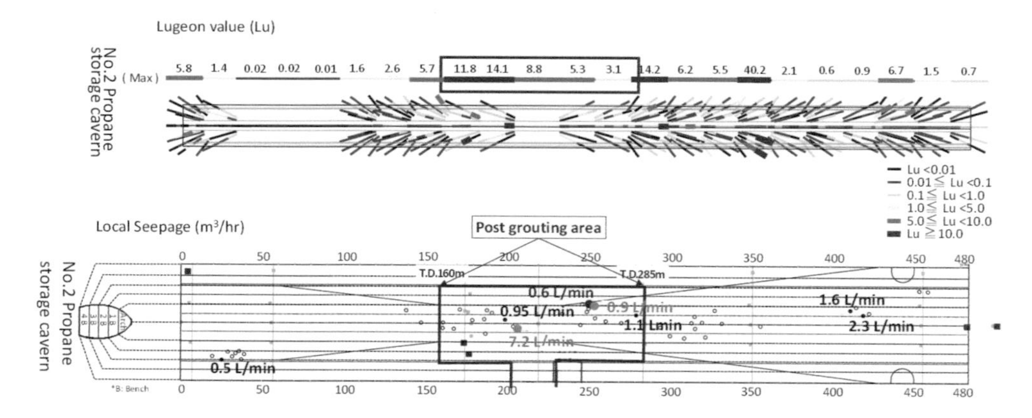

Figure 4.84 Map of No. 2 propane storage cavern with Lugeon values before post-grouting.

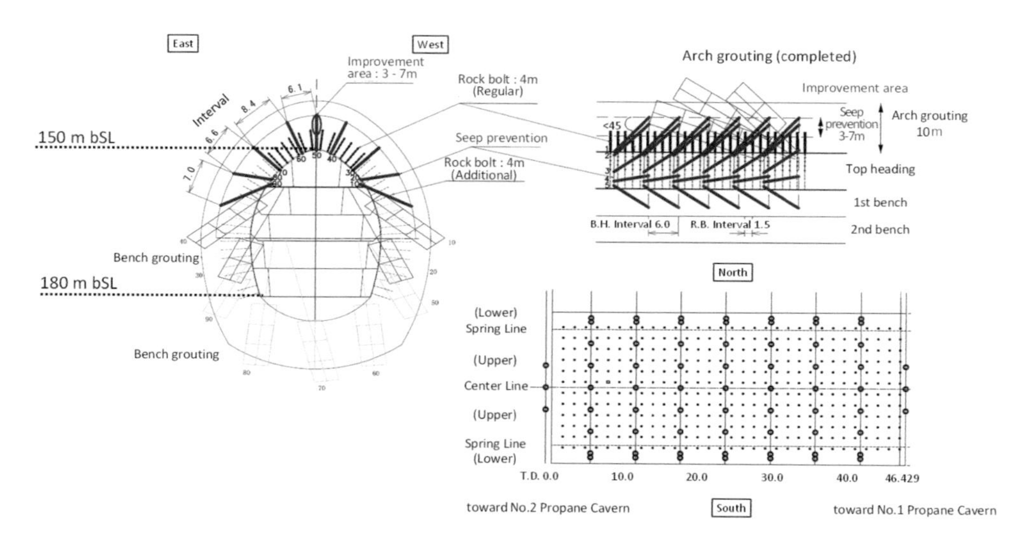

Figure 4.85 Layout of the post-grouting drillholes of No. 2 propane storage cavern in the Namikata facility.

was increased from 0.4 to 0.8 MPa. A seepage rate over 0.5 L/min was distributed in the centre of the cavern and the maximum flow rate was 5.8 L/min. This area had been classified as medium permeability at the time of arch pre-grouting, and one cycle of pre-grouting was applied. An analysis of the data at pre-grouting found a grouting borehole with a maximum 10 Lu in each cycle of boreholes, which implied the presence of local highly permeable cracks. In these zones, post-grouting was planned with the specification in Figure 4.85.

Figure 4.85 shows the specification of the post-grouting. The length of the boreholes was 8 m; the "cover rock" was 3 m; the injection length per stage was 5 m as a base; the injection pressure was 2.0 MPa above the seep water pressure; injection was for 30 min at 0.1 L/min/m; and the completion criterion was 1 Lu. The boreholes were obliquely arranged at 45° to avoid crossing the rock bolts.

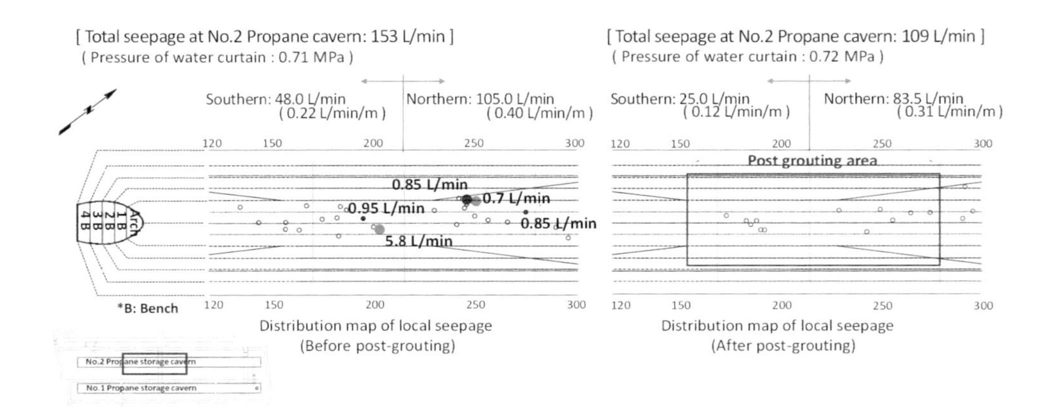

Figure 4.86 Map of seeps in the cavern before and after post-grouting in No. 2 storage cavern in the Namikata facility.

Figure 4.86 is a map of seeps in the cavern before and after post-grouting in this area. Post-grouting resulted in a reduction in the seepage rate from 153 to 109 L/min, local seep of less than 0.5 L/min and overall seepage less than the prediction. Thus, post-grouting was confirmed to be effective.

4.3 Evaluation of the Groundwater Control System

On completion of excavation of the caverns but before the air-tightness test, a water sealing test was carried out by pressurising the system to check the functionality of the groundwater controlling system. Once air leaks, causing an unsaturated zone in a crack, it is extremely difficult to re-saturate it. Therefore, it is very important to constantly maintain the water sealing function. On completion of cavern excavation, water sealing walls were constructed at the joints of water curtain tunnels and access tunnels, and then the water curtain tunnel was inundated. Water-tightness was confirmed by increasing the injection pressure into the water curtain system of water curtain tunnels and drillholes. Then, the pressure of the water curtain drillholes and tunnels was increased to the highest structural water level. The maximum water pressure was thus applied to the caverns against atmospheric pressure. The water sealing functionality and kinetic stability were confirmed in this manner.

4.3.1 The Kurashiki Facility

Figure 4.87 shows a conceptual diagram of the water curtain at the time of the pressurised water sealing test and the temporal variation in the water curtain pressure at the Kurashiki facility. Pressurised water was injected into the water supply shaft to increase the water pressure of the water curtains: pressure in the water curtain drillholes was from 1.1 to 1.35 MPa and the water immersing level in the water curtain tunnel was from 134 to 15 m below sea level. The increased immersing level of the water curtain tunnels corresponds to 0.06–1.225 MPa in water pressure.

The contour maps at the top of Figure 4.88 show the distribution of the piezometric head in cross section B on completion of cavern excavation and at the maximum sealing water

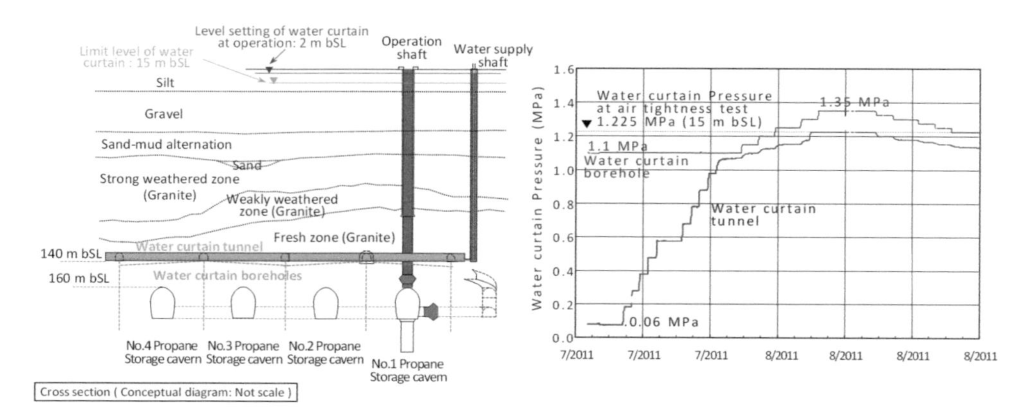

Figure 4.87 Conceptual diagram of the water curtain at the time of the pressurised water sealing test and the temporal variation in the water curtain pressure at the Kurashiki facility.

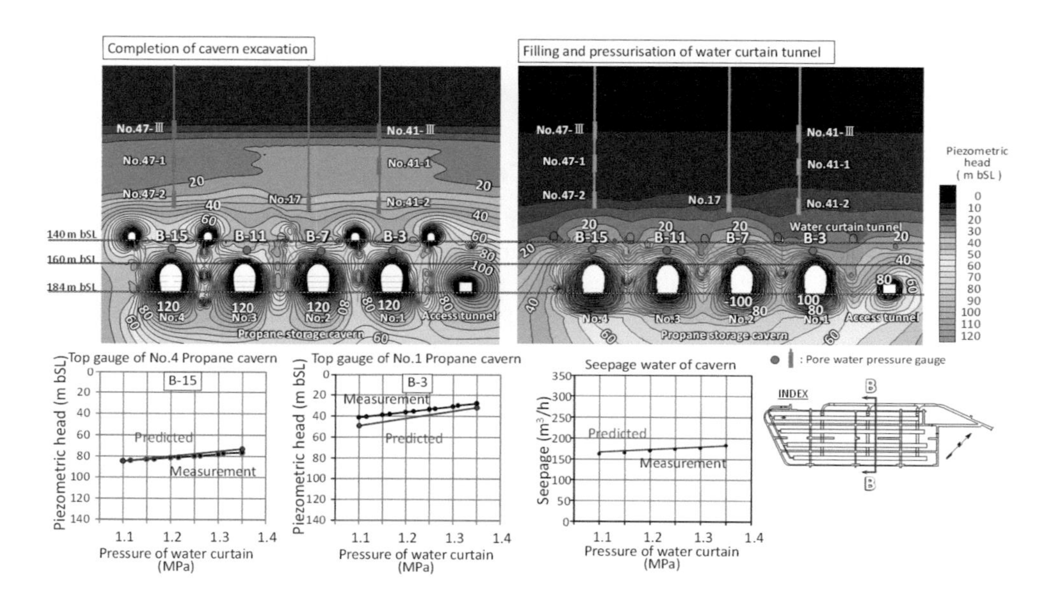

Figure 4.88 Distributions of the pore water pressure in cross section B on completion of the cavern and at pressurising water (top). Comparison between the measured and predicted values of the pore water pressure and seepage rate at pressurised water sealing in the Kurashiki facility (bottom).

pressure predicted by the 3D hydrogeological model. The graphs at the bottom of Figure 4.88 show a comparison between the measured and predicted seepage rate at pressurised water sealing. When the water curtain tunnels were immersed with water and the water pressure of the water curtain increased from 1.1 to 1.35 MPa, the pore water pressure increased by 10–15 m and the seepage rate increased by about 20 m^3/h. This was generally as predicted.

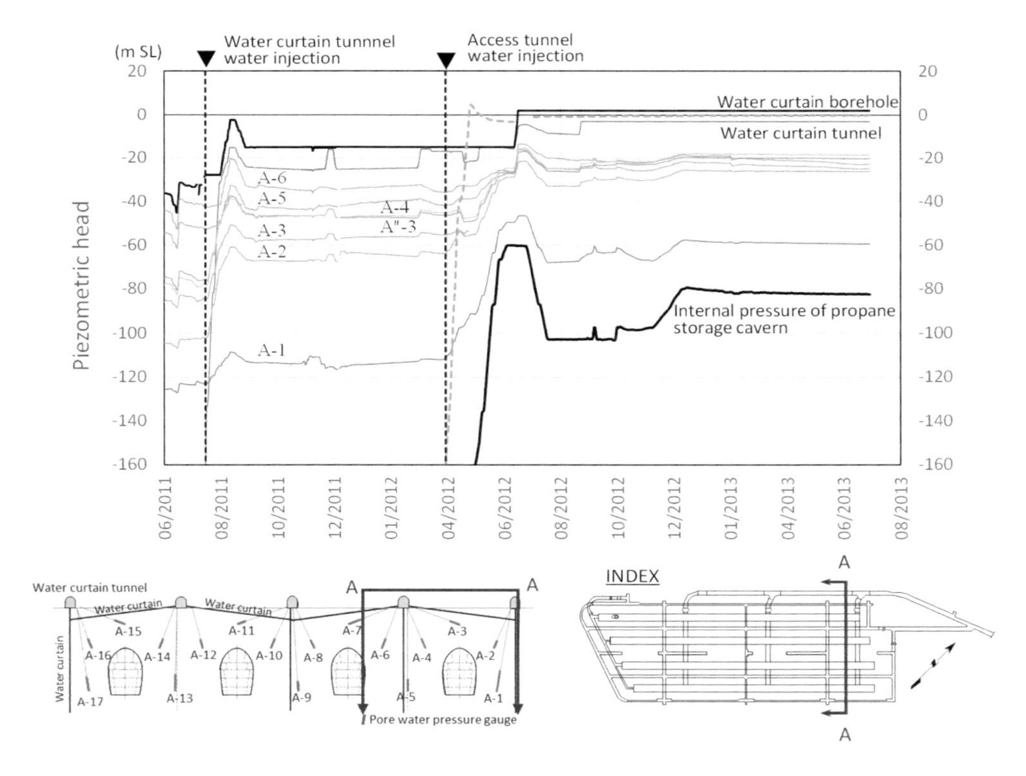

Figure 4.89 Temporal variation of the pore water pressure at the time of increasing the sealing water pressure (cross section A) in the Kurashiki facility.

The seepage rate at the maximum water pressure in the pressurised test was 182.3 m^3/h, which was lower than the flow rate specified in the design: 312 m^3/h.

Figure 4.89 shows the temporal variation in the pore water pressure at the time of increasing the sealing water pressure inside the water sealing curtains at the top (148 m below sea level) and shoulder (160 m below sea level) of cross section A as an example. When the water curtain tunnels were immersed with water and the water curtain pressure was increased from 1.1 to 1.35 MPa, an increase in the pore water pressure was recognised and its magnitude was 10–30 m.

Figure 4.90 shows the temporal variation in the pore water pressure from before cavern excavation to the increasing sealing water pressure test in the fresh granite above the water sealing curtains. The regional pore water pressure dropped at excavation of the water curtain tunnel, immersing and pressurising the water curtain tunnels to their original pre-excavation state. From this observation, the effectiveness of grouting improvement around the caverns and the water curtain system by pressurised water into the water curtain boreholes operated at excavation was confirmed.

4.3.2 The Namikata Facility

Figure 4.91 shows a conceptual diagram of the water sealing situation at the time of the pressurised water sealing test and the temporal variation in the water curtain pressure at the

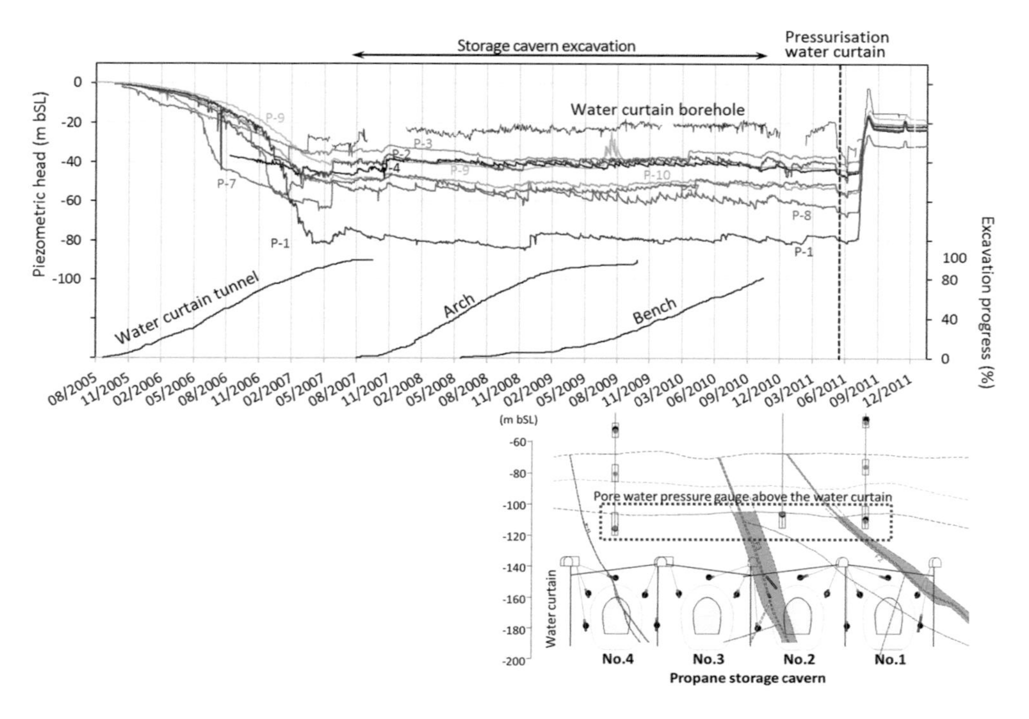

Figure 4.90 Temporal variation of the pore water pressure in the fresh granite above the water sealing curtains in the Kurashiki facility.

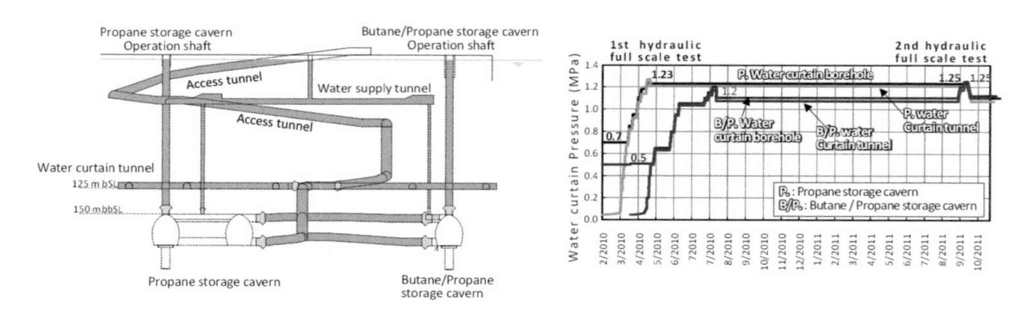

Figure 4.91 Conceptual diagram of the water curtain at the time of the pressurised water sealing test and the temporal variation in the water curtain pressure in the Namikata facility.

Namikata facility. The water curtain tunnels were pressurised to 0.5 MPa after immersing, then the water curtain boreholes were pressurised from 0.5 to 1.20 MPa matched by a subsequent increase in the pressure in the water curtain tunnel. The combined butane/propane storage cavern and propane storage cavern were separately pressurised.

The contour maps at the top of Figure 4.92 show the distribution of the piezometric head in cross section B on completion of cavern excavation and at the maximum sealing water pressure predicted by the 3D hydrogeological model. The graphs at the bottom of Figure 4.92

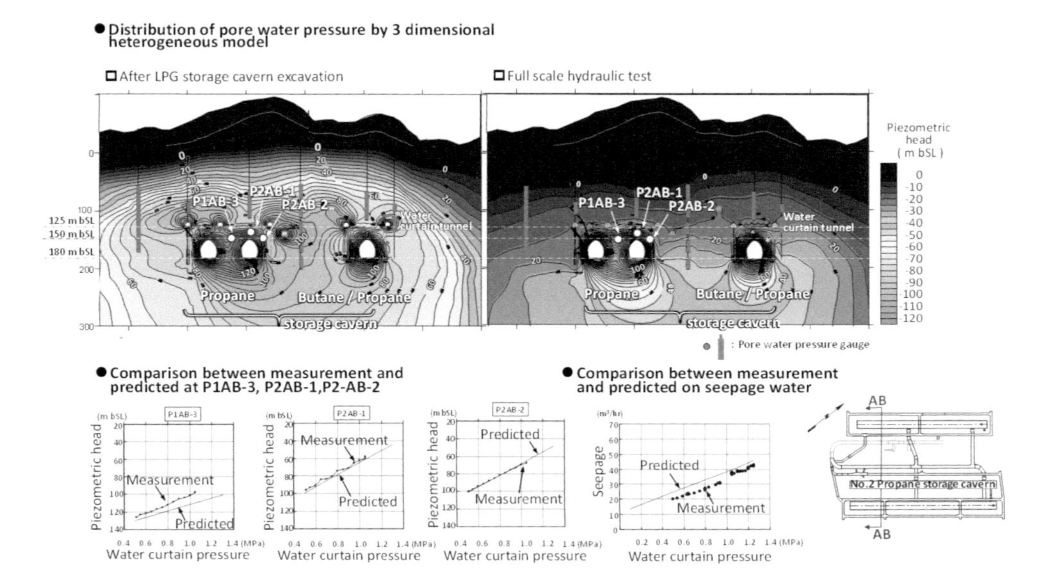

Figure 4.92 Distributions of the pore water pressure in the cross section AB on completion of the cavern and at pressurised water (top). Comparison between the measured and predicted values of the pore water pressure and seepage rate at the pressurised water sealing in the Namikata facility (bottom).

show a comparison between the measured and predicted seepage rate at pressurised water sealing. When the water curtain tunnels were immersed with water, the water curtain pressure was increased from 0.52 to 1.2 MPa, the pore water pressure increased by 40–50 m and by average rate increased about 25 m^3/h. This was generally as predicted. The seepage rate at the maximum water pressure in the pressurised test was 43 m^3/h, which was lower than the flow rate specified in the design: 48 m^3/h.

The regional pore water pressure above the water sealing curtain dropped at excavation of the water curtain tunnel, but immersing and pressurising the water curtain tunnels brought it back to its original pre-excavation state. This observation confirmed the effectiveness of grouting improvement around the caverns and the effectiveness of the groundwater control of the water curtain system with pressurised water in the water curtain boreholes at excavation.

4.3.3 Addition of Post-Grouting Based on Pressurised Water Sealing Test

Figure 4.93 shows the temporal variation in the pore water pressure and the seepage rate at the time of increasing the sealing water pressure inside the water sealing. At the pressurised water sealing test, the water curtain pressure was increased from 0.7 to 1.23 MPa in the combined butane/propane storage cavern and 0.5–1.20 MPa in the propane storage cavern. As a result (Figure 4.94), the seepage rate changed as predicted in the combined butane/propane storage cavern and propane storage No. 2 cavern. On the other hand, in propane storage No. 1, the seepage rate in the southern side of the cavern increased when the water curtain pressure reached 1.2 MPa, and the internal displacement of the cavern also increased. As a

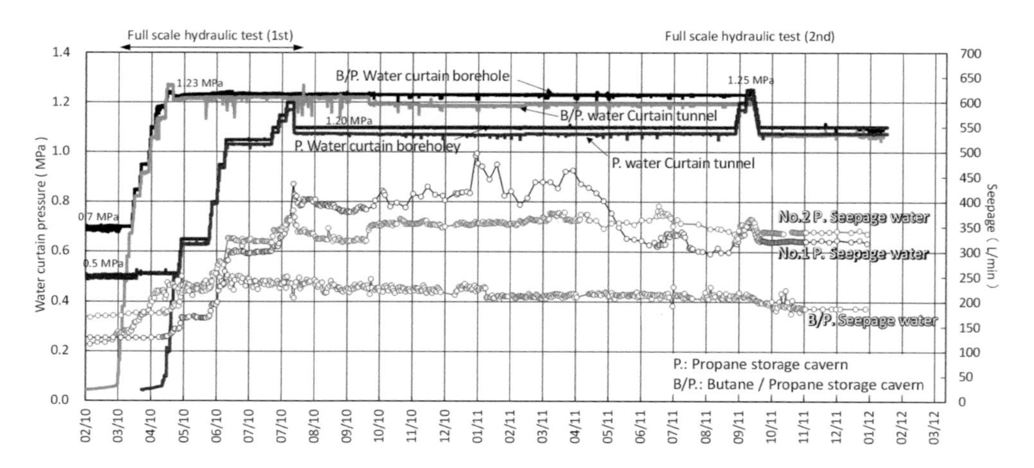

Figure 4.93 Temporal variation of the pore water pressure and the seepage rate at the time of increasing the sealing water pressure (cross section A) in the Namikata facility.

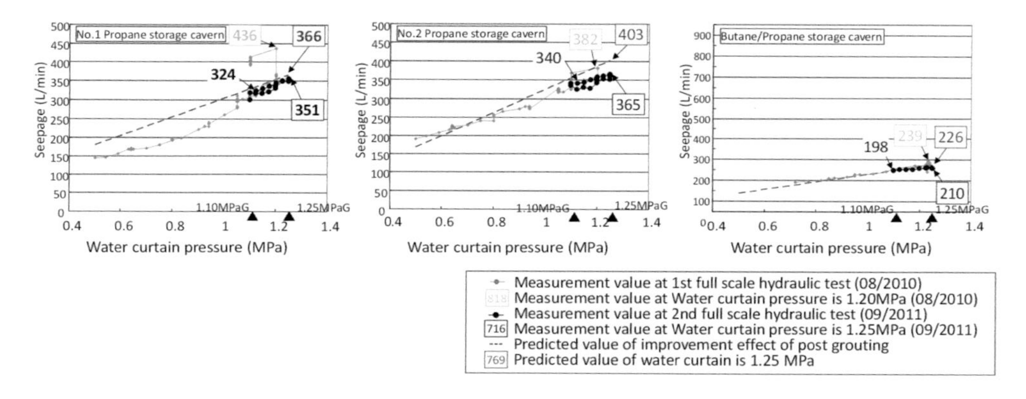

Figure 4.94 Correlation between the water curtain pressure and the seepage rate by cavern in the Namikata facility when the water sealing was pressurised.

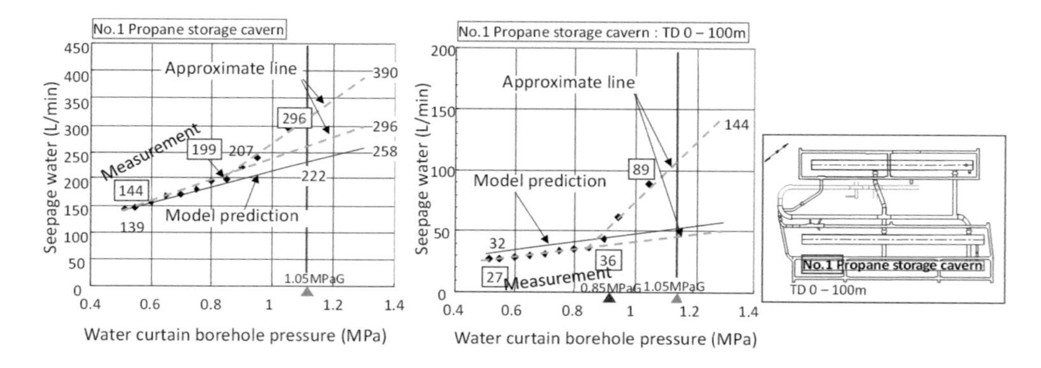

Figure 4.95 Correlation between the water curtain pressure and the seepage rate in propane storage cavern No. 1 of the Namikata facility when the water sealing was pressurised.

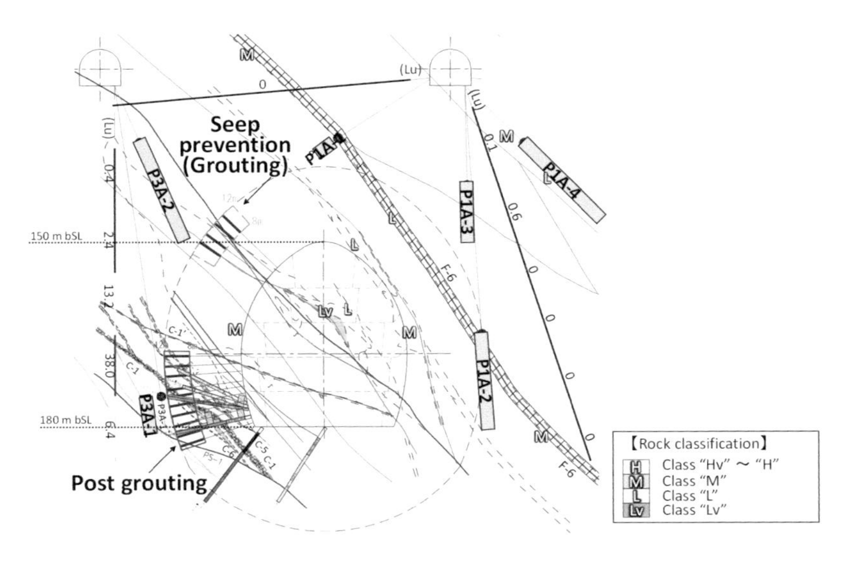

Figure 4.96 Result of a re-examination of the high seepage area of propane storage cavern No. 1 in the Namikata facility.

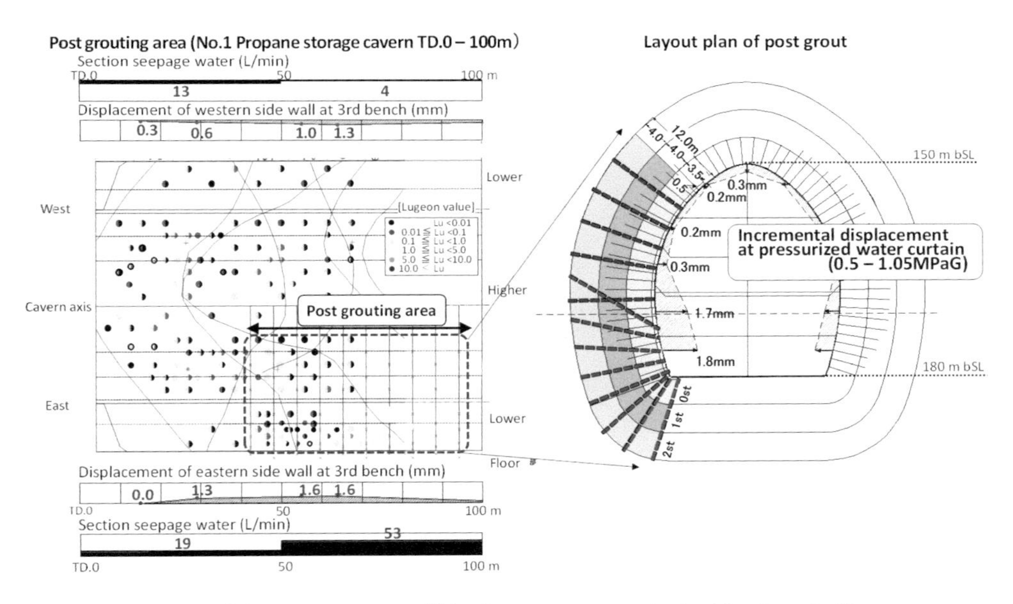

Figure 4.97 Outline of the post-grouting of No. 1 storage cavern in the Namikata facility.

local remedy for this, the sealing water pressure was reduced and post-grouting, additional rock bolts and shotcrete were applied.

Figure 4.95 shows the correlation between the water curtain pressure and the seepage rate in propane storage cavern No. 1 when the water sealing was pressurised. The flow rate

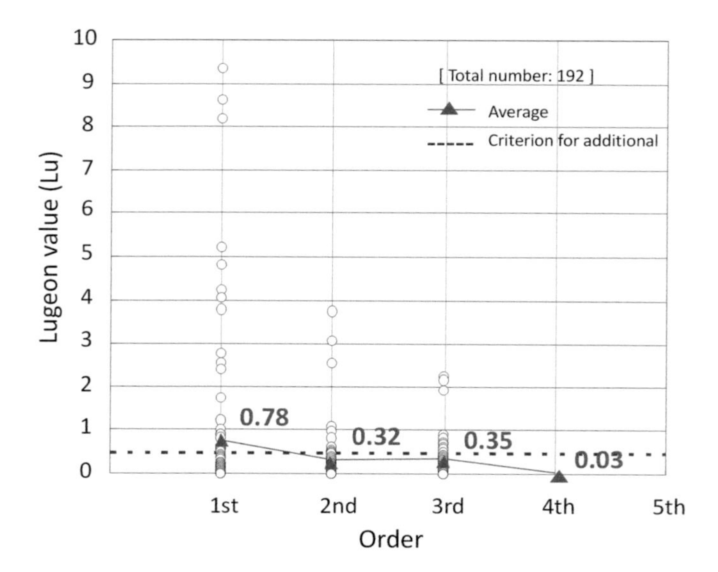

Figure 4.98 Reduction in the Lugeon value by stages of post-grouting of No. 1 propane storage cavern in the Namikata facility.

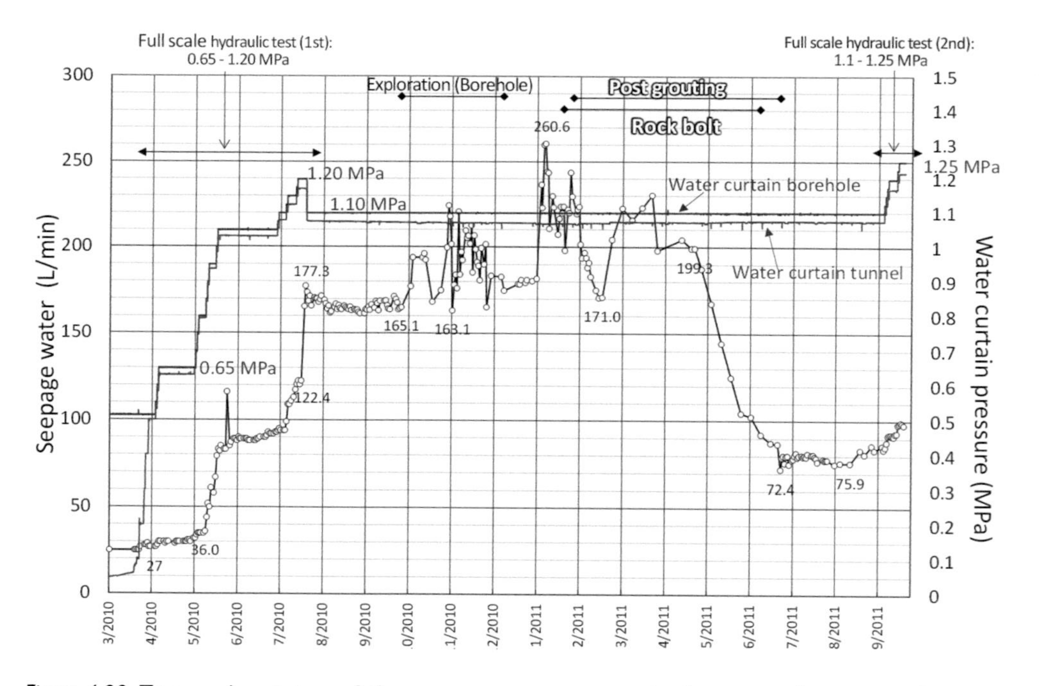

Figure 4.99 Temporal variation of the seepage rate at pressurised water sealing test in 0–100 m intervals of the No. 1 propane storage cavern in the Namikata facility.

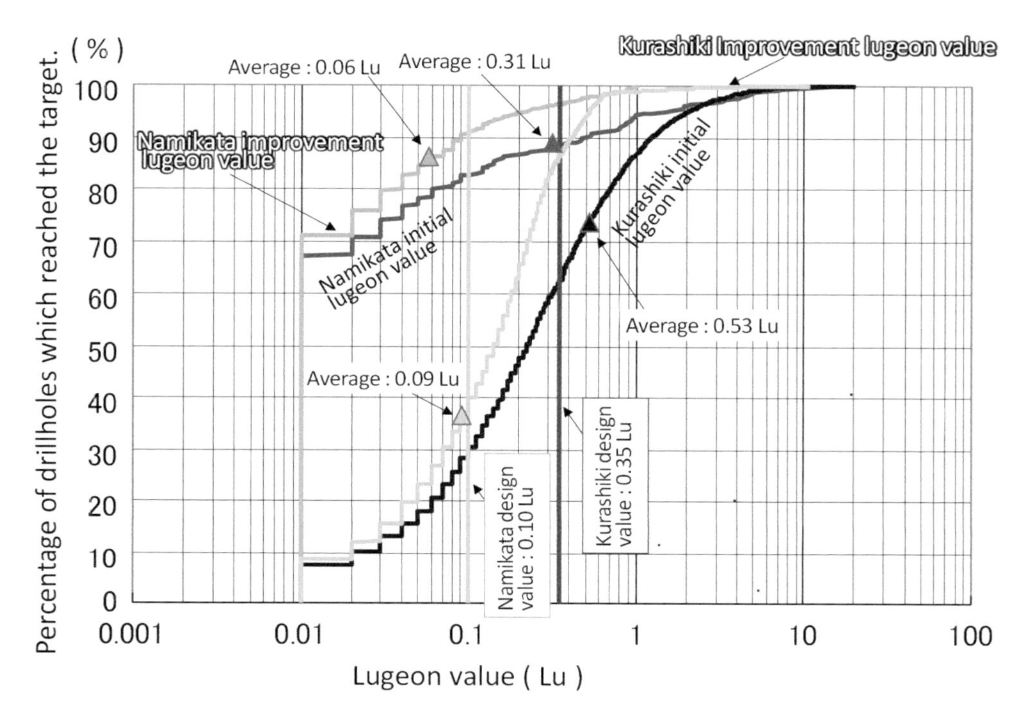

Figure 4.100 Distribution of permeability before and after grouting improvement in the Kurashiki and Namikata facilities.

linearly increased with the water pressure up to 0.85 MPa as predicted, but it exceeded the linear trend from 0.85 to 1.05 MPa. The section of this increase was 100 m in the southern part and other parts were as predicted. This section was re-examined by cores and BTV of additional boreholes.

Figure 4.96 shows the hydrogeological structure from this re-examination. Highly permeable low-angle cracks were detected in the lower part of the side walls where a low permeability zone had been expected. These cracks were considered the cause of the high water flow. These were locally treated by additional post-grouting to reduce permeability.

Local post-grouting was carried out as shown in Figure 4.97. The post-grouting water sealing injection was carried out to prevent leakage into the cavern and deformation of the wall of the cavern. The result of the post-grouting is shown in Figure 4.98. Using post-grouting, the Lugeon value was reduced to under 0.1 Lu.

Figure 4.99 shows the temporal variation of the increasing pressure of water sealing before and after post-grouting and the seepage rate in the 0–100 m interval of propane storage cavern No. 1. The seepage rate in the 0–100 m interval in the southern part of propane storage cavern No. 1 decreased from 150 to 75 L/min, which was within the prediction. The effectiveness of grouting improvement was thus confirmed.

Figure 4.100 shows the distribution of permeability in the areas of grouting improvement in the Kurashiki and Namikata facilities.

Among the grouting boreholes in the original bedrock of the Kurashiki facility area, less than 10% had Lugeon values lower than 0.01 Lu, and 15% of them were over 1 Lu. The average was 0.53 Lu. The grouting improvement brought the average Lugeon value around the cavern down to 0.15 Lu.

The average Lugeon value of the bedrock of the Namikata facility area was 0.3 Lu, and the grouting improvement reduced the average to 0.05 Lu around the cavern. As seen, the grouting improvement target was achieved in both the Kurashiki and Namikata facilities, and the indicators of the behaviour of the groundwater, such as the pore water pressure and the seepage rate, were brought to values within the management criteria.

References and Further Readings

Aberg B. (1977) Prevention of gas leakage from unlined reservoir in rock. *1st International Symposium on Storage in Excavated Rock Caverns, Rockstore77*, Stockholm, 399–414.

Aberg B. (1989) Pressure distribution around stationary gas bubbles in water-saturated rock fractures. *Proceedings of the International Conference on Storage of Gases in Rock Caverns*, Trondheim, 77–86.

Aoki K., Mito Y., Kurokawa Y., Yamamoto T., Date K., Wakita S. (2003) Development of dynamic grouting technique for the improvement of low-permeable rock masses. *Proceeding of the First Kyoto International Symposium on Underground Environment (UE-KYOTO)*, Kyoto, 345–352.

Aoki K., Mito Y., Nakamura Y., Yamamoto T., Date K., Watita S. (2005) Dynamic grouting technique for the improvement of low-permeable rock masses. *Proceedings of the 10th ACCUS Conference Underground Space, Economy and Environment, ISRM Regional Symposium 2005*, Moscow.

Aoki K., Mito Y., Yamamoto T., Date K., Wakita S. (2003) Evaluation of dynamic grouting effect for low-permeable rock. *Proceedings of the 10th ISRM Congress*, Johannesburg, 39–43.

Bergman M. (1977) Storage in excavated rock caverns. *Proceedings of the First International Symposium*, Stockholm, 832.

Date K., Wakita S., Yamamoto T., Nakajima Y., Hoshino Y., Aoki K., ProceedingMito Y. (2003) Development of dynamic grouting technique for the ground improvement. Proceedings of the ITA World Tunneling Congress, (Re)claiming the Underground Spase volume2, Amsterdam 2003, 929–935.

Fujii K., Chang, C.S., Okazaki, Y., Kaneto, T., Maejima, T. (2014) Observational groundwater control method for construction of hydraulic containment type LPG storage cavern. *International Symposium - 8th Asian Rock Mechanics Symposium, Sapporo, Japan.*

Goodall D.C., Kjorholt H. (2014) Estimation of gas loss from pressurised unlined cavern without water curtains. *Proceedings of the International Conference on Storage of Gases in Rock Caverns*, Trondheim, 49–55.

Gustafson G., Rhen I., Boden A. (1989) Storage of gases in rock caverns - Fieldtest in boreholes at Stripa, Sweden. *Proceedings of the International Conference on Storage of Gases in Rock Caverns*, Trondheim, 43–48.

Kjorholt H., Goodall D.C., Johansen P.P., Stokkebo O. (1989) Water curtain performance at the Kvilldal air cushion. *Proceedings of the International Conference on Storage of Gases in Rock Caverns*, Trondheim, 87–94.

Kobayashi S., Soya, M., Takeuchi, N., Onishi, M., Ymamoto, H., Kaneto, T., Maejima, T. (2014) Rock grouting with super-fine cement and supplementarily used colloidal silica at Kurashiki underground LPG storage base. *International Symposium - 8th Asian Rock Mechanics Symposium, Sapporo.*

Lindblom U.E. (1989) The performance of a water curtain during 10 years of operation. *Proceedings of the International Conference on Storage of Gases in Rock Caverns*, Trondheim, 347–354.

Nermoen B., Blindheim O.T. (1989) Ground water maintenance and leakage control during construction of unlined rock caverns for pressurised gas storage, Mongstad. *Proceedings of the International Conference on Storage of Gases in Rock Caverns*, Trondheim, 317–322.

Nilsen B., Olsen J. (1989) Storage of gases in rock caverns. *Proceedings of the International Conference on Storage of Gases in Rock Caverns*, Trondheim, 398.

Okazaki Y., Kurose H., Okubo S., Maejima T., Tezuka Y., Soya M., Aoki K. (2014) Construction of water curtain system for the hydraulic containment type LPG storage cavern. *International Journal of the JCRM*, 10(2), 32–41.

Shamoto Y., Aoki K. (1989) Groundwater control technique around underground crude oil storage caverns. *International Congress on Progress and Innovation in Tunneling*, Toronto, 165–172.

Wakita S., Date K., Yamamoto T., Aoki K., Mito Y. (2005) Development of dynamic grouting method effective for construction of underground storage caverns. *Proceedings of the International Symposium on Designs, Construction and Operation of Long Tunnels*, Taipei, 225–232.

Wakita S., Date K., Yamamoto T., Nakamura Y., Mito Y., Aoki K. (2004) Development of dynamic grouting technique for the low-permeable rock mass. *Proceedings of the ISRM International Symposium 2004 (3rd ARMS, KYOTO)*, Kyoto, 437–443.

Chapter 5

Mechanical Stability of Storage Cavern

The pressure of the groundwater surrounding a liquefied petroleum gas (LPG) storage cavern needs to be maintained higher than its internal pressure. If seepage increases during excavation of a storage cavern, the groundwater pressure around the cavern drops and this drop may continue in the highly permeable faults and cracks. This risks forming unsaturated zones during construction. If such unsaturated zones are formed, re-saturating the complex cracks in the bedrock is extremely difficult, and maintaining the groundwater pressure higher than the internal pressure of the cavern may not be possible. Therefore, the bedrock around the cavern should be kept saturated.

To prevent an unsaturated zone during excavation, pressurised water was injected into water curtain boreholes drilled from the water curtain tunnels on the top of the caverns, and the faults and cracks crossing the caverns were grouted. The mechanical stability of the cavern was ensured during excavation by installing support structures designed for the mechanical condition of the bedrock, the groundwater pressure and the effectiveness of grouting while preventing the development of loosened zones.

Safety was monitored during cavern excavation by analysing the data of displacement of the cavern and bedrock. The flow rate around the faults with clays increases when additional water injection is applied. Here, the cavern was stabilised by an additional amendment of the support pattern as necessitated by an analysis of the improvement in the grouted zones and geological surveys.

The functionality of the water curtains was tested by pressurising the water immersing the water curtain tunnels. The test assessed the stability of the cavern during pressurisation of the water using the measurements of bedrock displacement and management standard. If the result necessitated a redesign of the supports, supports were added to ensure the stability of the cavern.

5.1 Excavation of Storage Cavern

5.1.1 Excavation Procedure

The cavern was excavated in five stages (Figures 5.1 and 5.2). First, the arch was excavated after constructing the water curtains, and the bedrock of the arch was improved by pre-grouting. Then, the rocks of the side walls and floor were treated by pre-grouting and the first bench was excavated; this was subsequently repeated for the second, third and fourth benches.

DOI: 10.1201/9780367822163-5

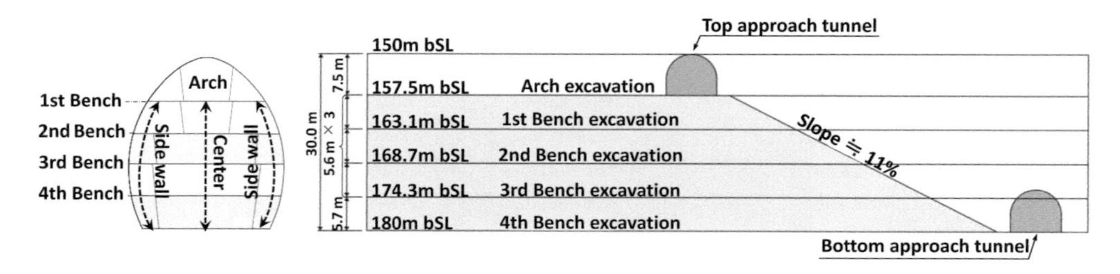

Figure 5.1 Procedure of cavern excavation (example from the Namikata facility).

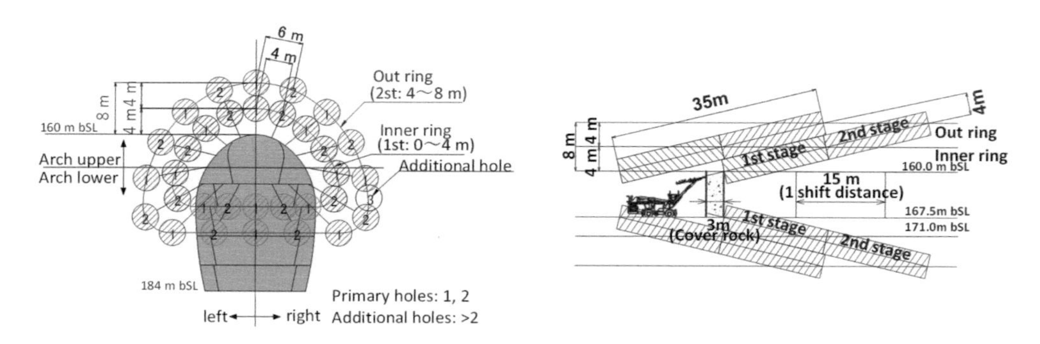

Figure 5.2 Procedure of cavern excavation (example from the Kurashiki facility).

To stop the loosened zone expanding during excavation of the arch, the crest of the arch was drilled first, extending to the walls of the arch to suppress the release of the stress caused by excavation. The walls were reinforced with rock bolts and shotcrete immediately after excavation. The smooth blasting method and breakers were used to minimise damage to the cavern walls and provide a smooth finish.

The optimum size and number of cross sections were selected to suppress looseness and ensure the stability of the cavern: four benches of 5.6 m each for the Namikata facility, where the rocks are hard, and four benches of 4.0 m each for the Kurashiki facility, where faults and cracks develop (Figure 5.3).

5.1.2 Designing Supports

5.1.2.1 Rock Properties

5.1.2.1.1 THE KURASHIKI FACILITY SITE

Figure 5.4 shows the lithological classification of the Kurashiki facility site. The bedrocks around the caverns are predominantly of hard Class H. However, five faults are identified across the cavern: F2 and F3 faults obliquely crossing the cavern with a dip of 50°–70°; F4–6 faults orthogonally crossing the cavern with a near-vertical dip. The faults are subject to intense hydrothermal alteration and are of Classes M and L bedrock with some clays sandwiched (Figure 5.5). Microfractures with very fine cracks develop in some parts of this area.

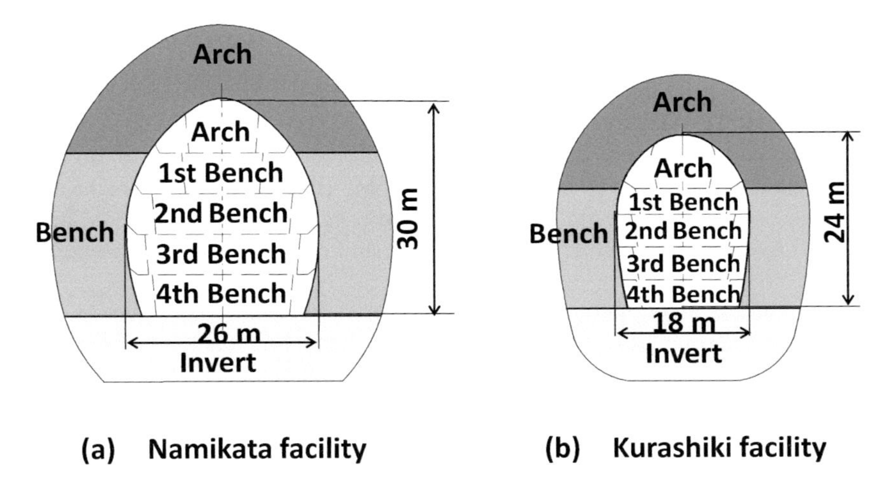

Figure 5.3 Section division of cavern excavation and range of pre-grouting improvement at the Namikata and Kurashiki facilities.

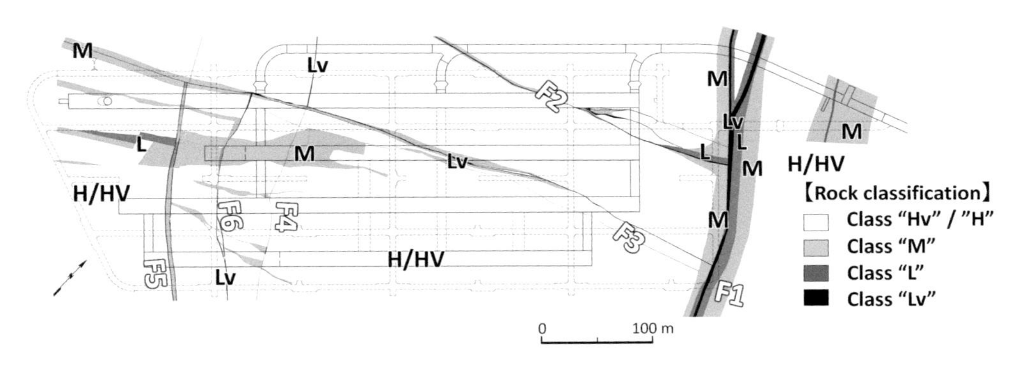

Figure 5.4 Major faults and rock classes in Kurashiki (plan view at 167.5 m bSL).

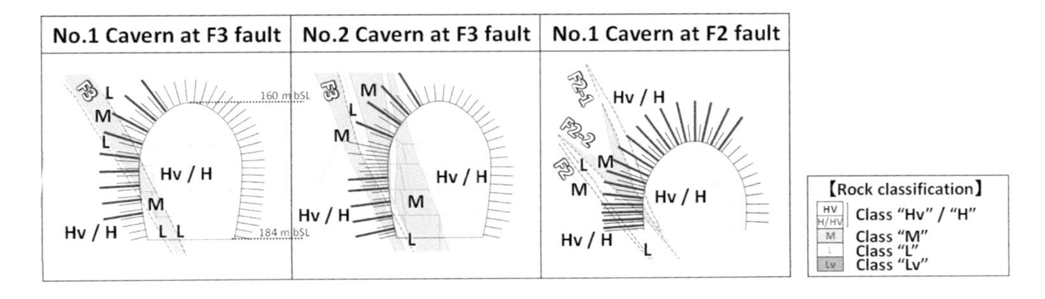

Figure 5.5 Examples of a cross section of the storage cavern (F2 and F3 faults).

The strike and dip of the F2 fault are N80E and 50NW, respectively. The F2 fault is continuous and crosses storage cavern No. 1 at an acute angle. The width of the crush zone is 1–2 m containing clay some 20–30 cm thick.

The strike and dip of the F3 fault are N70E and 70NW, respectively. The F3 is also continuous and crosses the storage caverns at an acute angle. The faults are crushed along the mylonite zones and contain green alterations and white clay veins 1–15 cm thick. The rocks along the crushed mylonite are Classes L and M. The widths of the cracks are 1–5 m at intervals of 5–10 cm and the dominant orientation is similar to the mylonite. The F3 fault is surrounded by crushed mylonite 20–30 cm wide and clay veins a few centimeters thick (Figure 5.6).

A microfracture is a continuous band of a rock mass with minute fractures with widths varying between a few centimeters to a few meters and a few meters to tens of meters long. Under a microscope, its structure is characterised by continuous microfractures less than 1 mm wide intersecting grains of quartz, feldspar and hornblende. The fractures may develop to a few microns wide along a cleavage of biotite. Some fractures have large openings from expansion and contraction, and others have small connected openings presenting the structure of a discrete jigsaw puzzle (Figure 5.7).

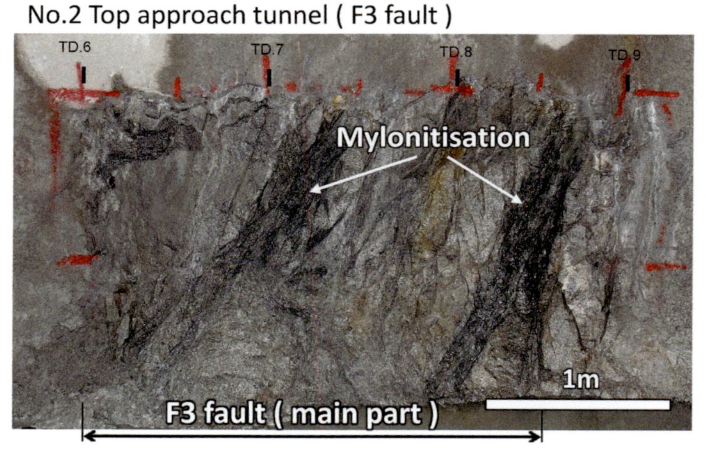

Figure 5.6 Photo of the F3 fault.

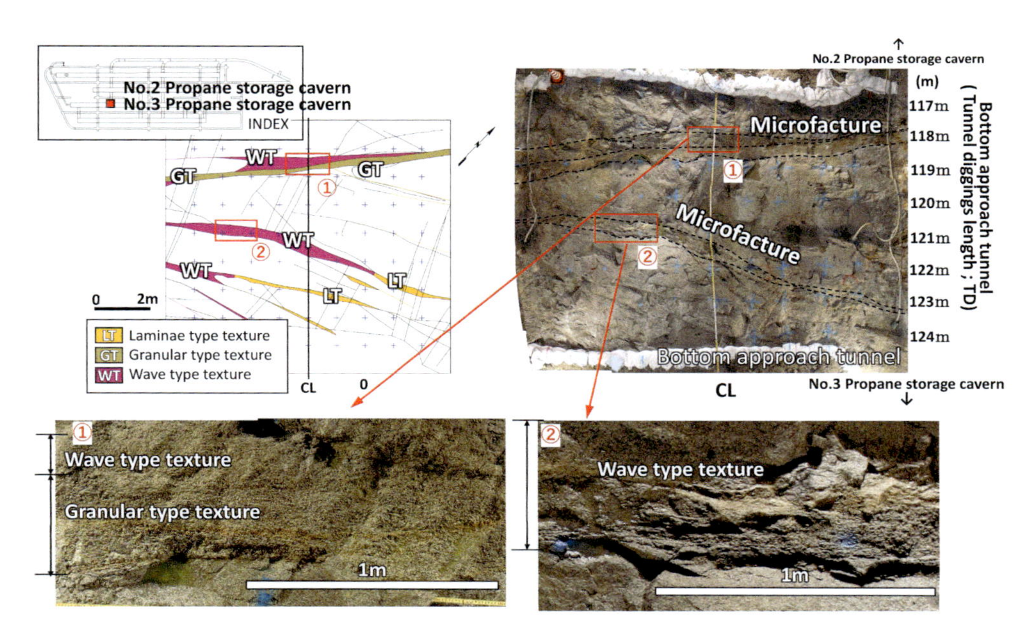

Figure 5.7 Photos of the microfracture.

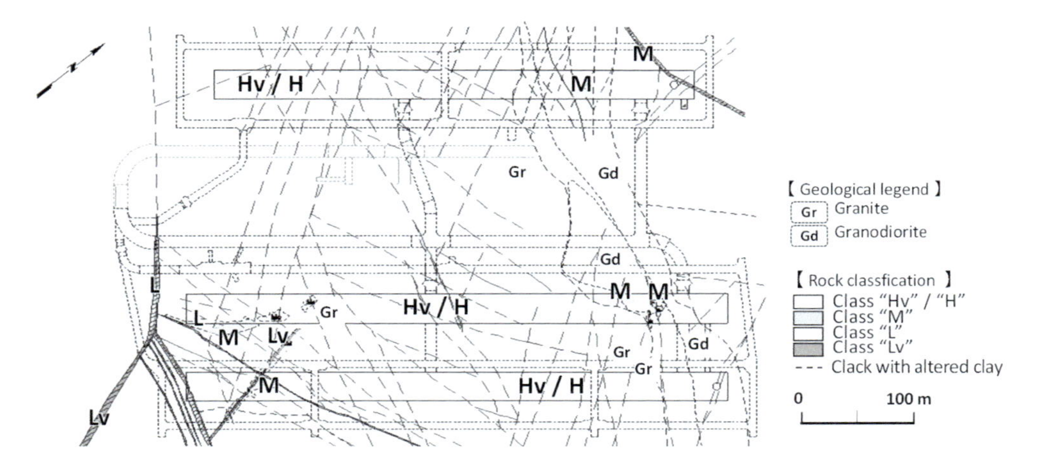

Figure 5.8 Major shatter zones and rock classification (plan view at 157.5 m bSL).

5.1.2.1.2 THE NAMIKATA FACILITY SITE

Figure 5.8 shows the lithological classification of the Namikata facility site. The bedrocks around the caverns are mainly of hard Class H but the south-western part of the propane storage cavern includes Class M rocks. The crack density is 1.5–3 m. Locally, there are continuous 3 mm cracks with clays obliquely crossing the storage caverns (Figures 5.8 and 5.9).

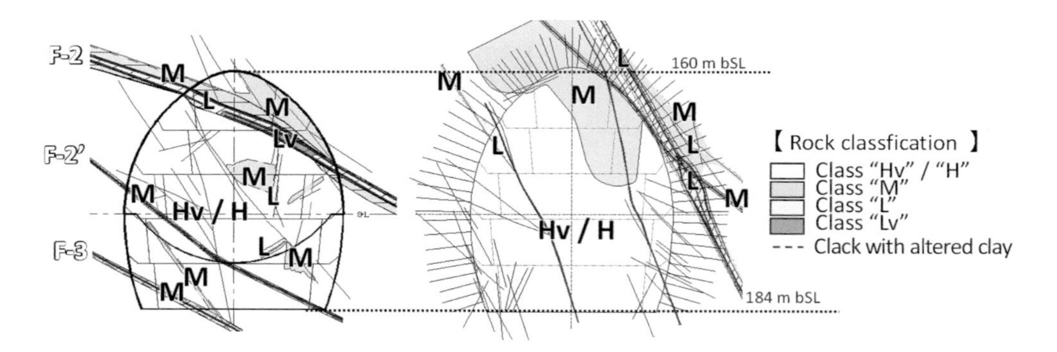

Figure 5.9 Example of a cross section of the storage cavern at the Namikata facility.

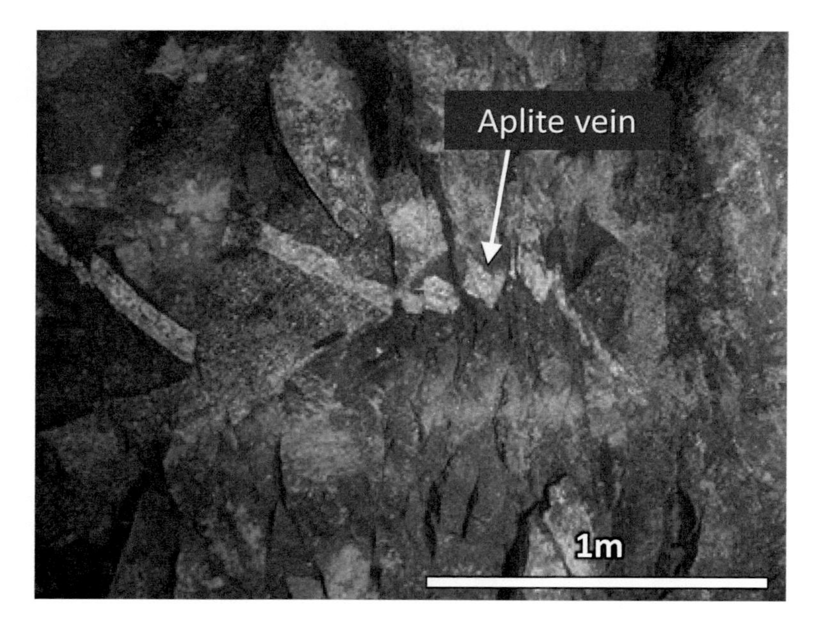

Figure 5.10 Aplite vein in the boundary zone between rock types.

Granites in the Namikata facility area are divided into Namikata Granite and Takanawa Granodiorite. The cracks are fine near the boundary of these rock types with intervals of 5–30 cm and are predominantly in the N60W direction across the storage cavern. Away from the boundary, the interval between the cracks tends to widen to 30–70 cm. In the granites in the boundary zone, aplite veins are subject to shear failure (Figure 5.10).

Failure zones F-2 and F-6 and calcite alteration zones are present to the southwest of the propane storage cavern. The strike and dip of the F-2 failure zone are N75W and 50N, respectively. This failure zone is 0.5–2 m wide and contains fragments of rocks mixed with clay. The clay is green and pink and is considered to be hydrothermally altered calcite. Other failure zones parallel to F-2 are also subject to hydrothermal alteration resulting in green and pink clays. Cracks around the F-2 failure zone are also affected by hydrothermal alteration and alteration clay is present.

The F-6 failure zone is in Class L rocks and it has a strike N65E and a dip 60N with a width of 1 m. The calcite alteration zone is about 8 m wide adjoining the granite with an irregular boundary. The rock class is mostly M with minor L, and its permeability is low. It is characterised by irregular calcite veins in the granite, which is considered to have replaced minerals such as quartz and feldspar (Figure 5.9). The storage caverns were excavated with knowledge of the above geological structure and the character of the rocks.

5.1.2.2 Designing Support in the Storage Cavern

5.1.2.2.1 SUPPORT DESIGN

In excavating a rock storage cavern, the mechanical stability of the cavern must be ensured considering the mechanical character of heterogeneous rocks, the condition of the groundwater pressure, the effect and pressure of the groundwater and the improvement by grouting. An analysis of the mechanical behaviouur of the cavern during excavation was checked considering the distribution of faults, the improvement by grouting and the pressure of the groundwater (Figure 5.11).

5.1.2.2.2 SUPPORT PATTERN

In designing supports, a behaviour prediction analysis checked the stress that occurs in the support material (rock bolts and shotcrete) and the stability of the rocks in the loosened zones. The stability of the rock mass was checked by measuring its resistance to the groundwater pressure which satisfies the water-tightness requirement applied to the side walls of the arches. Figure 5.12 shows the workflow of the support design on excavation of the storage cavern. The stability was checked by suspension assessment of the arches, a sliding stability assessment of the side walls and a stability assessment of the key blocks.

The supports of the Kurashiki facility's caverns were generally composed of rock bolts and shotcrete. For Class H rocks, the rock bolts used were 3–4 m long and the thickness of the concrete was 8–12 cm. For Classes L and M rocks in the fault zone, high yield strength bolts 4–8 m long were used with steel fibre reinforced shotcrete (Figure 5.13).

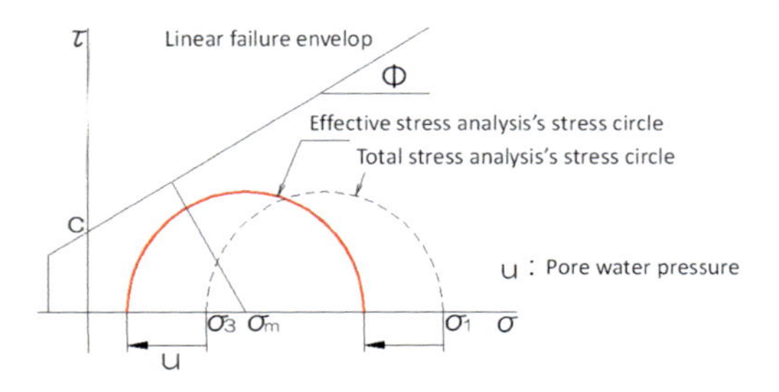

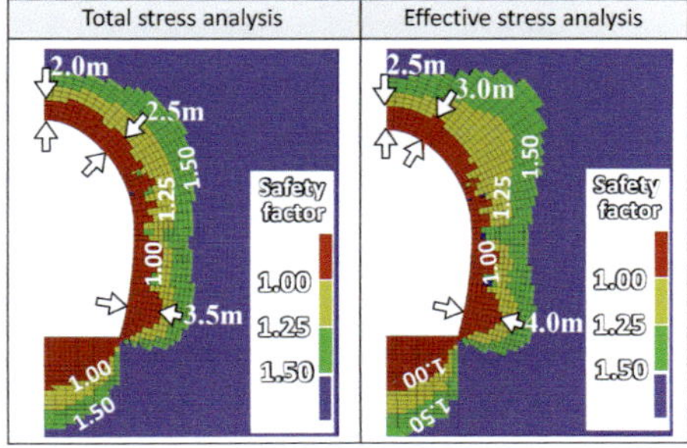

Figure 5.11 Stability assessment of the cavern considering the groundwater pressure.

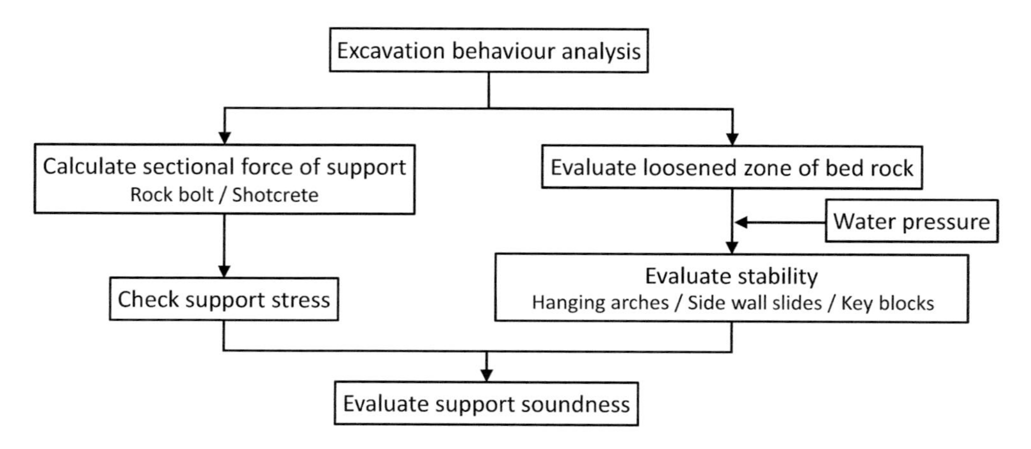

Figure 5.12 Workflow of support design at excavation of the storage cavern.

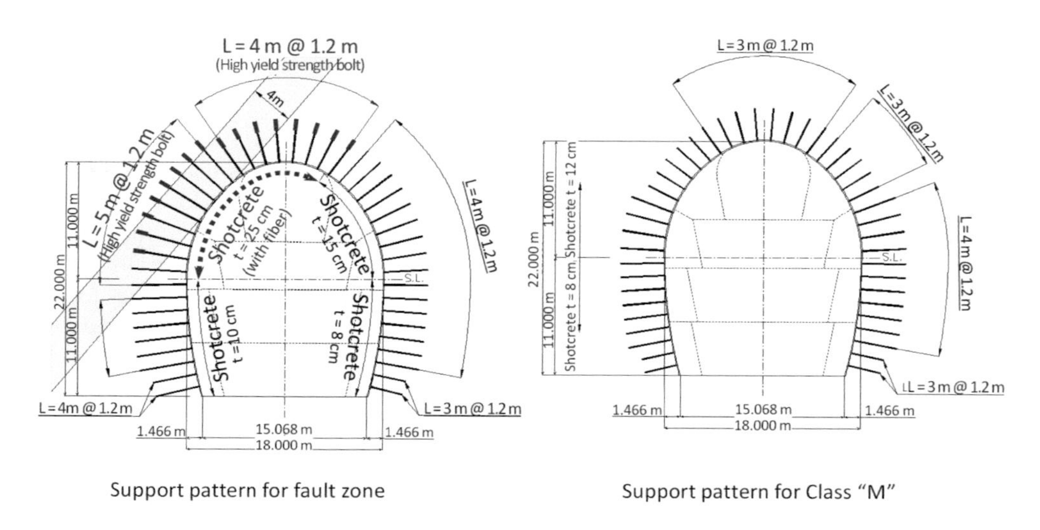

Figure 5.13 Support pattern for fault zone and for Class M rocks in the Kurashiki facility.

The supports of the Namikata facility's caverns were, similarly to Kurashiki's, generally composed of rock bolts and shotcrete. For Class H rocks, the rock bolts used were 4–6 m long at 1.5 m intervals and the thickness of the concrete was 8–12 cm. For Class M rocks with high crack density, bolts of 4–7 m long were used at 1.2 m intervals (Figure 5.14).

5.1.2.3 Measurement Management

In the construction of a rock cavern with a large cross section, the design needs to monitor and provide feedback on the behaviour of the cavern and the surrounding rocks to cover the uncertainty of the geology and the ground pressure. For this purpose, monitoring points were

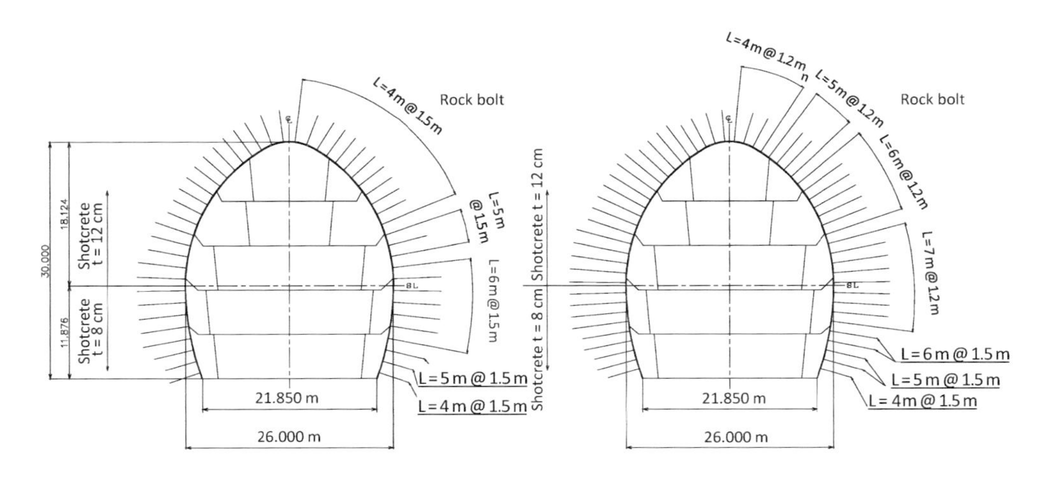

Figure 5.14 Support pattern for Classes H and M in the Namikata facility.

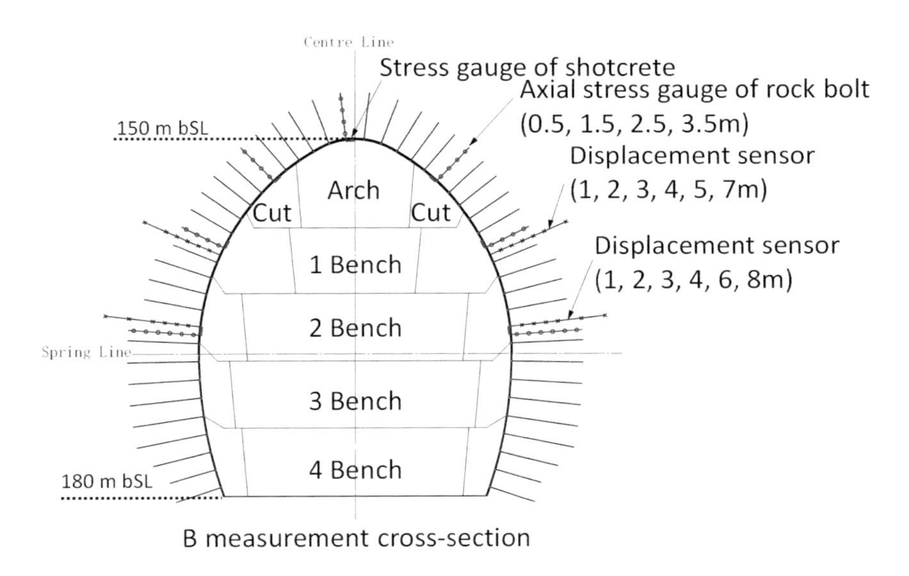

Figure 5.15 Outline of measurements for cavern stability at the Namikata cavern.

systematically arranged in the entire cavern, and inner space displacement was estimated from surveys with electro-optical distance measuring instruments. The displacements were measured particularly densely around the geologically weak parts including the fault zones, failure zones and intersections of faults. Additionally, the axial stress of the rock bolts, stress in shotcrete and displacement using a rock displacement gauge were measured in these zones (Figure 5.15).

5.1.3 Assessment of Cavern Stability at Cavern Excavation

Table 5.1 shows the assessment criteria of cavern stability at the time of cavern excavation. Using these criteria, the behaviour of the rocks was measured and confirmed satisfactory.

5.1.3.1 Excavation of Arches

This section describes the assessment of the behaviour of the F3 fault of the Kurashiki storage at the time of excavation of the arch. The strike and dip of the F3 fault were N70E and 70NW, respectively. Class L rocks were distributed 0.5–1 m; Classes M and L were 0–2 m in both the upper wall and lower walls. Microfracture was mixed within the fault-affected zone in the lower wall of the F3 fault.

The F3 fault was close to No. 2 storage at a high dip angle with layered rocks of Classes ML and L, both 1 m thick. The rocks of Class L were abundant with dense high angle cracks. Therefore, the stability of the cavern was closely monitored through the excavation of the benches. The result of the predictive analysis considering the anisotropy of the cracks in Class L rocks showed that the maximum loosened zone was under 2 m in the arch and 3 m on the benches, which led to an expectation that the planned support pattern would maintain the stability of the caverns (Figure 5.16).

5.1.3.2 Excavation of Benches

5.1.3.2.1 KURASHIKI FACILITY

Figure 5.17 shows the inner space displacement of Bench 1 on completion of cavern excavation. The maximum inner space displacement in Class H rocks was about 8 mm, which is within the management standard of 15 mm. The inner space displacements were 46 mm at the F2 fault obliquely crossing the cavern and 21 mm at the F3 fault crossing storage cavern No. 2. However, it was 5–7 mm in the microfracture zones, which is comparable to Class H rocks.

Figure 5.18 shows the support pattern on completion of cavern excavation for various rock properties. The basic support pattern was designed for Class H rocks, and more supports were added in the fault and microfracture zones. The support pattern at No. 1 storage cavern crossing the F2 and F3 faults is shown in Figure 5.18 as an example.

Representative temporal variations of the inner space displacement of a cross section are shown in Figure 5.19. The change in the inner space displacement occurred at the time of

Table 5.1 The assessment criteria of cavern stability

Assessment criteria	*Measurement items*
Inner space displacements should be convergent. (The amount of displacement should be predictable.)	Arch subsidence Inner space displacements
Slack zones should be stable in the range of rock bolts' length. (Safety factor criteria of cavern stability should be satisfied.)	Axial stress of rock bolts
Cavern timbering and lining should be staunch. (Applied stress of shotcrete and axial stress of rock bolts should be less than safety acceptance criteria values.)	Stress of shotcrete

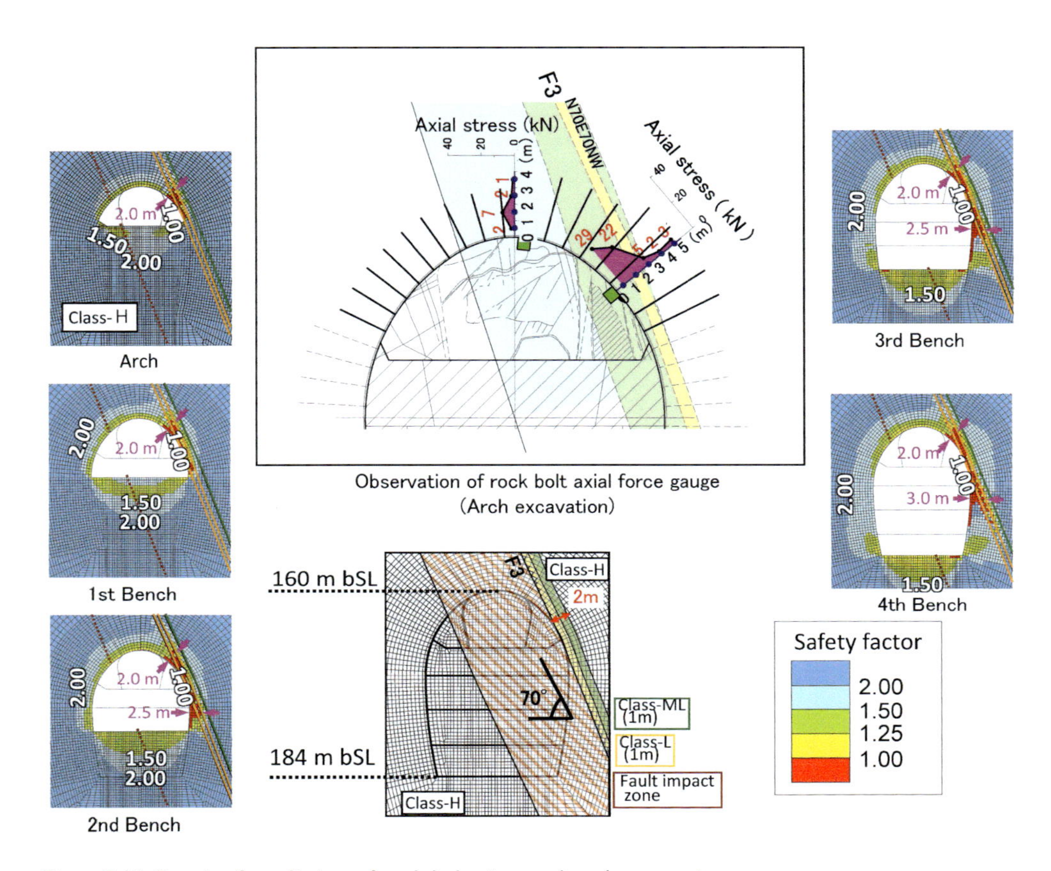

Figure 5.16 Result of prediction of rock behaviour at bench excavation.

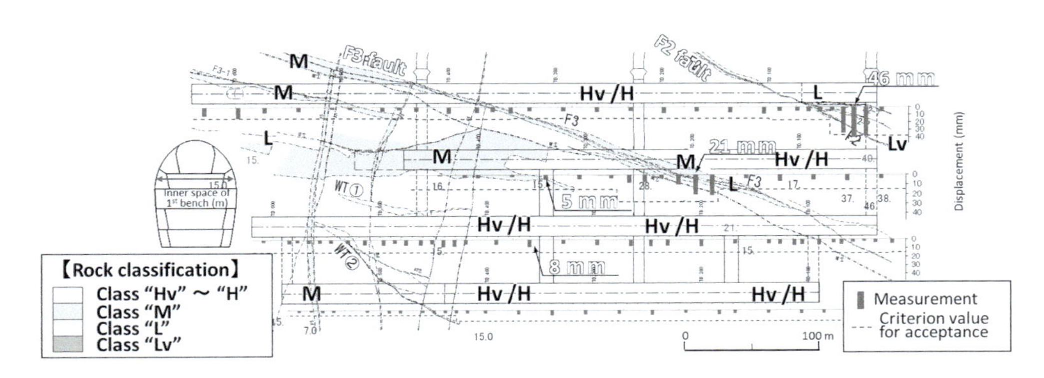

Figure 5.17 Result of measurement of inner space displacement on completion of cavern excavation (inner space displacement of Bench 1).

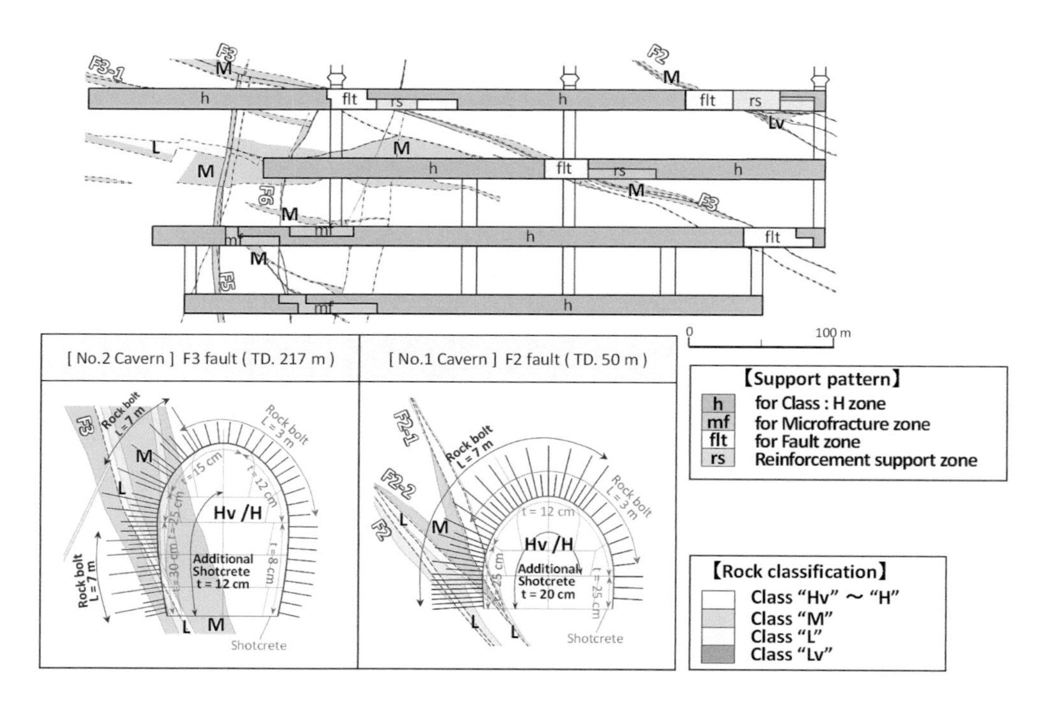

Figure 5.18 Plan view of rock classification and support pattern in the fault zones.

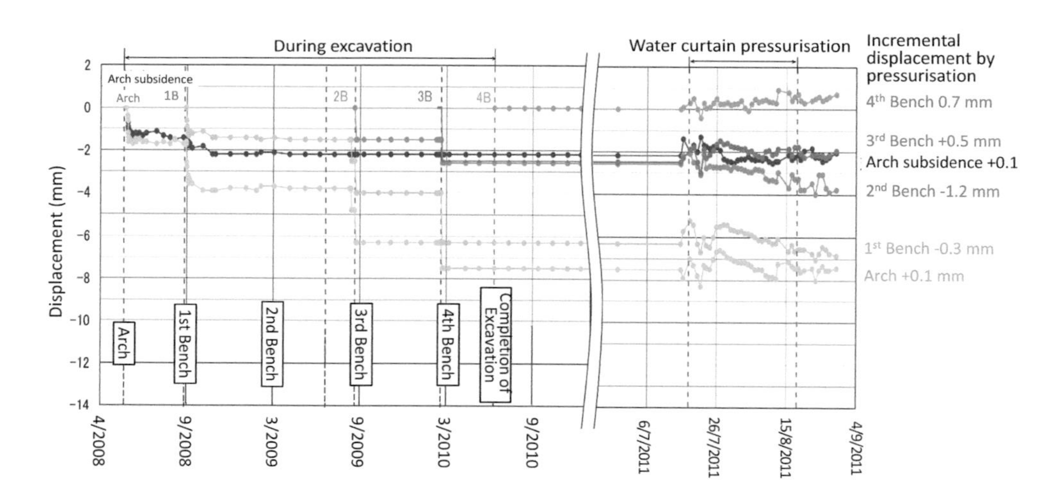

Figure 5.19 Temporal variations of sink of the roof and inner space displacement.

excavation of the arch and benches, and stabilised after the excavations. The magnitude of distortion was within the standard and the stability of the cavern was assured.

Representative temporal variations of the inner space displacement are shown in Figure 5.20. The change in the inner space displacement occurred at the time of excavation of

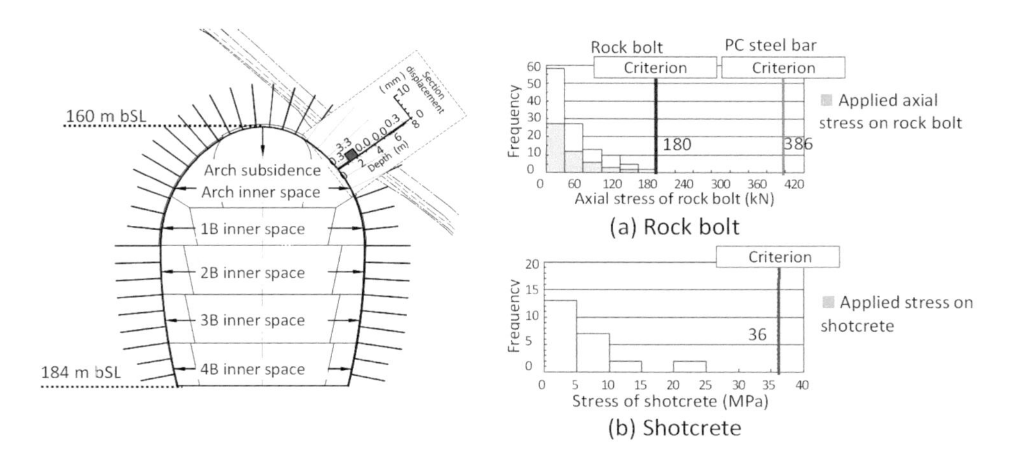

Figure 5.20 Distribution of underground displacement in F2 fault and the result of stress measurements of rock bolt axis and shotcrete.

the arch and benches and converged after the excavations. Displacement did not occur after completion of excavation. The magnitude of displacement was within the standard criteria and the excavation was carried out ensuring the stability of the cavern.

Gauges for displacement, the axial stress of the rock bolts and shotcrete stress were installed in the areas near the fault zone where the stability of the cavern was affected. Figure 5.20 shows the distribution of segment displacement by an underground displacement measurement, the stress of the rock bolt axis and stress in shotcrete around the F2 fault. The displacement occurred within the fault about 2 m from the wall, which is less than the length of the rock bolts of 5 m. The stress both on the rock bolts and in shotcrete was within the standard design strength of the management criteria.

5.1.3.2.2 NAMIKATA FACILITY

Figure 5.21 shows the inner space displacement of the two benches on completion of cavern excavation. The inner space displacement during excavation was generally about 9 mm, which is within the range expected by the model for Class H rocks. An exception was the southern part of propane storage No. 1 (TD 0–130 m), which is discussed later. The stress on the supports was caused by the cavern excavation, but it was within the standard management criteria and the health of the support material was maintained.

A representative temporal variation of the inner space displacement is shown in Figure 5.22. The change in the inner space displacement occurred at the time of the arch and benches excavation and converged after the excavations. Displacement did not occur on completion of excavation. The magnitude of displacement was within the standard and the excavation was carried out ensuring the stability of the cavern.

The inner space displacement reached 20–25 mm during cavern excavation in the F6 crush zone near the southern part of propane storage cavern No. 1. In this area, the F-6 faults developed and associated cracks with clays obliquely crossed the cavern. The cracks obliquely crossing the cavern caused safety concerns as they could move along the cavern

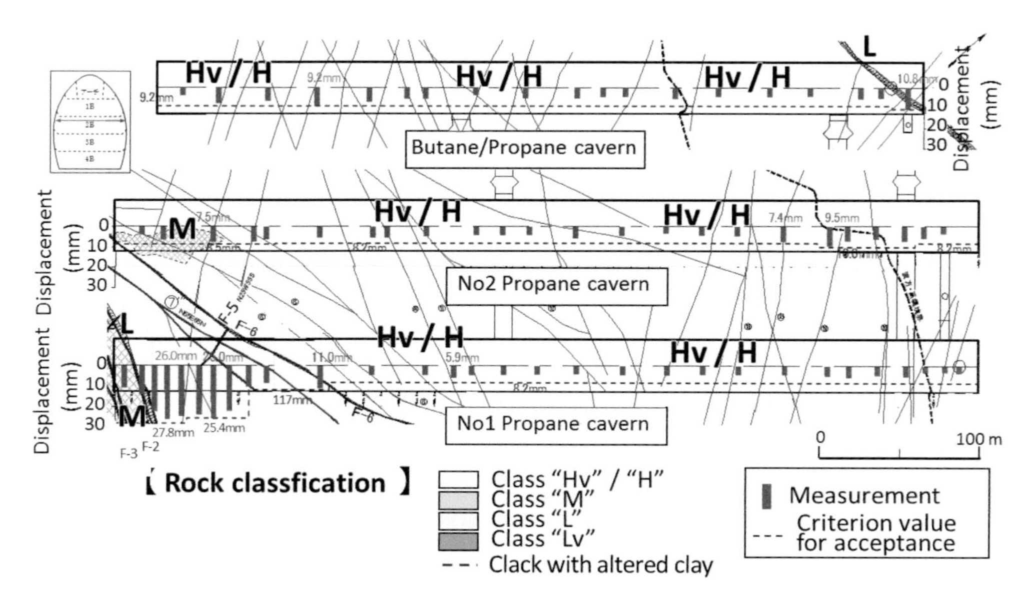

Figure 5.21 Result of measurement of inner space displacement on completion of cavern excavation (inner space displacement of Bench 1).

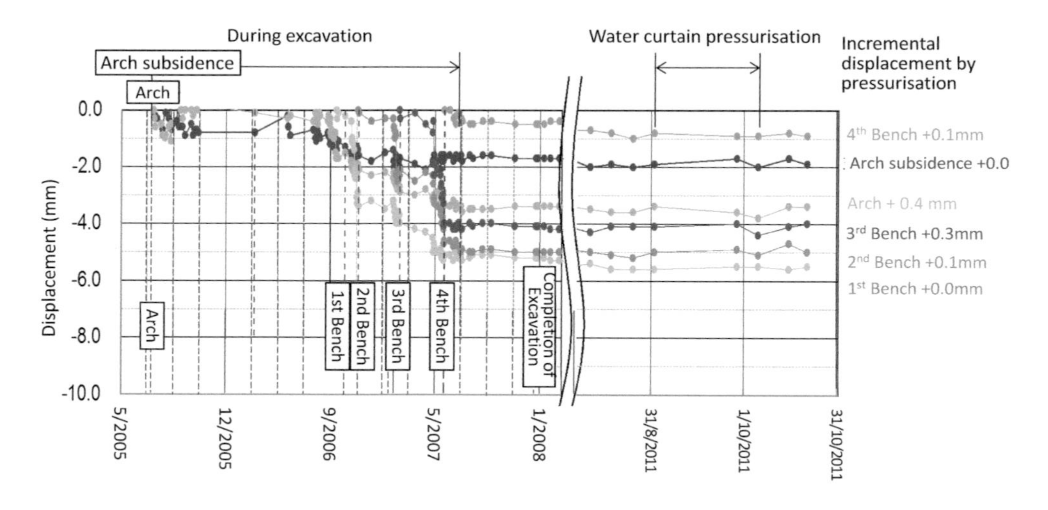

Figure 5.22 Temporal variation of sink of the roof and inner space displacement.

axis. Therefore, the influence of the F-6 crush zone on the stability of the cavern was closely examined. It was found that the distortion behaviour of the rocks could be explained by the analytical model with rock property parameters set based on the in-situ shear stress test and the tri-axial test of cores. According to the results of the modelling, the maximum size of the loosened zone induced by excavation was about 6 m away from the cavern wall, and the

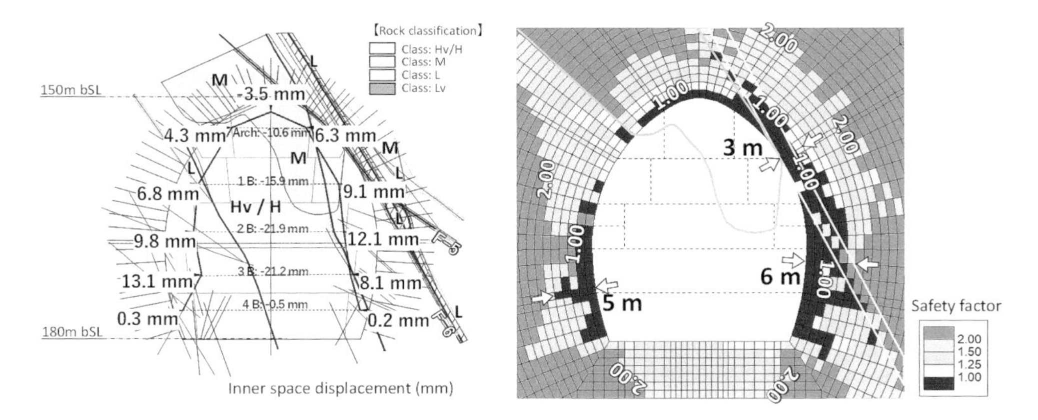

Figure 5.23 Analytical model and the distribution of the local safety factor of the F-6 crush zone across propane storage No. 1.

current support pattern ensured the stability of the cavern. The magnitude of distortion was within the standard management criteria considering the F-6 crush zone, and the cavern was considered secure (Figure 5.23).

In addition, the supports were designed with consideration of the response of the groundwater pressure to pressurised water injection. The existing support pattern was confirmed to be sufficiently secure with this design and the sealing water was pressurised. As a result, the rock behaviour was within the standard management criteria and the stability of the cavern was assured (Figure 5.24).

5.2 Mechanical Stability of Water Curtain at Pressurisation

Prior to the air-tightness test, water barriers (concrete plug) were constructed at the joints of the water curtain tunnels and access tunnels on completion of excavation of the caverns. Then, the water tunnels were filled with water and the water curtain, composed of the water curtain tunnels and the water curtain boreholes, were pressurised. Thus, the mechanical stability of the caverns and the piping shaft plugs were tested and confirmed under the highest water curtain pressure against the cavern which was under atmospheric pressure.

Figure 5.25 shows the pressurising pattern of the Kurashiki facility.

The total stress analysis for the stability of the cavern continued from the time of excavation. Using 3D hydrogeological modelling, this analysis calculates the variation in the groundwater pressure caused by an increase in the water curtain pressure. The increased groundwater pressure was applied from the peripheral surface of the cavern (8 m from the cavern wall in Kurashiki) to the peripheral surface of the grouting as an external force. The stability of the rock behaviour was examined by this pressurising (Figure 5.26). The stress on the supports was checked based on the information about the loosened zone from this analysis for planning final grouting and additional supports.

According to the above design, the water curtain pressure was increased and the mechanical stability of the cavern was confirmed by measurements of the behaviour of the rocks after post-grouting and the installation of additional supports.

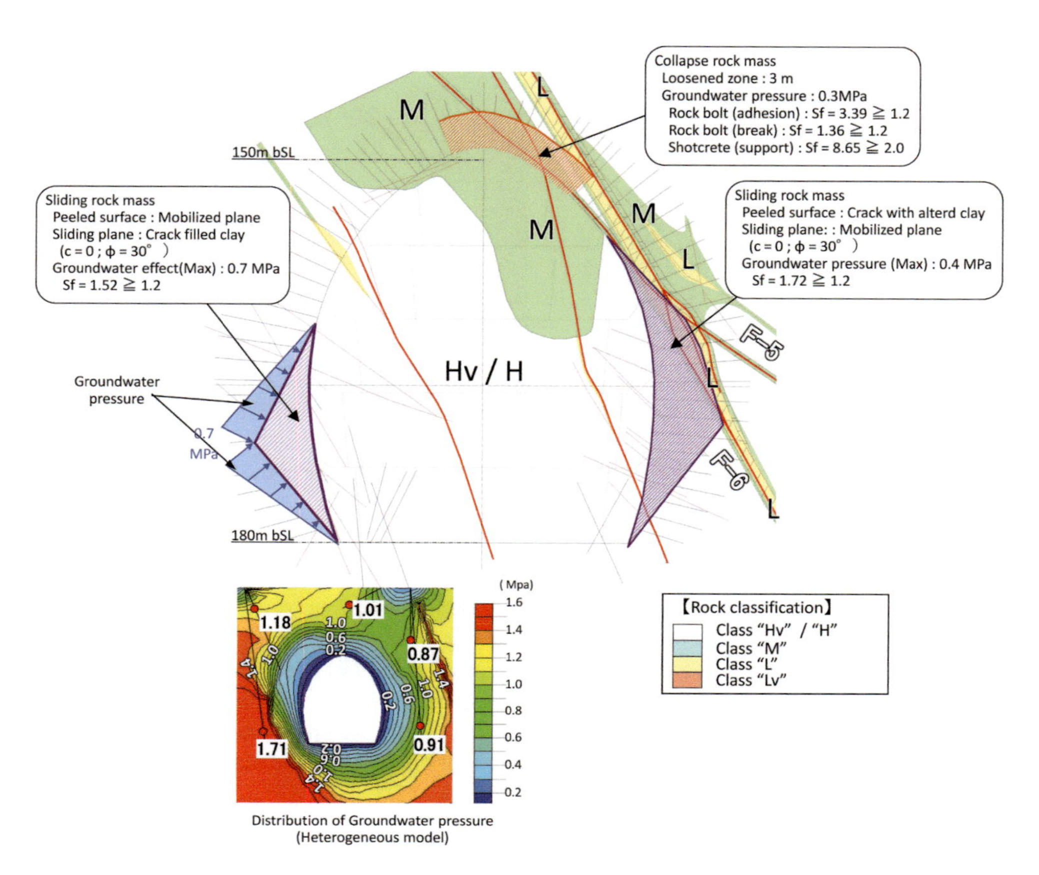

Figure 5.24 Result of cavern stability analysis and support design around the F-6 crush zone in propane storage No. 1.

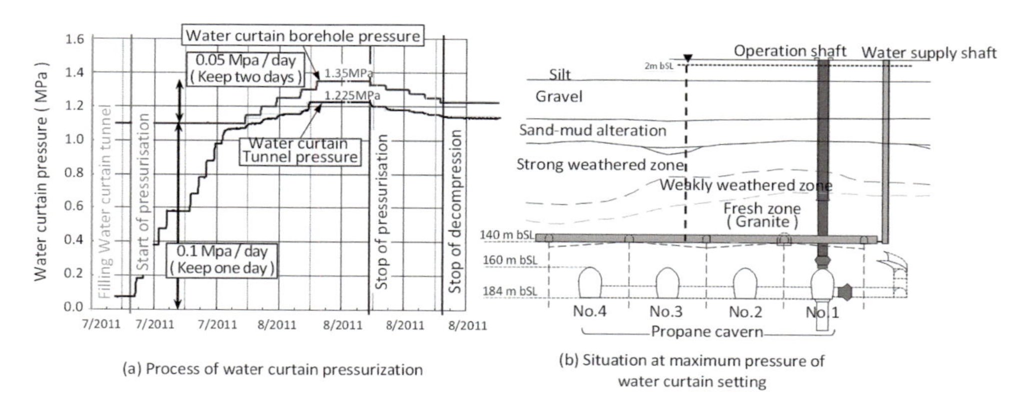

Figure 5.25 Pressurising pattern and the situation of water curtains under the maximum pressurise.

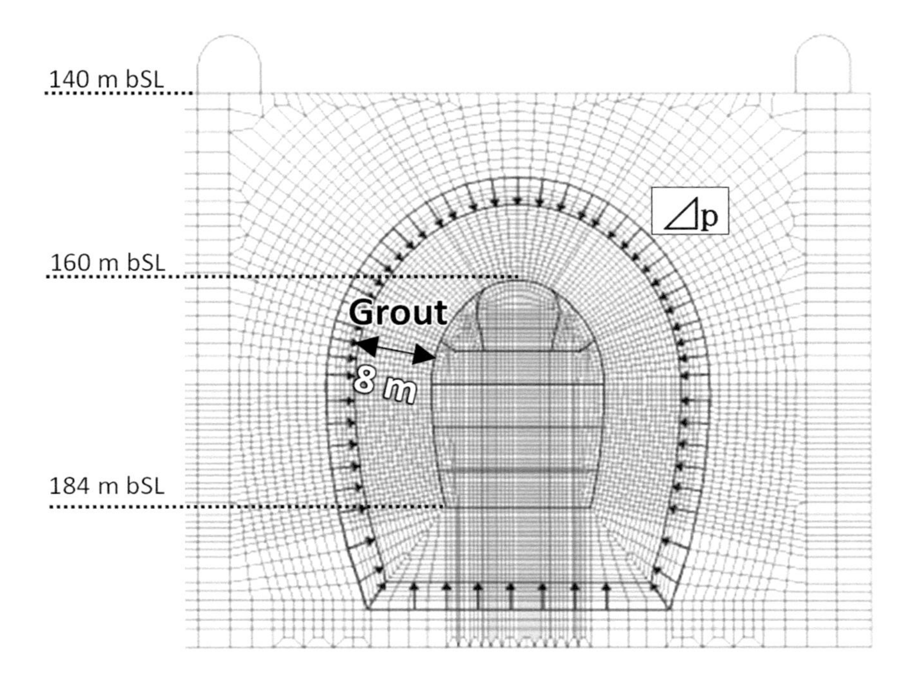

Figure 5.26 Schematic diagram of analysis at the pressurised water sealing at the Kurashiki facility.

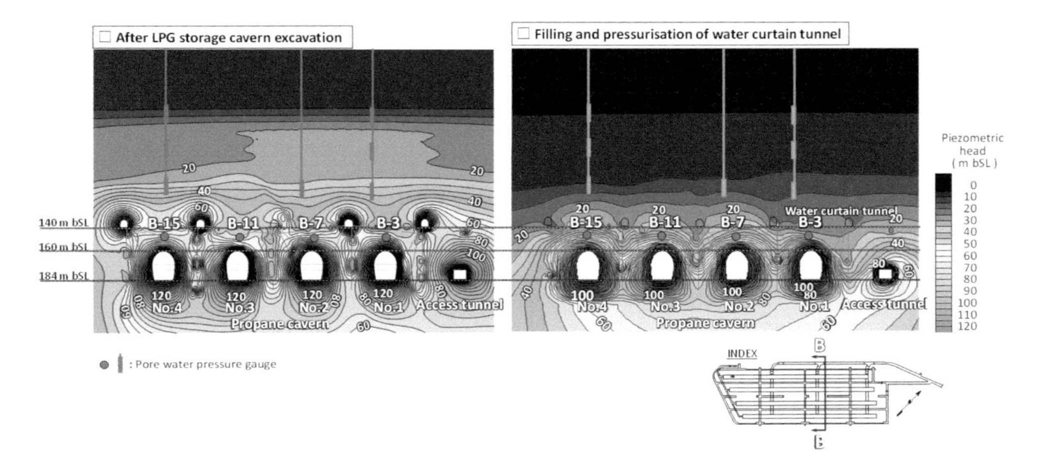

Figure 5.27 Piezo water head contours on completion of excavation and at pressurising cross section B.

5.2.1 The Kurashiki Facility

Among the predictions using 3D hydrogeological modelling, the pore water pressure contours of cross section B on completion of excavation and the highest water pressure are shown in Figure 5.27. The pore water pressure of the large area over the water curtains dropped during excavation of the water curtain tunnels and the caverns, but it recovered almost to pre-excavation pressure by pressurising the water curtain boreholes and the water curtain tunnels to the structural maximum water level of operation despite the fact that the cavern and access tunnels were under atmospheric pressure.

Figure 5.28 shows examples of comparisons between actual and predicted values for the pore water pressure and the flow rate of seepage at the time of pressurising. While the water curtain pressure increased from 1.1 to 1.35 MPa, the pore water pressure increased by about 10–15 m and the flow rate of seepage increased about 20 m^3/h, which were roughly as predicted. The maximum flow rate during pressurisation was 182.3 m^3/h.

Figure 5.29 shows subsidence of the cavern ceiling, inner space displacement and the distribution of ground displacement during pressurising of the water curtain pressure from 1.1 to 1.35 MPa. The increase in the inner space displacement of Class H rocks was around 1 mm, within management standard, thus the stability of the caverns by grouting and support under

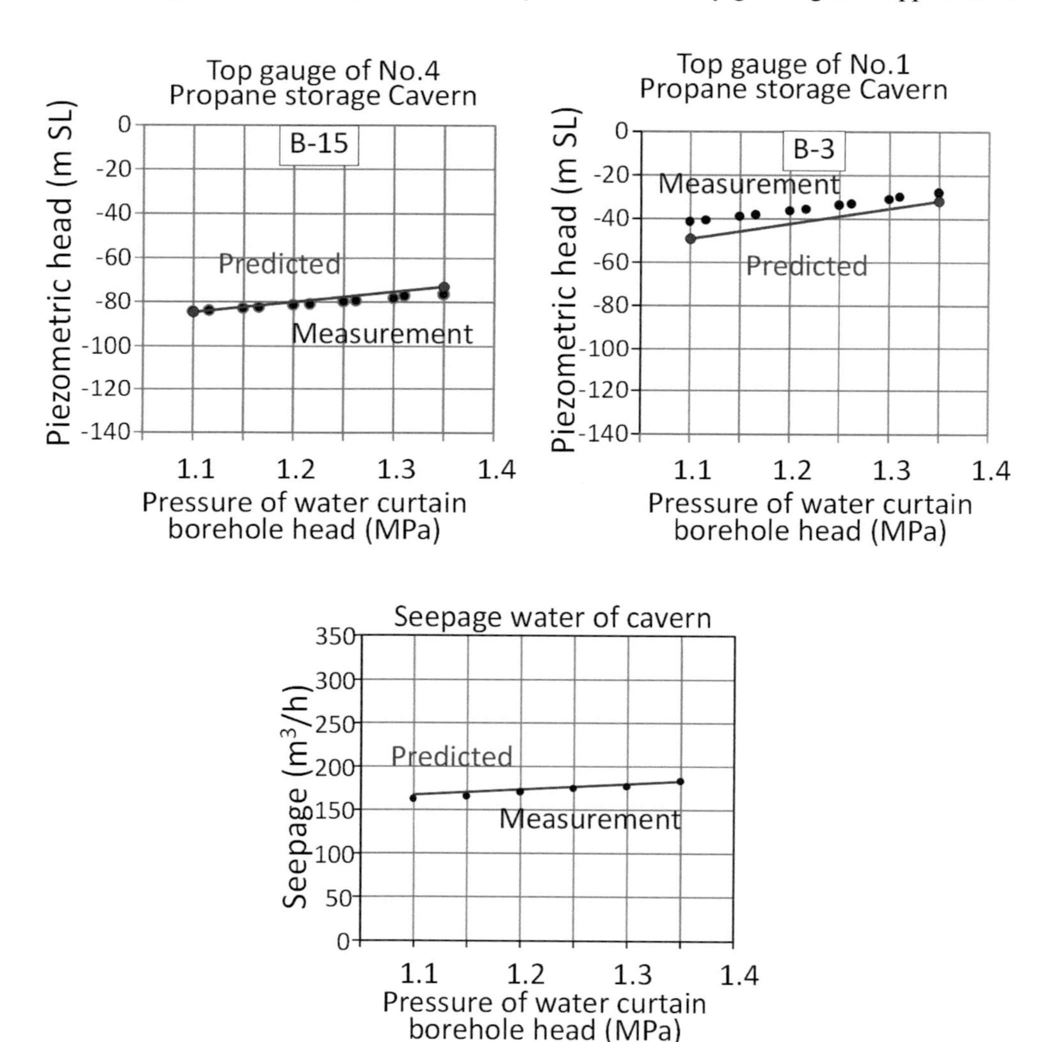

Figure 5.28 Comparison between actual and predicted values of pore water pressure and the flow rate of seepage at the time of pressurising.

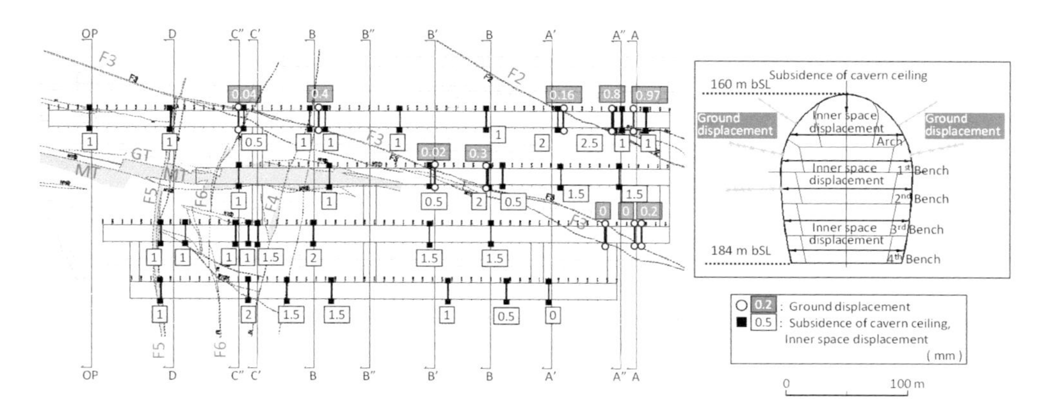

Figure 5.29 Subsidence of the cavern ceiling, inner space displacement, and distribution of the ground displacement in the Kurashiki facility.

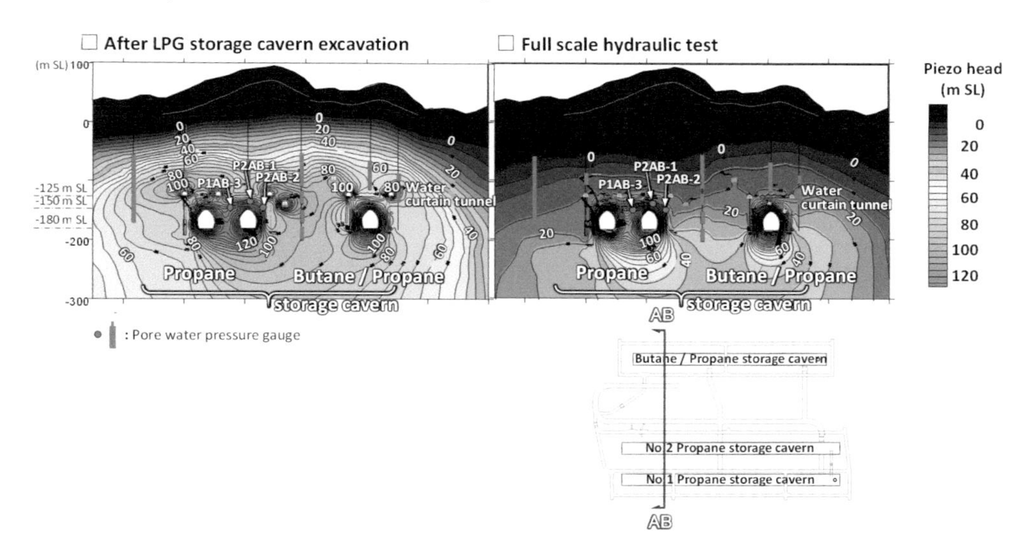

Figure 5.30 Piezo water head contours on completion of excavation and at pressurising cross section AB.

high pressure was secured. In the F3 fault zone, the increase in displacement was around 2 mm, which is within the standard management criteria, because of additional supports and post-grouting during the increase in the water curtain pressure. There, the displacement by an underground displacement measurement only increased less than 0.5 mm, also within the standard management criteria.

Thus, the required mechanical stability of the caverns under 1.35 MPa was ensured.

5.2.2 The Namikata Facility

Among the predictions using 3D hydrogeological modelling, the pore water pressure contours of cross section AB on completion of excavation and the highest water pressure are shown in Figure 5.30. The pore water pressure of the large area over the water curtains dropped during excavation of the water curtain tunnels and the caverns, but it recovered almost to pre-excavation pressure by pressurising the water curtain boreholes and the water curtain tunnels to the structural maximum water level of operation despite the fact that the cavern and access tunnels were under atmospheric pressure.

The pore water pressure increased by 40–50 mm and the flow rate of seepage about 25 m³/h by immersing the water curtain tunnels and increasing the water curtain pressure from 0.5 to 1.2 MPa. These were largely as expected by the 3D hydrogeological model (Figure 5.31).

To the south of storage cavern No. 1, inner space displacement increased and the flow rate of seepage also increased when the water curtain pressure was increased to 1.2 MPa. Detailed investigations by borehole televiewer observation and tests on the permeability and seepage pressure were carried out in the reconnaissance bore. The investigation revealed the presence of three highly permeable open cracks behind the eastern floor, where the seepage pressure was as high as 1.55 MPa similar to the surroundings (Figure 5.32).

The highly permeable crack zones adjacent to the cavern were considered the main cause of the displacement at the pressurising test. Accordingly, its behaviour was predicted by analysis and a countermeasure of reinforcing the rocks and post-grouting was designed.

● Comparison between measurement and predicted value of P1AB-3, P2AB-1,P2-AB-2

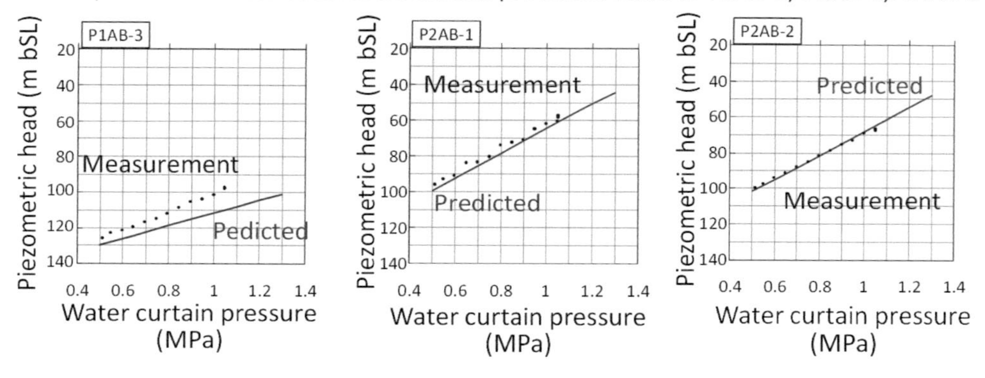

● Comparison between measurement and Predicted value of seepage water

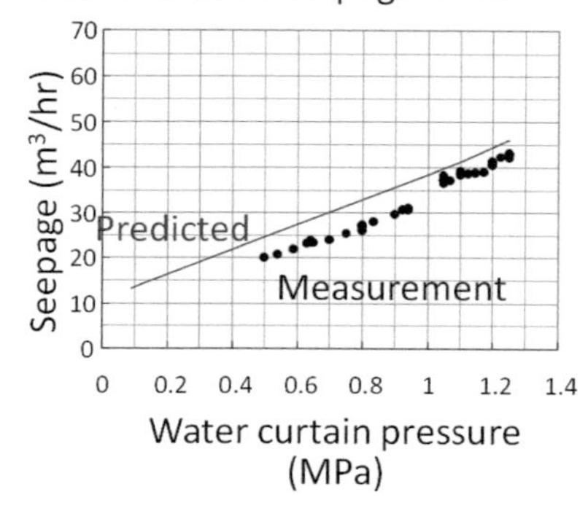

Figure 5.31 Comparison between actual and predicted values of pore water pressure and the flow rate of seepage at the time of pressurising.

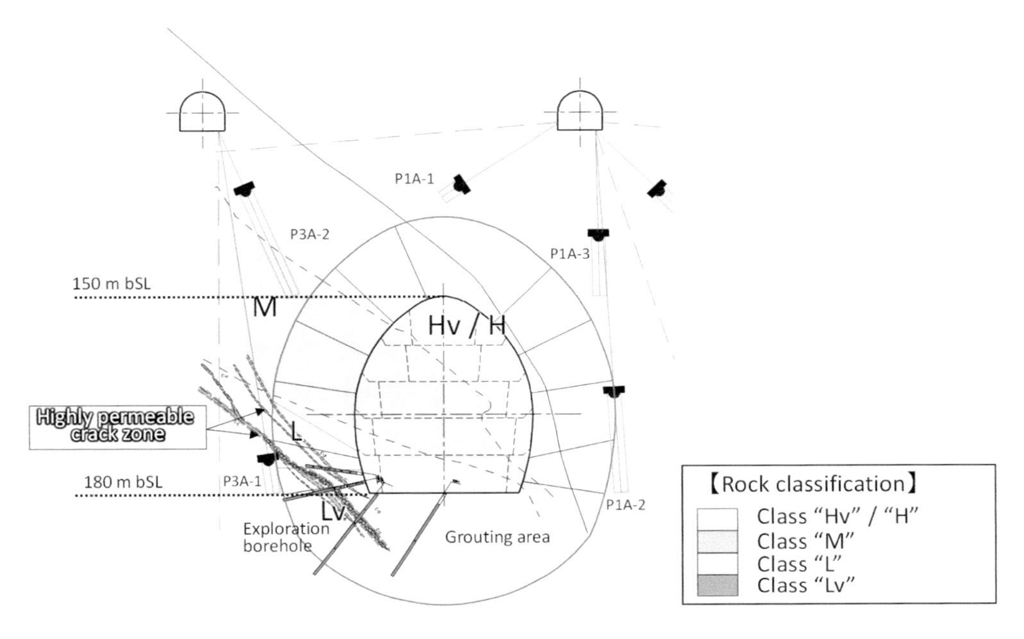

Figure 5.32 Geological cross section incorporating the borehole investigation.

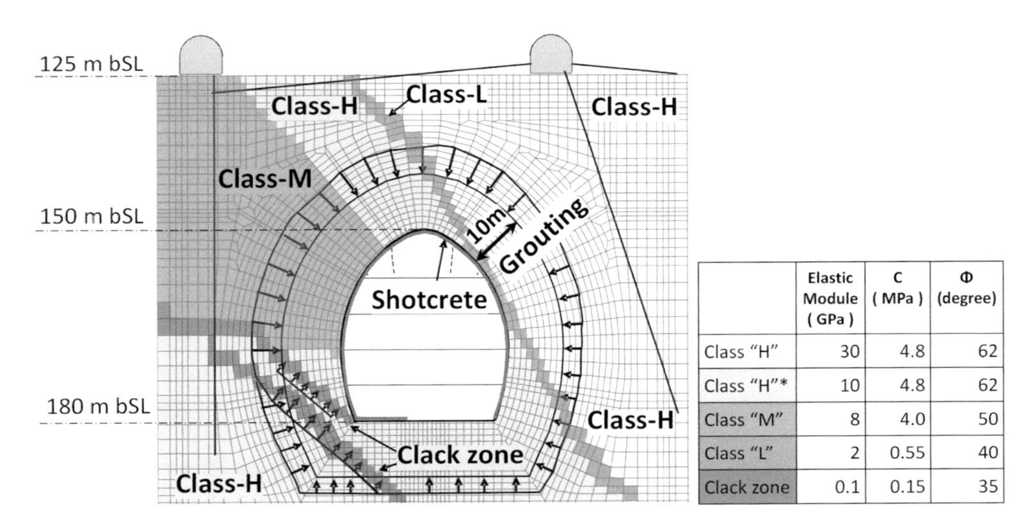

	Elastic Module (GPa)	C (MPa)	Φ (degree)
Class "H"	30	4.8	62
Class "H"*	10	4.8	62
Class "M"	8	4.0	50
Class "L"	2	0.55	40
Clack zone	0.1	0.15	35

Figure 5.33 Model for the cavern stability analysis at pressurising sealing water.

The analysis included an inversion to reproduce the rapid increase in displacement at the water curtain pressure of 1.2 MPa during excavation and the step analysis for water curtain pressure of 1.25 MPa, which was applied at post-grouting and the final stage, as a predictive analysis using the model from the inversion.

Figure 5.33 shows the model of rocks and physical properties at a water curtain pressure of 1.2 MPa. The water pressure to apply was set on the basis of an osmotic flow analysis with

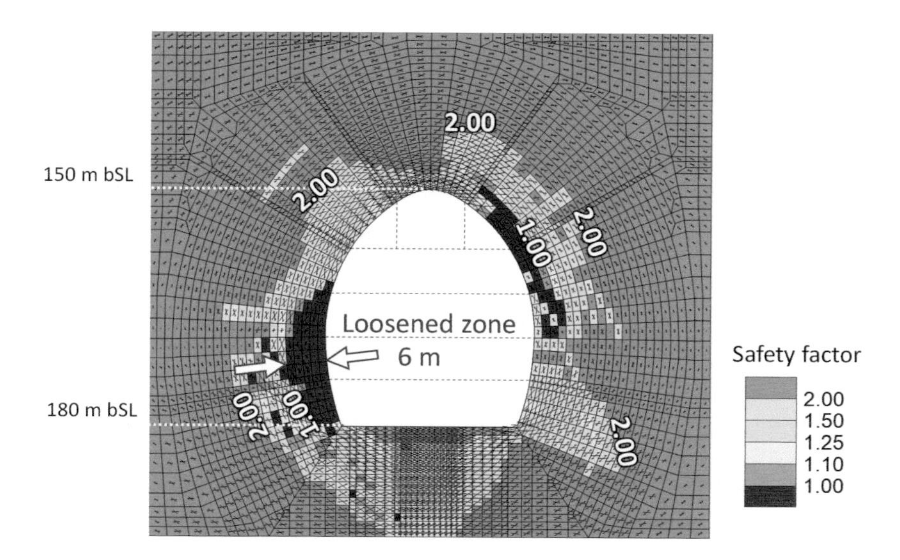

Figure 5.34 Distribution of the local safety factor at pressurising water sealing by predictive analysis.

the seepage pressure measurement and the 3D hydrogeological model, and applied to the highly permeable crack zones and the peripheral surface of grouting.

This method is considered to precisely predict the variation in the inner space displacement as in the comparison between the predicted and measured inner space displacement from excavation through to pressurising (Figure 5.32). Effect of the planned post-grouting to the highly permeable cracks at the bottom was predicted by the model. In the modelling process, the permeability coefficient of the highly permeable crack zone was substituted by the expected permeability coefficient after the grouting, and model calculation predicted the internal displacement and distribution of relaxation.

Figure 5.34 shows a comparison of predicted and measured internal displacement from excavation to pressurisation and prediction of relaxation zone. This method provided an accurate prediction of the internal displacement. The extent of the relaxation at pressurisation was predicted to be about six metres. Additional rock bolts and shotcrete reinforced by steel fibre were installed on the side walls (Figure 5.35). Post-grouting was also applied to the highly permeable crack zone on the floor.

The result showed that the extent of the loosened zone was about 6 m. Accordingly, rock bolts and steel fibre reinforced shotcrete were added as necessary support (Figure 5.35). The highly permeable crack zones on the cavern floor were also treated by post-grouting.

Figure 5.36 shows the subsidence of the roof, the inner space displacement and the distribution of ground displacement at the time of increasing the water curtain pressure from 1.1 to 1.25 MPa after the safety countermeasure work. The maximum increase in the inner space displacement was 0.5 mm, which was within the prediction. The displacement, flow rate of seepage and pore water pressure were also stable.

The excavation was completed with the assurance of the mechanical stability of the whole cavern by the above measures during excavation.

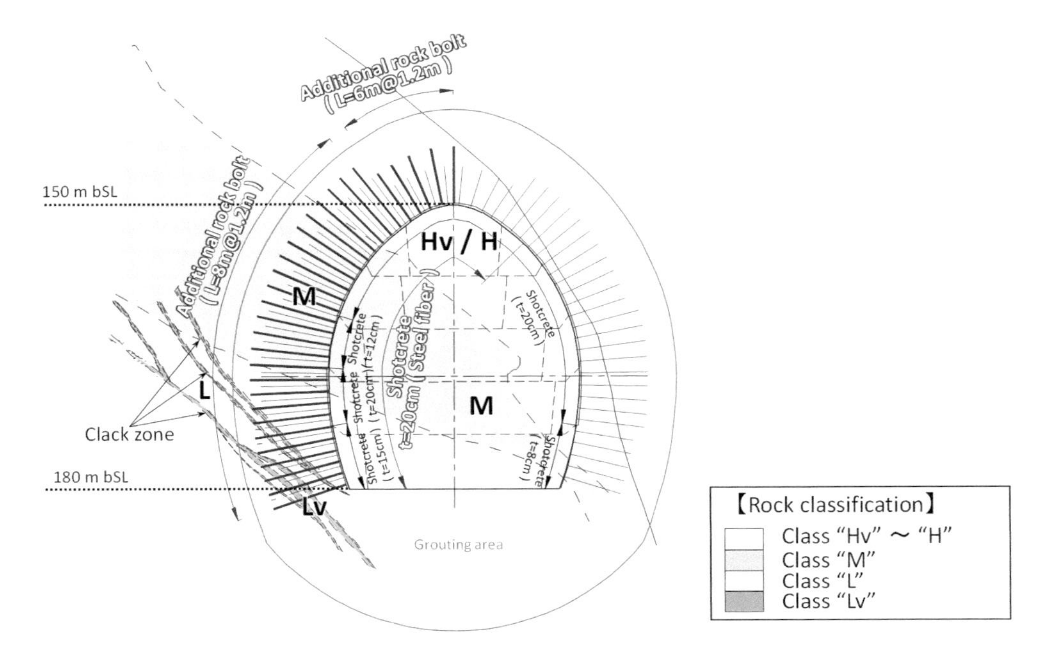

Figure 5.35 Additional supports in the crack zone of propane storage No. 1.

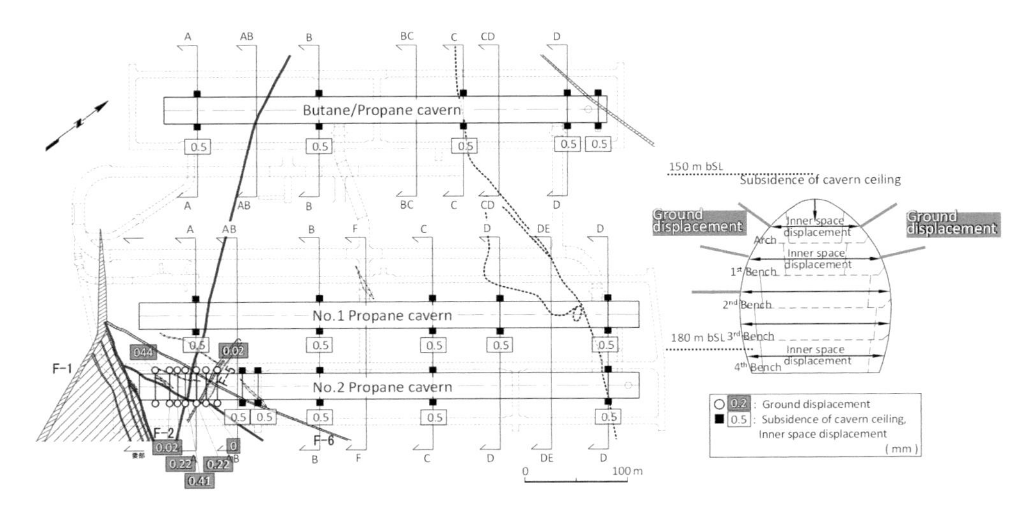

Figure 5.36 Subsidence of the roof, inner space displacement and distribution of ground displacement at the time of increasing the water curtain pressure at the Namikata facility.

References and Further Readings

Aoki K., Hibiya K. (1994) Groundwater control during the construction of large underground crude oil storage caverns. Integral approach to applied rock mechanics. *ISRM International Symposium*, Vol.1, 567–577.

Aoki K., Maejima T., Morioka H., Mori T., Tanaka M., Kanagawa T. (2003) Estimation of rock stress around cavern by CCBO and AE method. *Proceedings of the Third International Symposium of Rock Stress RS Kumamoto*, Kumamoto, Japan, 203–209.

Aoki K., Mito Y., Chang C.S., Maejima T. (2006) Numerical analysis of rock fracturing process by DEM using bonded particles model. *ISRM International Symposium 2006 (4th ARMS)*, Singapore, 398.

Aoki K., Mito Y., Chang C.S., Tasaka Y., Maejima T. (2007) Hydro mechanical coupled discrete modelling for the assessment of airtightness of unlined large rock cavern. *11th Congress of the International Society for Rock Mechanics*, Lisbon, 925–929.

Aoki K., Mito Y., Matsuoka T., Kondoh D. (2004) Design of gas storage rock cavern by the hydro-mechanical coupled discrete model. *Proceedings of the 2nd International PFC Symposium*, Kyoto, 289–299.

Aoki K., Mito Y., Mori T., Maejima T. (2005) Evaluation of EDZ around highly stressed rock cavern by AE measurements. *Eurock 2005 - Impact of Human Activity on the Geological Environment*, Brno, 23–29.

Aoki K., Mito Y., Yamamoto T., Shirasagi S. (2006) Geostatistical evaluation of the mechanical properties of rock mass for TBM tunnelling by seismic reflection method. *Rock Mechanics and Rock Engineering*, Springer nature, Austria, 591–602.

Aoki K., Toida M., Koshizuka K. (1989) In situ investigation of loosened zone around deep underground opening by acoustic emission monitoring technique. *Rock at Great Depth*, Pau, 313–319.

Aoki K., Toida M., Koshizuka K. (1990) The behavior of the loosened zone around rock caverns monitored by acoustic emission. In *Static and Dynamic Considerations in Rock Engineering*, Swaziland, 23–28.

Aoki K., Toida M., Koshizuka K. (1991) Monitoring of loosened zone around rock cavern by acoustic emission measurement. *3rd International Symposium on Field Measurements in Geomechanics*, Vol.1, 395–403.

Bergman M. (1977) Storage in excavated rock caverns. *Proceedings of the First International Symposium*, Stockholm, 832.

Ito M., Maejima T., Aoki K. (1995) Observational construction management system for the excavation of a large underground cavern in heavily jointed rock. *International Congress on Rock Mechanics*, Vol.2, 889–892.

Mito Y., Chang C.S., Aoki K., Matushi H., Niunoya S., Minami M. (2007) Evaluation of fracturing process of soft rocks at the great depth by AE measurement and ADEM simulation. *11th Congress of the International Society for Rock Mechanics*, Lisbon, 273–276.

Mito Y., Yamamoto T., Shirasagi S., Aoki K. (2003) Prediction of the geological condition ahead of the tunnel face in TBM tunnels by geostatistical simulation technique. *Proceedings of the 10th ISRM Congress*, Johannesburg, 833–836.

Mori T., Aoki K., Morioka H., Iwano K., Tanaka K., Kanagawa T. (2003) Application of borehole seismic and AE monitoring technique in the rock cavern. *Proceedings of the 10th ISRM Congress*, Johannesburg, 845–848.

Mori T., Iwano K., Morioka H., Aoki K. (2004) Application of micro seismic monitoring technique in the rock cavern. *Proceedings of the ISRM International Symposium 2004 (3rd ARMS, KYOTO)*, Kyoto, 379–383.

Nilsen B., Olsen J. (1989) Storage of gases in rock caverns. *Proceedings of the International Conference on Storage of Gases in Rock Caverns*, Trondheim, 398.

Nishioka K., Aoki K. (1998) Rapid tunnel excavation by hard rock TBM's in urban areas. *Proceedings of the World Tunnel Congress '98 on Tunnels and Metropolises*, Vol.2, 655–661.

Shirasagi S., Aoki K. (2001) Development of intelligent TBM excavation control system. *Proceeding of the ISRM Regional Symposium Eurock 2001 Espoo*, Finland, 615–620.

Shirasagi S., Yamamoto T., Murakami K., Ogura E., Mito Y., Aoki K. (2005) Evaluation system of tunnel excavation by geostatistics applying for seismic reflective surveys and TBM. S driving date. *Proceedings of the International Symposium on Design, Construction and Operation of Long Tunnels*, Taipei, 1255–1262.

Takewaki H., Yoshimura T., Aoki K., Hanamura T., Tajima T. (1988) Seismic observation and analysis of an underground crude oil storage facility. *Rock Mechanics and Power Plants, ISRM Symposium*, Madrid, 599–606.

Chapter 6

Air-tightness Test

The air-tightness of a rock storage cavern is tested by measuring the variation in the internal pressure ΔP over a 72 h period after compressed air has been injected. This test is designed to confirm the suitability of the cavern for the storage of liquefied petroleum gas (LPG). Several factors affect the pressure of the contained gas including temperature, change in the space for the gas in the storage cavern due to water invasion and the dissolution of the gas into the water. The effects of these factors must be accurately measured and appropriately compensated for in evaluating the air-tightness of the rock storage.

The pore water pressure around the storage cavern increases considerably in response to the injection pressure. Its response varies due to the heterogeneity of the hydrogeological properties of the surrounding rock. At the time of injection, the pore water pressure variation must be closely monitored to ensure that the hydrodynamic containment is functioning.

Air-tightness was examined in a small-scale tunnel before testing the actual sites in the Kurashiki and Namikata facilities. This experiment was aimed to analyse for

1. the effect of the variation of these factors and the accuracy of their measurements of the parameters on the evaluation of the air-tightness; and
2. the relationship between the behaviour of the pore water pressure around the cavern at pressurised injection and the hydrogeological properties.

According to the results of this experiment, a high-precision measurement system was developed and a standard guideline for evaluating air-tightness was drawn up for the Kurashiki and Namikata facilities. The measurement of the temperature distribution in the storage cavern is one issue among many influential factors. To precisely estimate the spatial distribution of the temperature in the storage cavern, high-precision thermometers were used at optimum locations as predicted by a 3-dimensional (3D) gas flow analysis by the computed fluid dynamics (CFD) method.

A delay in the response of the pore water pressure to the injection was measured and confirmed: it was due to the non-stationary behaviour of groundwater. The standard rate of increase in air pressure was determined by a non-stationary analysis to maintain the difference between the pore water pressure and the internal pressure of the storage cavern positive throughout the duration of the pressurisation. For this, an analysis of the unstable state using the results of the sealing experiment in the small tunnel was carried out before the event. At the time of actual pressurisation, the difference between the internal pressure and the piezometric head was constantly monitored at pressure measurement points and the rate of pressure change was adjusted accordingly; sometimes pressurisation was halted. The behaviour

DOI: 10.1201/9780367822163-6

of groundwater at pressurisation was evaluated by comparing the prediction with an analysis by a 3D hydrogeological model and actual measurements.

The variation in the internal pressure of the cavern, ΔP, did not decrease during air-tightness tests of the storage cavern after pressurisation. Using the high-precision monitoring system, excellent air-tightness was confirmed.

6.1 Air-tightness Test in a Small Reconnaissance Tunnel

6.1.1 Preliminary Test Using Small Tunnel

An air-tightness test in a small-scale tunnel was carried out as preliminary confirmation prior to the real test at the Kurashiki and Namikata facilities. For this test, the tunnel had been excavated for geological investigation at the Namikata facility. This section describes two aspects of this test: (1) evaluation and confirmation of the applicability of the air-tightness test using a variation in the internal pressure; and (2) evaluation of the behaviour of the pore water pressure at the time of injection into the storage cavern.

6.1.1.1 Applicability of Evaluation Method of Air-tightness Using Variation of Internal Pressure of the Cavern

The air-tightness of an underground storage cavern is evaluated by the variation in the internal pressure, ΔP, corrected for the influence of a variation in temperature and volume (Eq. 6.1). The measurement of the variation of internal pressure ΔP has inevitable uncertainty due to the precision of the instruments used for the measurement pressure, temperature and water level. Hence, the acceptable standard for air-tightness set as the absolute value of ΔP is smaller than its uncertainty $\varepsilon\Delta P$ (Eq. 6.2) throughout the test period:

$$\Delta P = P_0 - P_t \cdot \frac{T_0}{T_t} \cdot \frac{V_t}{V_0} \tag{6.1}$$

$$|\Delta P| \leq \varepsilon\Delta P \tag{6.2}$$

where the subscript is the time parameter ($t = 0$ at the start of the test and $t = t$ is its arbitrary time from the start)

P: internal pressure in Pa

T: temperature inside the storage cavern in K

V: its volume in m^3

The main parameters for ΔP are the measured values of pressure, temperature and volume of the storage cavern. The uncertainty associated with these measurements is regarded as the precision of the instruments, and $\varepsilon\Delta P$ is calculated as

$$\varepsilon\Delta P = P_t \cdot \left\{ \left(\frac{\sigma P_t}{P_t}\right)^2 + \left(\frac{\sigma P_0}{P_0}\right)^2 + \left(\frac{\sigma T_t}{T_t}\right)^2 + \left(\frac{\sigma T_0}{T_0}\right)^2 + \left(\frac{\sigma V_t}{V_t}\right)^2 + \left(\frac{\sigma V_0}{V_0}\right)^2 \right\}^{1/2} \tag{6.3}$$

where

σP_t, σT_t, σV_t: uncertainties of the measurement of pressure, temperature and volume at time $t = t$

σP_0, σT_0, σV_0: uncertainties of the measurement of pressure, temperature and volume at time $t = 0$.

Table 6.1 shows the calculation of uncertainty of the variation in the internal pressure of cavern $\varepsilon\Delta P$ using the precision parameters of instruments from preceding LPG rock storages in Eq. 6.3. Although the testing environments were not the same, $\varepsilon\Delta P$ in the last column is about 0.5 kPa at all the sites.

Figure 6.1 is a comparison of the influence of each measurement item of the air-tightness test on the uncertainty of the variation in the internal pressure of cavern $\varepsilon\Delta P$.

It was noted that the influence of the thermometer with ±0.1°C precision was considerably high. Following this, in order to make it comparable to other instruments, the precision of the thermometer was calculated to be ±0.01°C. Using a thermometer with this precision, the new $\varepsilon\Delta P$ was 0.1 kPa, a comparable value.

The influence of the air-tightness standard $\varepsilon\Delta P$ on the variation in the quantity of gas in the storage cavern is calculated by:

$$V_r = \frac{\varepsilon\Delta P}{P_{\mathrm{STD}}} \cdot V \cdot \frac{T_{\mathrm{STD}}}{T} \cdot \frac{24}{t_{\mathrm{test}}} \tag{6.4}$$

Table 6.1 Standard air-tightness $\varepsilon\Delta P$ against the precision of the measurement instruments and cavern specifications from preceding overseas studies

Precision	*Facility*	*Test pressure P (kPa)*	*Volume V (10^3 m^3)*	*Temperature T (°C)*	*ε (ΔP) (kPa)*
Pressure: ±0.05 kPa	Site A	808	170	13	0.46
Temperature: ±0.1°C	Site B	885	310	15	0.45
Water level (volume):	Site C	852	340	13	0.48
±10 mm	Site D	885	260	19	0.49
	Site E	995	140	16	0.55

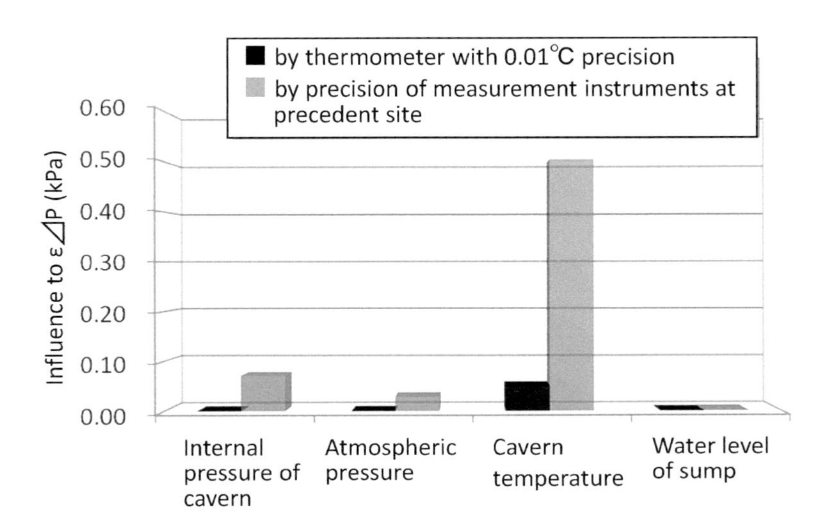

Figure 6.1 Contribution of measurement to $\varepsilon\Delta P$ based on the estimate of the Kurashiki facility.

V_r: variation in the quantity of the gas in the storage cavern (Nm^3/day)
V: gas capacity of the storage cavern (m^3)
P_{STD}: standard atmospheric pressure (Pa)
T: temperature of the storage cavern at the time of test (K)
T_{STD}: standard temperature (K)
t_{test}: duration of test (h)

Figure 6.2 shows V_r calculated for the precedents and the Kurashiki and Namikata facilities. As the capacity of the Kurashiki and Namikata facilities is greater than the overseas precedents, the value of V_r is two to six times larger.

Hence, a measurement system similar to the overseas precedents does not deliver a reliable evaluation of air-tightness comparable to the overseas precedents.

The reliability of an evaluation of the air-tightness of the large storage caverns at the Kurashiki and Namikata facilities needs to be comparable with the overseas precedents. It is evaluated by the value of $\varepsilon\Delta P$. The influence of the precision of the measuring instruments on $\varepsilon\Delta P$ was analysed based on the overseas precedents. The same precision parameters were used to simulate the effects on the Kurashiki and Namikata facilities for comparison. The result at the Kurashiki facility is shown in Figure 6.2 as an example.

On the other hand, there is no reported case history of using the thermometer with ±0.1°C precision to confirm the influence of the improvement in precision of the thermometer on ΔP. For example, some influence which was negligible in the measurement bya thermometer with ±0.1°C precision may have considerable influence on the system using an instrument with ±0.01°C precision. The applicability of the thermometer with ±0.01°C precision must be examined and confirmed by a test in the small tunnel before the air-tightness tests at the Kurashiki and Namikata facilities.

6.1.1.2 Variation of Pore Water Pressure at Pressurised Air Injection

A small-scale test was carried out in the test tunnel drilled for geological investigation at the Namikata facility site with the aim:

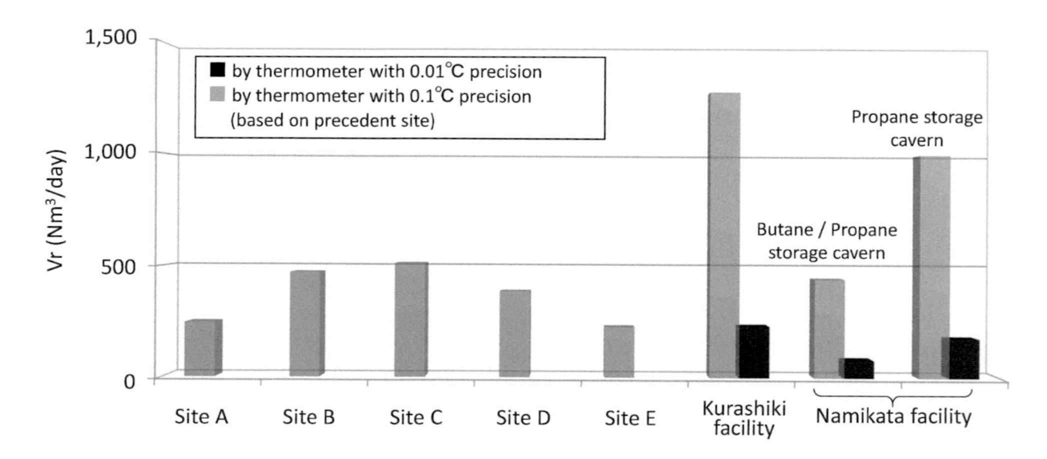

Figure 6.2 Variation in quantity of gas in the storage cavern, V_r, based on $\varepsilon\Delta P$.

1. To evaluate the air-tightness by a variation of the internal pressure:
   - To calculate the variation in the internal pressure of the cavern, ΔP, using thermometers with $\pm 0.01°C$ precision, and to confirm their applicability to sealing evaluation.
   - To analyse the temperature distribution of the cavern at the time of the test and to clarify the mechanism of its behaviour to contribute to designing the placement of thermometers considering the scale of the actual cavern and the distribution of the seep in the cavern.
2. To determine the behaviour of the pore water pressure at the time of injection:
   - To measure and analyse the pore water pressure around the small-scale model cavern at the time of injection to contribute to water-tightness management.

6.1.2 Outline of the Test

The test took place in the investigation tunnel which had been drilled for geological information at the Namikata site (Figure 6.3). Its depth at 125 m below sea level is the same level as the hydraulic sealing tunnel. The investigation tunnel was reinforced by lock bolts and concrete spray, but not waterproofed by grouting. Figure 6.4 shows the layout of the test cavern, and Figure 6.5 shows its plan view and cross section. The side tunnel for the test is horseshoe shaped, 3 m wide and 3.4 m high. The cavern for the air chamber is 16 m long at the front end of the tunnel, followed by No. 1 concrete plug to separate the air chamber, the water chamber and No. 2 concrete plug, then a wider area for test drilling for water-tightness and a measurement room separated by a passage tunnel. To maintain the water pressure in the rock formation surrounding the air chamber, six hydraulic sealing boreholes were placed 3 m from the tunnel wall.

The measuring instruments were pore water pressure gauges for water-tightness management around the air chamber cavern, and pressure gauges, thermometers and water level

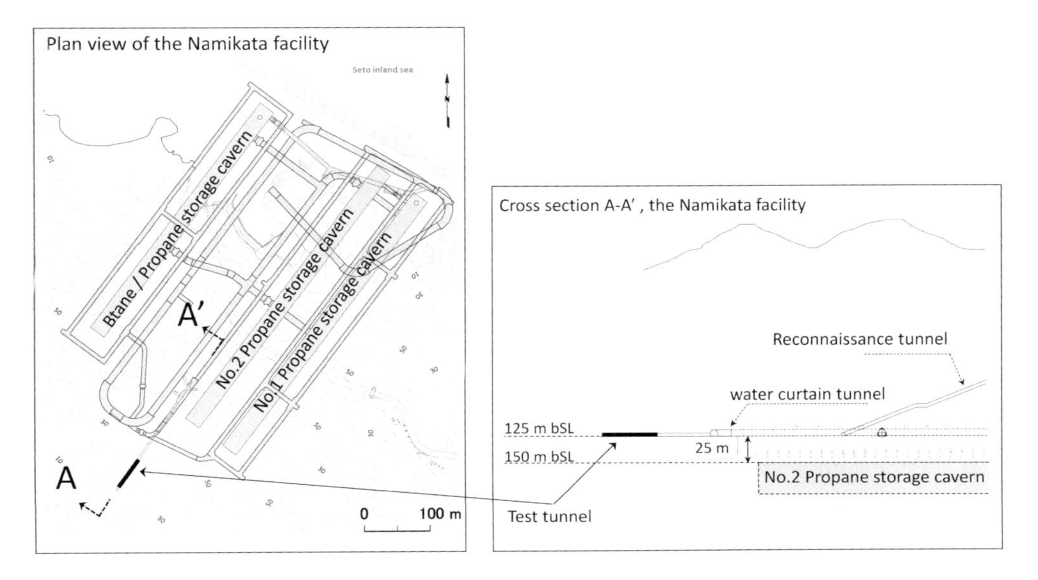

Figure 6.3 Location of the test cavern in the Namikata site.

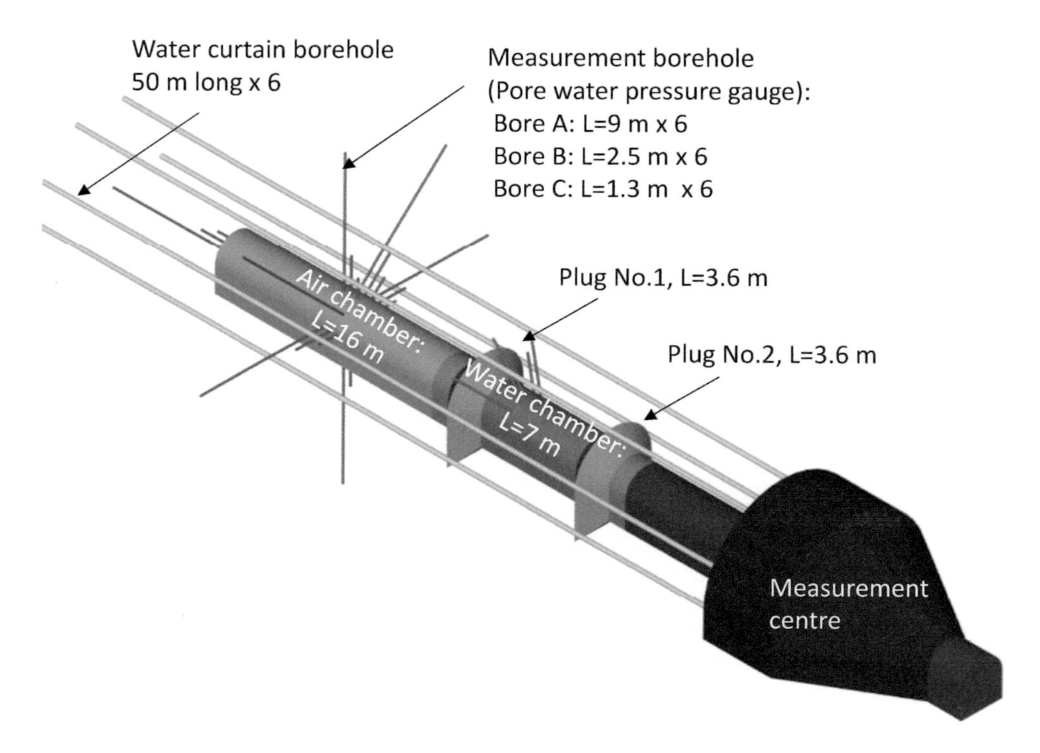

Figure 6.4 Bird's-eye view of the layout of the test cavern.

gauges in the air chamber. Three cross section planes for measurements A–C were set up 1 m apart near the centre of the air chamber (Figure 6.5); six boreholes were drilled in six directions along each of the A, B and C planes from the tunnel drilled 9, 2.5 and 1.3 m, respectively, in length. These were named A-1, A2,..., C-6. The last 30 cm of the boreholes was isolated by packers for the measurements interval. In addition, three boreholes were drilled above No. 1 concrete plug and water pressure gauges were installed at the centre and the bottom of the hole isolated by packers. Acoustic emission (AE) sensors were installed on the walls within pressure measurement intervals of the B-2, B-3, B-4, C-2, C-3, C-4 and hole P on the top of No. 1 plug. Figure 6.6 shows the arrangement of the pore water pressure gauges.

Figure 6.7 shows the arrangement of the measurement instruments in the air chamber. The air pressure inside the chamber was measured in the measurement station through a pressure-tolerant nylon tube. Nineteen thermometers with ±0.001°C precision were used, rather than the ±0.1°C models used in the overseas predecessors. A top cover was installed on each thermometer to keep dripping water off (Figure 6.8). A water level gauge was installed in the drainage pit to monitor the variation in the volume due to the fluctuation in the water level. The volume of the air chamber was measured with a 3D laser scanner prior to the experiment (Figure 6.9). Its volume was 165.78 m^3. A hygrometer was also installed in the air chamber.

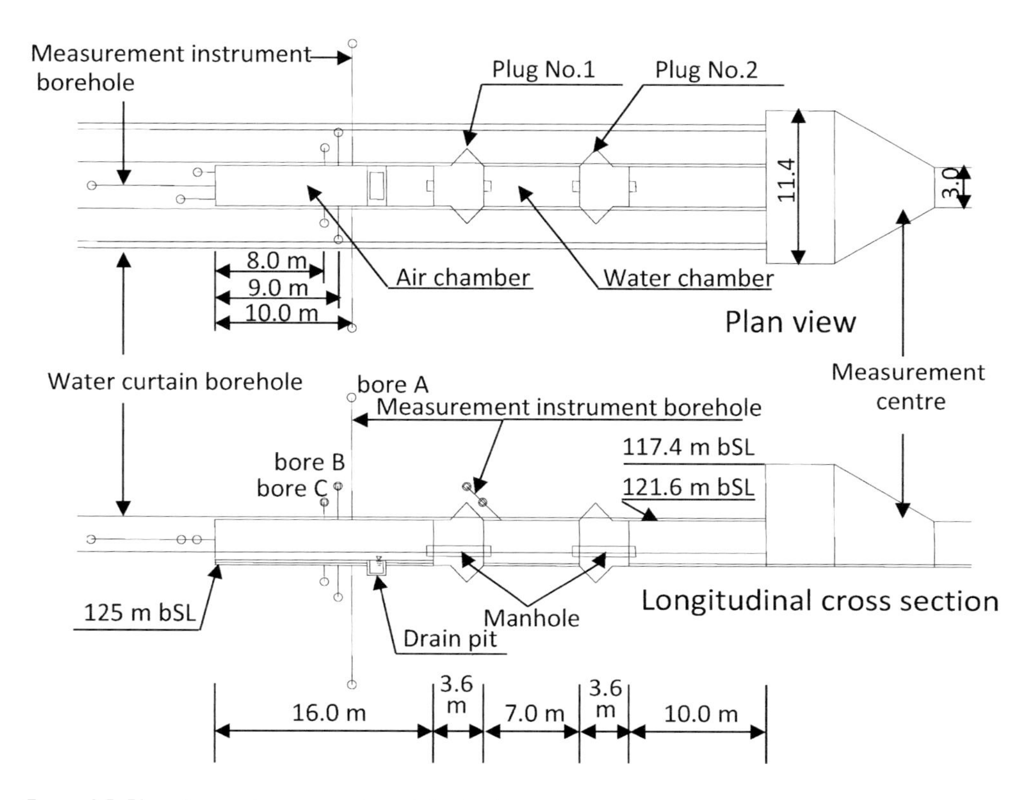

Figure 6.5 Plan view and cross section of the test cavern.

6.1.3 Hydrogeological Structure of the Test Tunnel

Cretaceous granite and granodiorite were present in the test site, as recognised in the cavern. Figure 6.10 shows the local geology around the test tunnel in plan view and the hydrogeological setting along cross section B. The dominant orientations of the cracks were strike between NS and NS20W (the orientation perpendicular to the tunnel) with a southerly dip and strike N30E to N70SE with a northerly dip. The permeability of the rocks was measured in the measurement holes and at the time of boring the hydraulic sealing holes. Their logarithmic average was 0.08 Lu, which was comparable to the average Lugeon value of the actual storage cavern of the Namikata facility. However, the geological zones with cracks connecting to the cavern with the N-S strike presented discrete Lugeon values between 0 and 34 Lu showing a heterogeneous hydrogeological structure similar to the actual cavern.

In observation hole A-2, where cracks in the N-S strike were dominant, borehole television (BTV) observation revealed seeping. The crack path from the air chamber to the water chamber (① in Figure 6.10) formed a geological boundary to the fracturing structure. The wall surface of the air chamber was generally wet and drips were observed along the southern seep line of the crack ①.

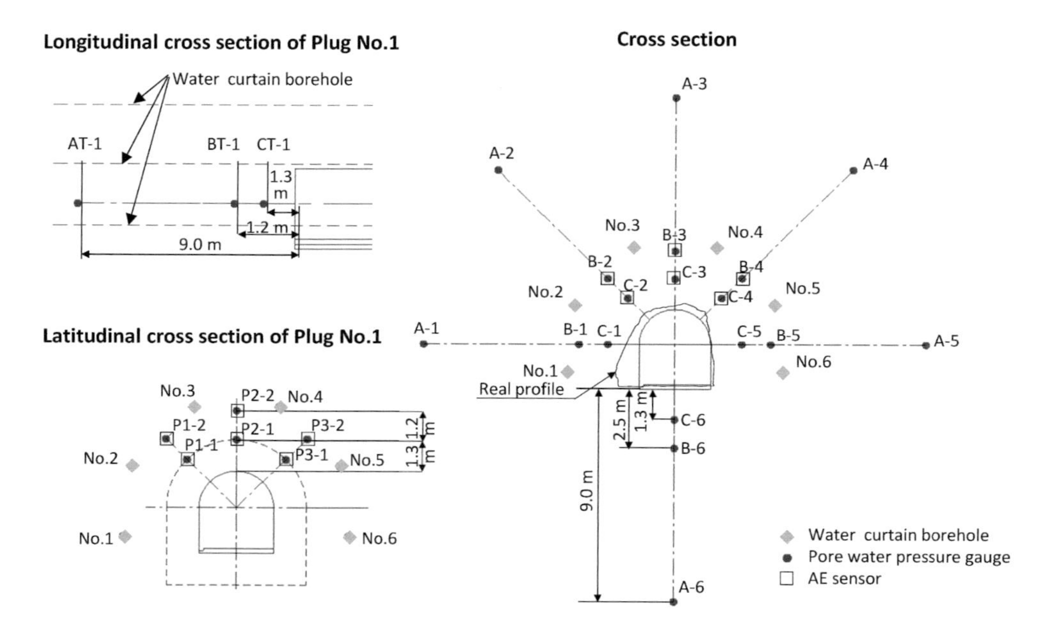

Figure 6.6 Layout of the sensors for gas sealing around the water-tightness management cavern in the test tunnel.

6.1.4 Test Procedure

Figure 6.11 shows the test procedure undertaken:

① Drilling a water sealing bore; measurement bore, start water injection.
② Installing the measurement instruments.
③ Building plug concrete No. 1 (air chamber side) and No. 2 (water chamber side).
④ Injecting pressurised water up to 1800 kPa into the water chamber to confirm the tolerance of the plug.
⑤ Injecting water into the air chamber as a preliminary test. This tested the increase in pressure by water and aimed to examine the response of the water pressure surrounding the tunnel and the speed of the increase in pressure at injection. The confined water pressure was kept at 1200 kPa during the test, and the internal pressure in the air chamber was increased from 400 to 900 kPa by controlling the water drainage from the air chamber after it filled up.
⑥ Following the water injection test in the air chamber, the water was drained from the air chamber and the water chamber. Then, measurement instruments (thermometers, hygrometers, water level gauges and tubing for pressure measurement) were installed in the air chamber. The water chamber was again filled with water and the pressure was kept at 1200 kPa.
⑦ The final test was carried out after confirming stable temperature in the air chamber and the surrounding water pressure.

The pressure for the air-tightness test ⑦ was set to 700 kPa according to the preliminary water injection test ⑤ where the pore water pressure outside the water curtain bores was stable up

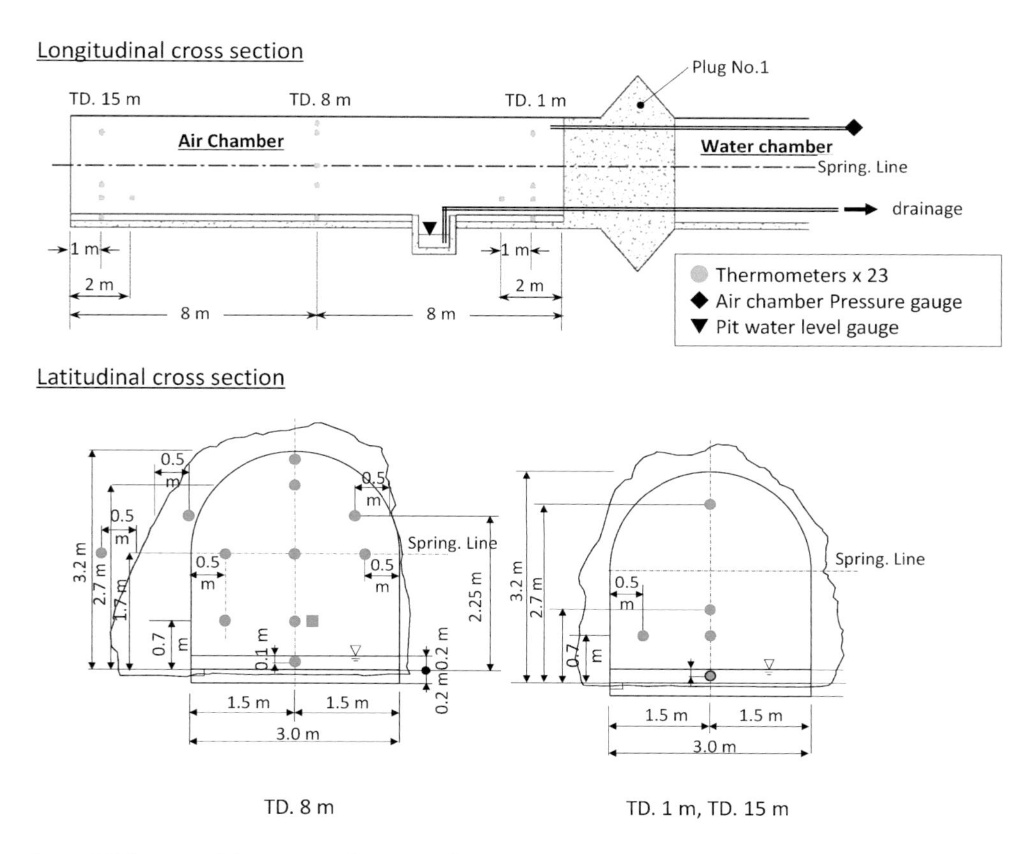

Figure 6.7 Layout of the sensors for gas sealing within the water-tightness management cavern in the test tunnel.

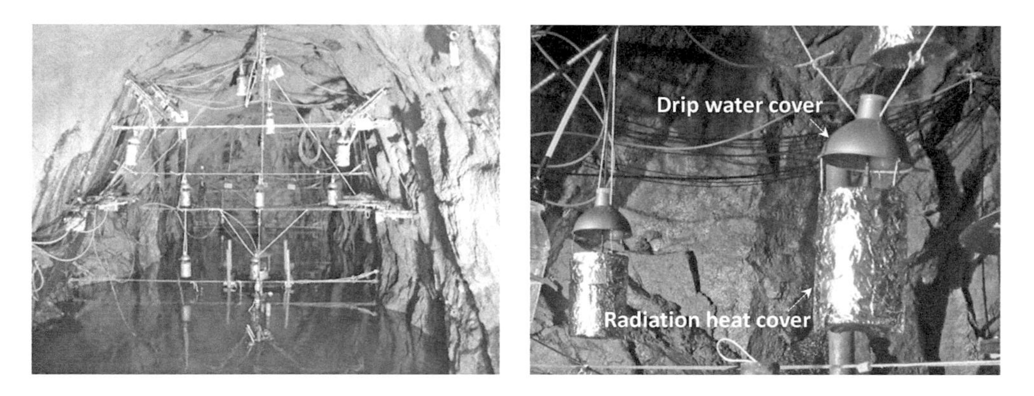

Figure 6.8 Installation of thermometers in the air chamber.

to 790 kPa of the internal pressure of the chamber. After the air-tightness test, the internal pressure of the chamber was gradually increased again to revaluate the air-tightness under higher internal pressure and to analyse the behaviour of the surrounding pore water pressure.

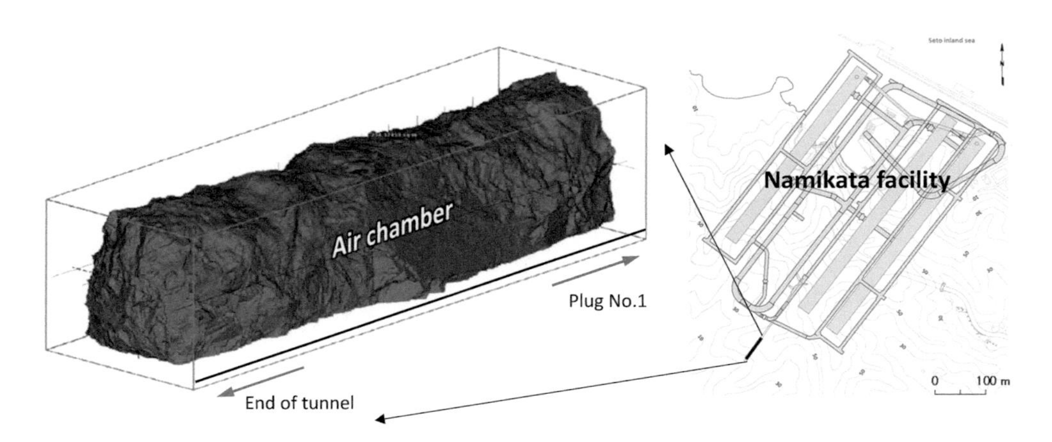

Figure 6.9 Shape of the air chamber by the 3D scanner.

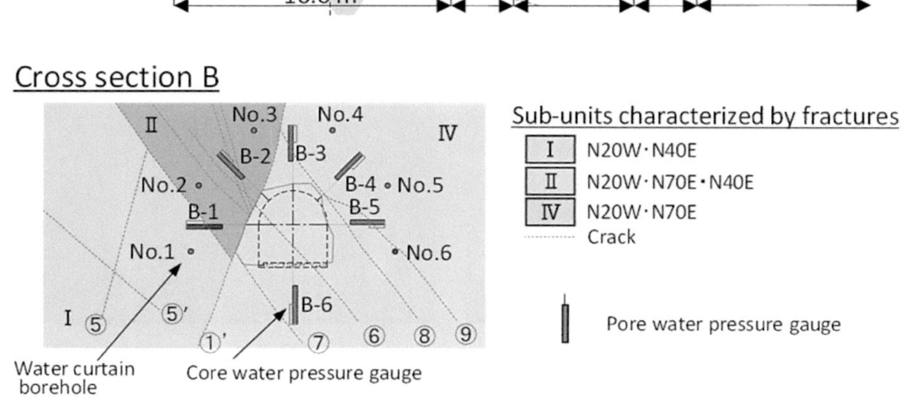

Figure 6.10 Geology around the test tunnel.

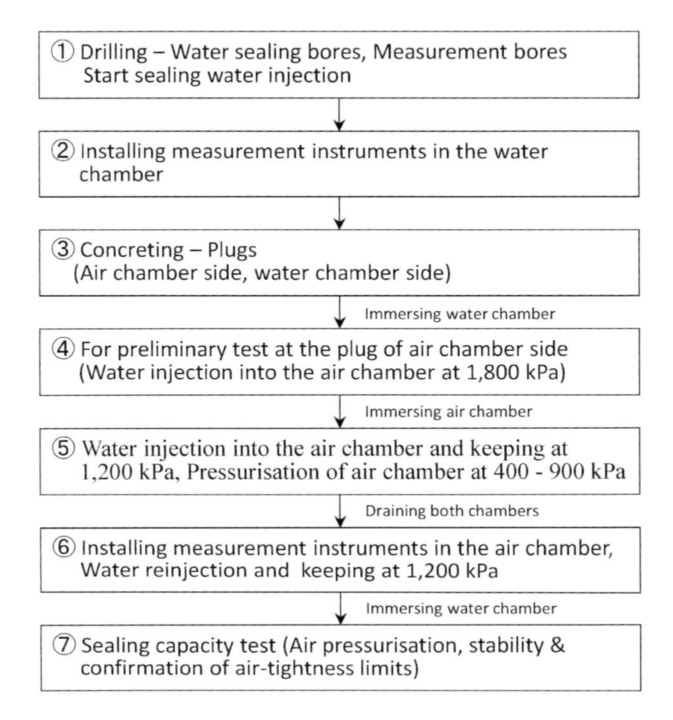

Figure 6.11 Procedure of the air-tightness test at the small tunnel.

Figure 6.12 shows the test procedure.

As stated in Section 6.1.1.1, the air-tightness in this test is evaluated by the variation in the internal pressure in the air chamber, ΔP, corrected for temperature and volume changes, and it is judged adequate if there is no loss of internal pressure in the storage cavern and if ΔP is smaller than its uncertainty $\varepsilon\Delta P$.

$$|\Delta P| \leq \varepsilon\Delta P \qquad (6.2) \text{ Repeat}$$

The value of $\varepsilon\Delta P$ for this test was 0.117 kPa calculated from the precision of the instruments and the test condition: test pressure 800 kPa abs (700 kPa); temperature 294 K; volume 165.78 m^3, test duration 1440 min; flow rate of seep 0.005 m^3/min; solubility of air 0.019 m^3/m^3 (at 294 K under atmospheric pressure, by Chronological Scientific Table, 2000). This is summarised in Table 6.2. From this calculation, the sealing is regarded as good if ΔP is less than 0.177 kPa for the duration of the test.

6.1.5 Test Result

6.1.5.1 Applicability of the Evaluation Criteria of Air-tightness Based On Variation of Internal Pressure of Cavern

Figure 6.13 shows the measurement of the temperature and pressure during the air-tightness test from the start of air injection through the test duration. The temperature values

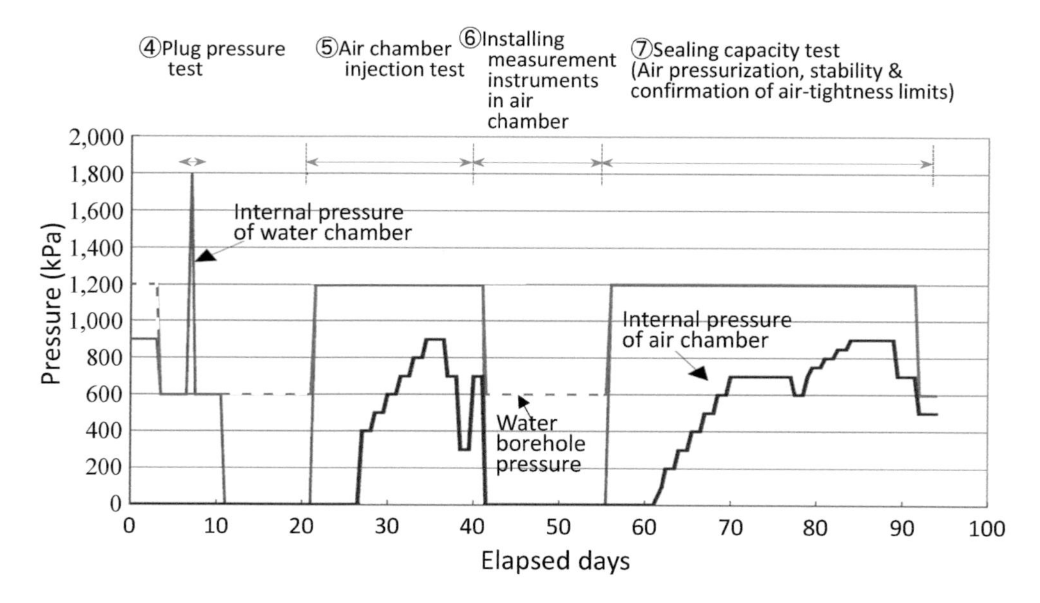

Figure 6.12 Plan of test procedure.

in Figure 6.13 are those measured at the nine points along the main test cross section. The variation in pressure over 100 h, ΔP, was calculated from the values measured by the instruments arranged in the cavern at 700 kPa internal chamber pressure. This section details the seal capacity test for the applicability of a high-precision thermometer over 100 h. In addition, the mechanism of the temperature changes in the cavern at the time of the air-tightness test was examined by analysing the variation in the temperature distribution from the start of injection to after injection, in order to design the placing of the thermometers considering the site characteristics including the size of the cavern and the distribution of seeps in the cavern.

Figure 6.14 shows temporal variations in the air chamber pressure, the average temperature in the air chamber and the water level and humidity in the pit during the air-tightness test under 700 kPa internal pressure. In this test, the internal pressure of the air chamber was raised to 700.1 kPa followed by a stabilising period of one day. Then, the air-tightness was evaluated for the next 100 h by calculating ΔP.

After stopping the injection, the temperature in the air chamber dropped sharply and slowly converged. Accordingly, the pressure in the air chamber showed lowering by 0.6 kPa over 100 h from the initial pressure of 698 kPa. Variations in pressure and temperature due to a decrease in the water level in the pit by draining were also recognised: the temperature change from 0.02°C to 0.04°C was clearly observed.

Figure 6.15 shows the pressure variation ΔP after all corrections were applied.

At the end of the 100-h test, the value of ΔP was 0.003 kPa. Its maximum value during the test was 0.051 kPa; both values well cleared the range of uncertainty, $\varepsilon\Delta P$, ±0.177 kPa, calculated for the precision of the instruments. During the test, ΔP did not drop and stayed within the range of $\varepsilon\Delta P$. This test concluded that thermometers with ±0.01°C precision were suitable under the site condition and that the evaluation of air-tightness using these thermometers was reliable.

Table 6.2 Calculation of air-tightness standard $\varepsilon \Delta P$

	Pressure		*Temperature*		*Volume*		*Vapor pressure*		*Dissolved air*	
Test condition	P	800 kPa	T	294.15 K	V	168.78 m^3	P_w	800 kPa	V_s	165.78 m^3
Instrument accuracy	εP	0.11 kPa	εT	0.01 K	εV	0.0002 m^{3}[a]	εP_w	0.045 kPa[b]	εV_s	0.0081 m^{3}[c]
	$2\times\left(\frac{P}{P}\right)^2$	3.8×10^{-8}	$2\times\left(\frac{T}{T}\right)^2$	2.3×10^{-9}	$2\times\left(\frac{V}{V}\right)^2$	2.9×10^{-12}	$2\times\left(\frac{P_w}{P}\right)^2$	6.3×10^{-9}	$\left(\frac{V_s}{V}\right)^2$	2.4×10^{-9}
										$\varepsilon \Delta P$ 0.177 kPa

Notes: [a] Area × thickness = 2.0 (m^2) × 0.0001 (m) = 0.0002 (m^3).

[b] Vapor pressure (at 21°C) × εH = 2.5 (kPa) × relative humidity 1.8%/100

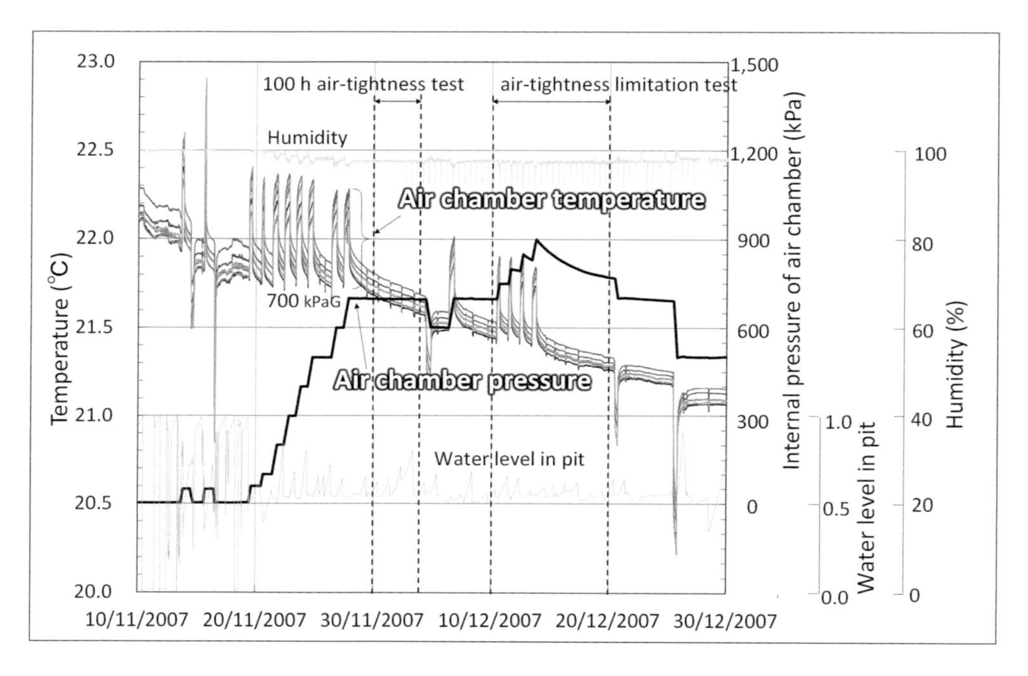

Figure 6.13 Measured values for the evaluation of air-tightness at the air-tightness test in the small tunnel. (Temperature at the main measurement cross section TD = 0.8 m).

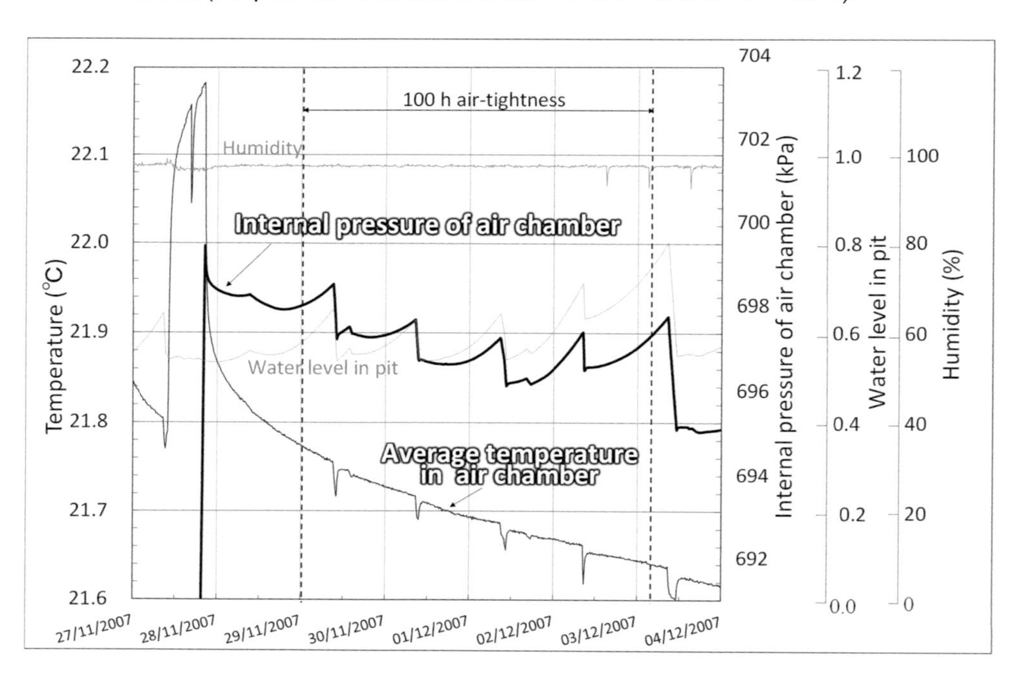

Figure 6.14 Result of measurements in the air chamber under 700 kPa in the air-tightness test in the small tunnel.

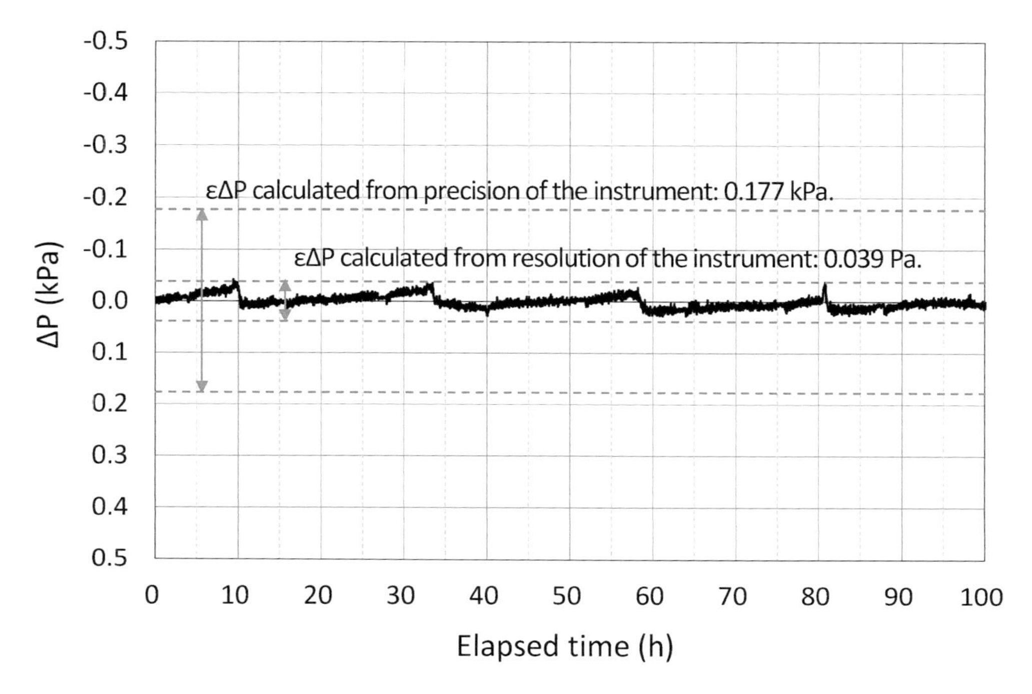

Figure 6.15 Pressure variation ΔP after all corrections (tested under 700 kPa).

A temporal variation in the temperature distribution during the test in the small tunnel was analysed.

Figure 6.16 shows the change in temperature for the duration of the 100-h test after stopping injection at each measurement point. Figure 6.17 shows the thermal contours, the average humidity of the latitudinal cross section and the variation in the water table temperature along the longitudinal axis of the air chamber. The distribution of the temperature in the air chamber presented a difference of around 0.2°C in places. The temperature was higher at the upper test positions (dark colours) than the lower positions (light colours). The temperature showed a layered distribution in the longitudinal cross section (Figure 6.17). The temperature was lowest in the deep part of the air chamber (TD 15 m) and increased toward the injection valve (TD 1 m). This trend was similar to the distribution of the water table temperature.

The temporal variation in the temperature of the air chamber showed a trend converging to the temperature of the surrounding rocks after stopping the injection and maintaining the same distribution.

The variation in the temperature was analysed by a heat flow analysis with a 3D model of the air chamber at the time of the air-tightness test (Figure 6.18).

This analysis used the ANSYS FLUENT 12 (reference to be added) method for the low-pressure injection (injection rate 100 kPa/h; confined air pressure 0 → 50 kPa) before the 100-h confined air pressure test. The model parameters of the material used in this analysis are shown in Table 6.3.

Figure 6.19 shows the temporal variation in temperature at the measurement points on the measurement cross sections at TD = 1, 8 and 15 m as an example of a comparison between

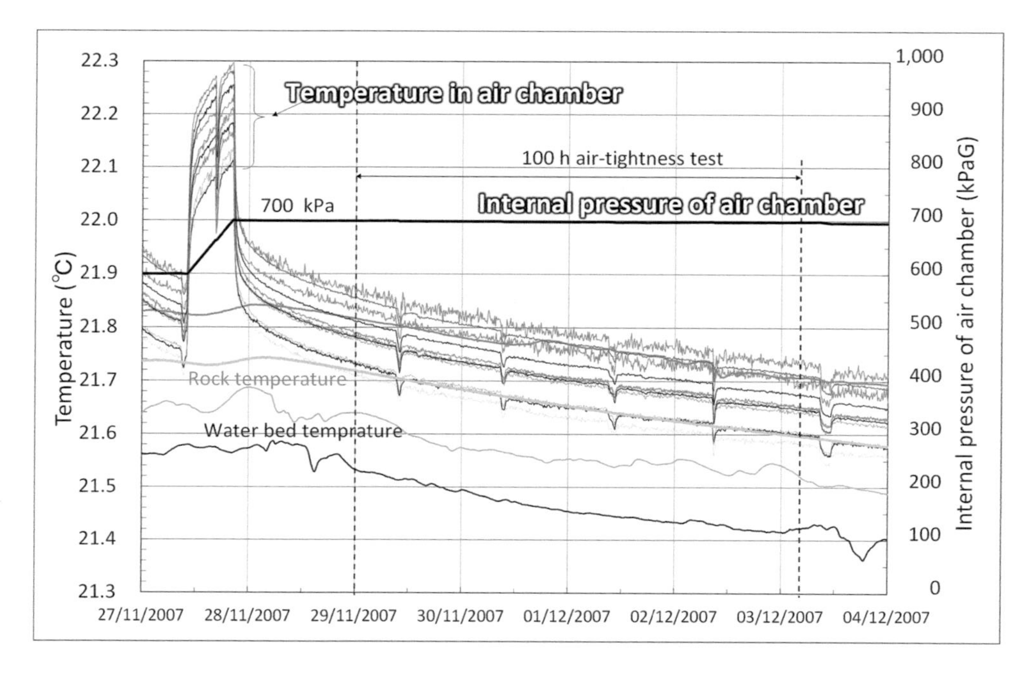

Figure 6.16 Temporal variation of the air chamber after injection in the test tunnel experiment.

the analysed and measured values. Figure 6.20 is the contour of the predicted values and its flow velocity vectors at 280 minutes after the start of injection. The predicted values mimic the layered structure of the measured values. However, the difference at TD = 1 m near the air intake is relatively large while the deeper part at TD = 15 m shows good agreement. One of the reasons for this discrepancy may be the omission of the gradient of the water table temperature which decreases toward the end of the tunnel and the water seeping from the wall surfaces in the model. This suggests that an analysis of the temperature distribution along the length of the tunnel should include these effects. Especially in the Kurashiki facility, where the quantity of seepage is large, the placement of the thermometers should be decided with consideration of the effect of the water.

From the flow velocity vectors in Figure 6.20, the convection of the air was considered to be the mechanism to form the layered thermal distribution in the latitudinal section: the air is heated by compression and its reduced density climbs in the middle of the air chamber, then it is cooled on the wall surface and moves down along the surface.

6.1.5.2 Behaviour of Pore Water Pressure at Injection

The pore water pressure around the cavern varies along with pressurised injection. Monitoring this behaviour contributes to designing the management system of water-tightness at the time of injection into the cavern.

Figure 6.21 shows the internal pressure of the air chamber, the water curtain pressure and the measured pore water pressure along cross section B in the preliminary test with pressurised water in the air chamber.

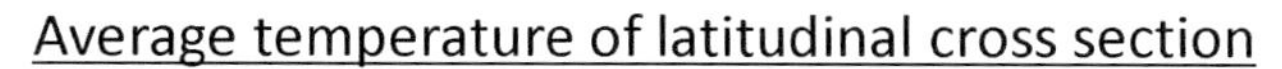

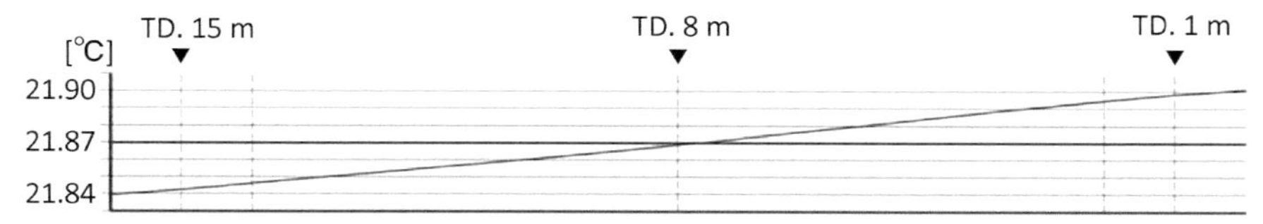

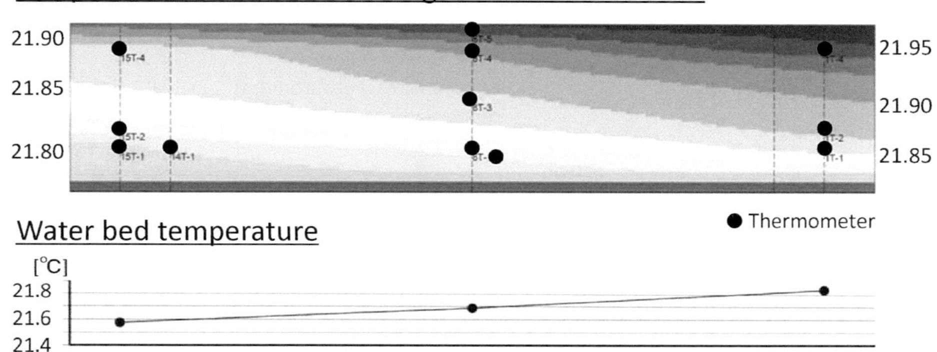

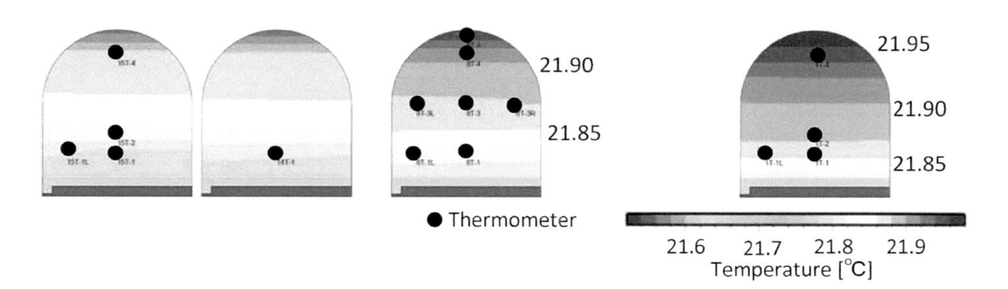

Figure 6.17 Temperature distribution immediately after stopping injection in the test tunnel experiment.

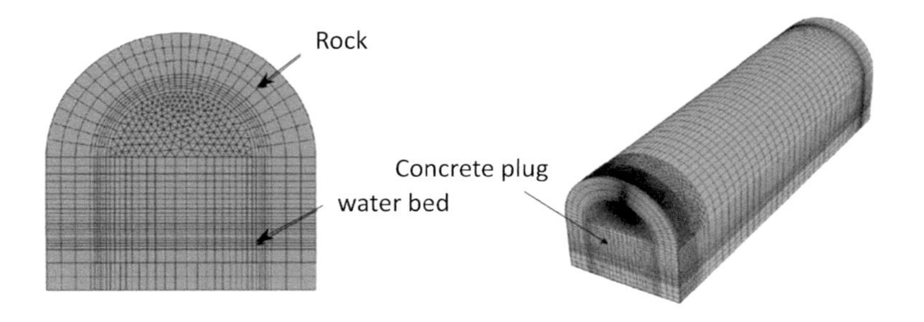

Figure 6.18 3D model of heat flow analysis in the test tunnel experiment.

Table 6.3 Model parameters for 3D heat flow analysis in the test tunnel experiment

Gas	The standard *k*-ε turbulent model is used in analysis taking compressibility into consideration.
Concrete	Internal heat conduction of concrete is considered. The external temperature is set to a constant at 23.8°C.
Rocks	Heat conduction in the rocks within 1 m around the air chamber is considered. The external temperature is set to a constant at 21.8°C.
Water bed	In the model, the water bed is treated as a solid with heat conduction. The cell is set to 20 cm thick.

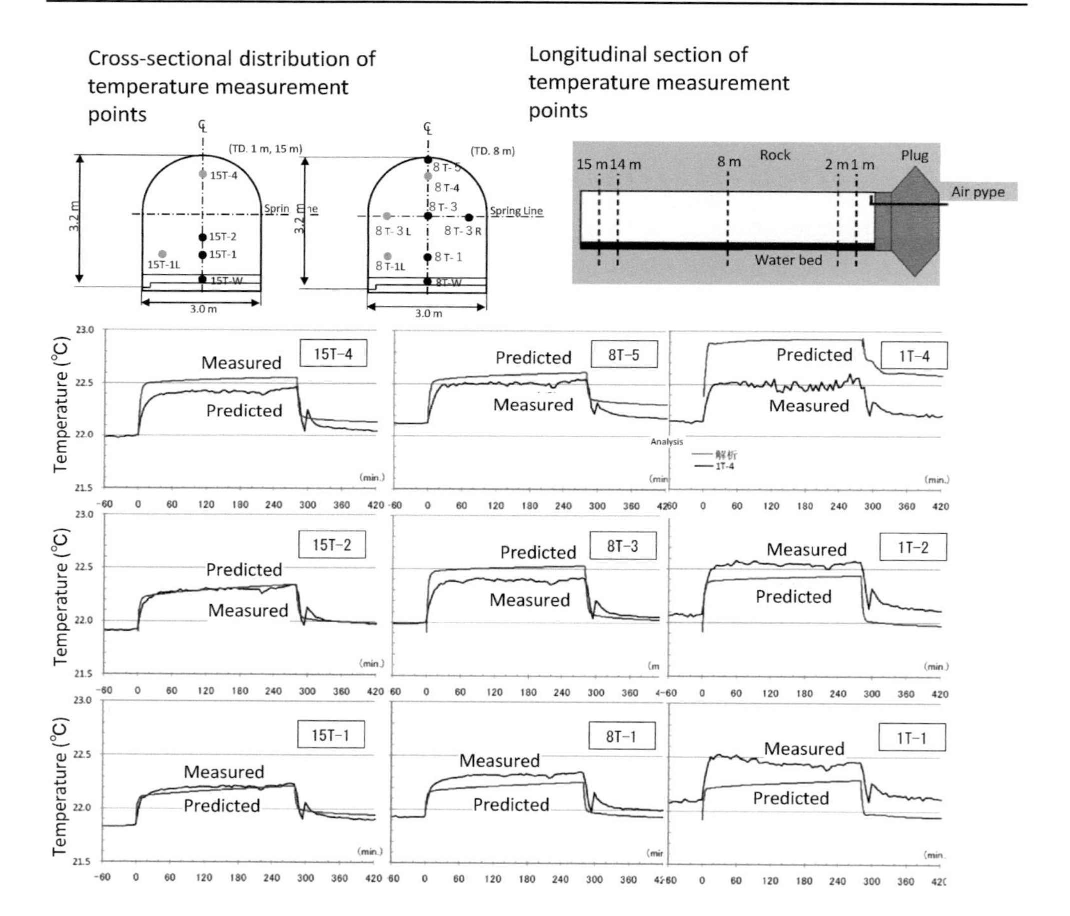

Figure 6.19 Analysed and measured temperature at the measurement points in the test tunnel experiment.

A small-scale air-tightness test was carried out in the investigation tunnel with the internal pressure of the chamber at 700 kPa for 100 h. The air chamber pressure was subsequently increased up to 900 kPa in 50 kPa steps, and the variation in the internal pressure was observed for 12 h to ensure air-tightness. The behaviour of the surrounding pore water pressure was also analysed at the same time (Figure 6.22).

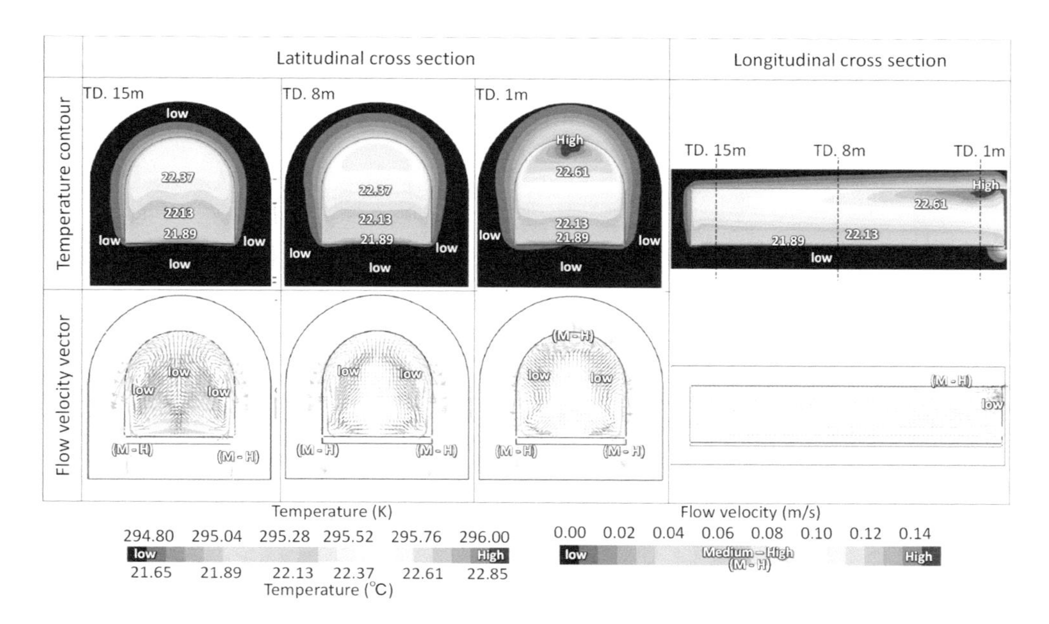

Figure 6.20 Temperature distribution and flow velocity vector at 280 min into the injection in the test tunnel experiment.

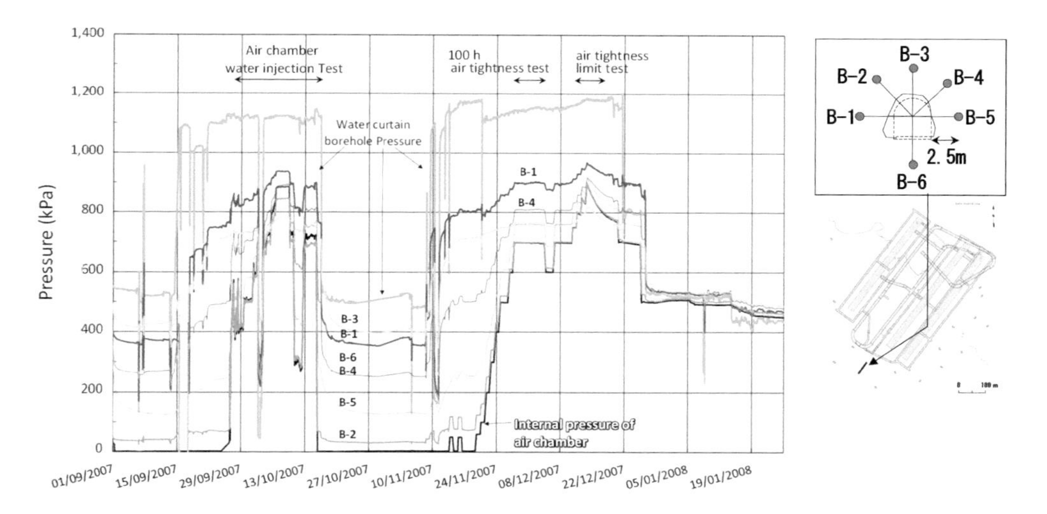

Figure 6.21 Temporal variation of the internal pressure of the air chamber and pore water pressure of cross section B.

The test found that ΔP maintained a constant trend up to 750 kPa but it clearly decreased over 800 kPa. Therefore, the critical sealing limit under the test condition was considered between the internal pressures of the air chamber at 750 and 800 kPa.

This section provides details of the behaviour of the pore water pressure around the air chamber in the air pressure range 750–800 kPa.

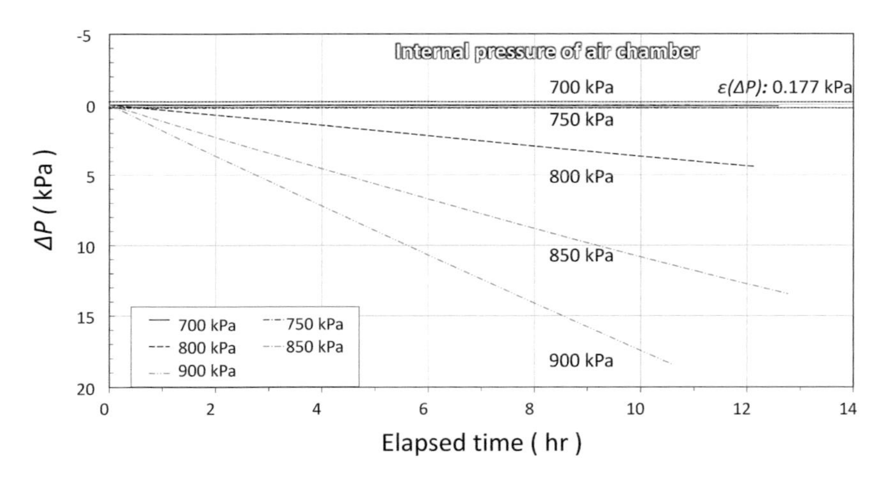

Figure 6.22 ΔP at each internal pressure of the air chamber during the critical sealing test.

Figure 6.23 shows the temporal variation of the pore water pressure at B-2, A-2 and C-2 in the geological subdivision with cracks in the N-S strike and the AE counts in B-2 and C-2 holes.

These points were selected because B-2 showed the lowest pore water pressure value in the B boreholes at the time of the injection from 750 to 800 kPa; A-2 and C-2 were on the same line; and B-3 was close to B-2. The bottom of Figure 6.23 shows the temporal variation in the difference between the pore water pressure and the internal pressure of the air chamber at B-2 and C-2.

Starting the pressure injection into the air chamber from 750 kPa, the pore water pressure of B-2 and C-2 holes increased in parallel to the internal pressure of the air chamber with similar value. When the internal pressure of the air chamber reached around 758 kPa, the piezometric head difference between the internal pressure of B-2 and the air chamber was nearly 0 m. Then, the gradient of the pore water pressure increase at B-2 suddenly changed and an AE was observed at B-2 and C-2 immediately thereafter. When the air chamber continued to be pressurised, AE stopped after 765 kPa of internal pressure in the air chamber, and the pore water pressure at B-2 again increased in parallel with the internal pressure.

The pore water pressures at A-2 and B-3 hardly showed any variation even when the pore water pressure at B-2 steeply increased. It gently increased from the time the internal pressure of the air chamber reached 773 kPa where the internal pressure of the air chamber exceeded the measured pore water pressure at these holes. At the same time, AEs were observed continuously in the C-2 hole. No other AE sensors showed such an observation.

Figure 6.24 shows hydrological field contour maps of the pore water pressure and the pressure difference between the pore water pressure and the internal pressure of the air chamber at the critical sealing pressure of the latitudinal cross section of the air chamber at internal pressure 758 kPa, immediately before which the discontinuous behaviour of the pore water pressure was observed. This figure demonstrates that the area with the smallest piezometric head difference occurred to the top left of the air chamber. This area coincides with the orientation of the continuous cracks from the air chamber. Pressurising the air chamber with continuous cracks needs special attention.

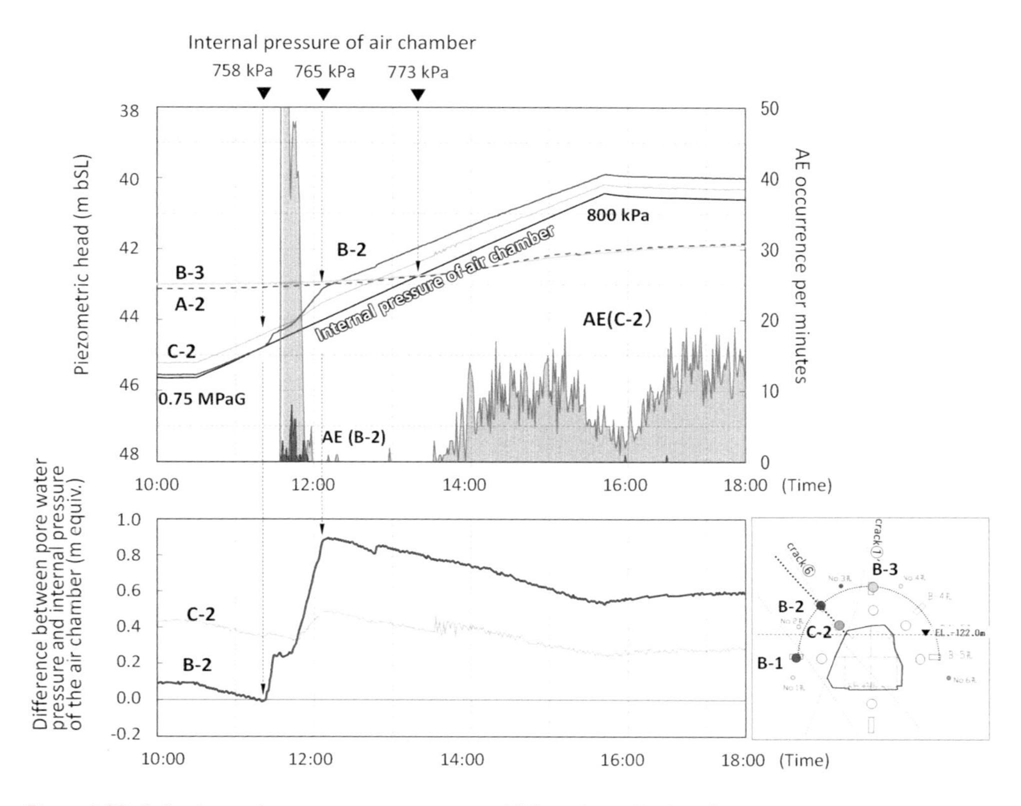

Figure 6.23 Behaviour of pore water pressure and AE at the critical sealing.

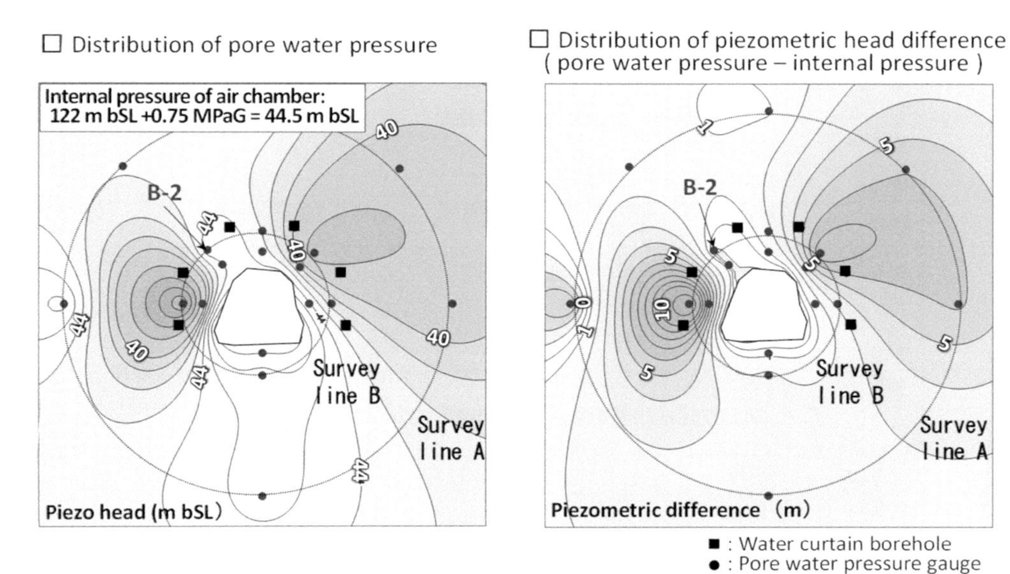

Figure 6.24 Pore water pressure contour and piezometric head difference contour at the critical sealing pressure (758 kPa internal pressure of the air chamber).

The behaviour of the pore water pressure was examined along the pressurised injection.

Figure 6.25 shows the temporal variation in the internal pressure of the cavern and the pore water pressure measured on cross section B. All the pore water pressure gauges show a stepwise increase along with stepwise pressurisation into the caverns. The measurement at B-5 shows a delayed response to the internal pressure of the cavern presenting an unstable behaviour, which resulted in the graph with rounded corners.

The response of the pore water pressures to the pressurisation of the caverns was influenced by two hydrological boundaries of the water curtain borehole and the cavern. Figure 6.26 shows the correlation between two piezometric head differences: the water curtain pressure minus the internal pressure of the cavern and the pore water pressure minus the internal pressure of the cavern. Most of the data points present a largely linear correlation between these piezometric head differences except for B-5 where unstable behaviour was observed. The correlation of the pressure differential of B-5 showed periodic irregularity, which confirms the influence of the unstable behaviour.

Figure 6.27 shows a comparison of the correlation plots between the tests of varying internal pressure and varying water curtain pressure. All the piezometric pressure differences showed positive linear correlations, largely indicating similar responses under the two tests. This suggests that where the hydrological boundary is perfect, the pore water pressure between the water curtain boreholes and the cavern is determined by the piezometric pressure difference.

From this observation, the management of the behaviour of the underground water was determined by considering the piezometric pressure differentials between the water curtain pressure minus the internal pressure of the cavern and the pore water pressure minus the internal pressure of the cavern as well as its temporal variation.

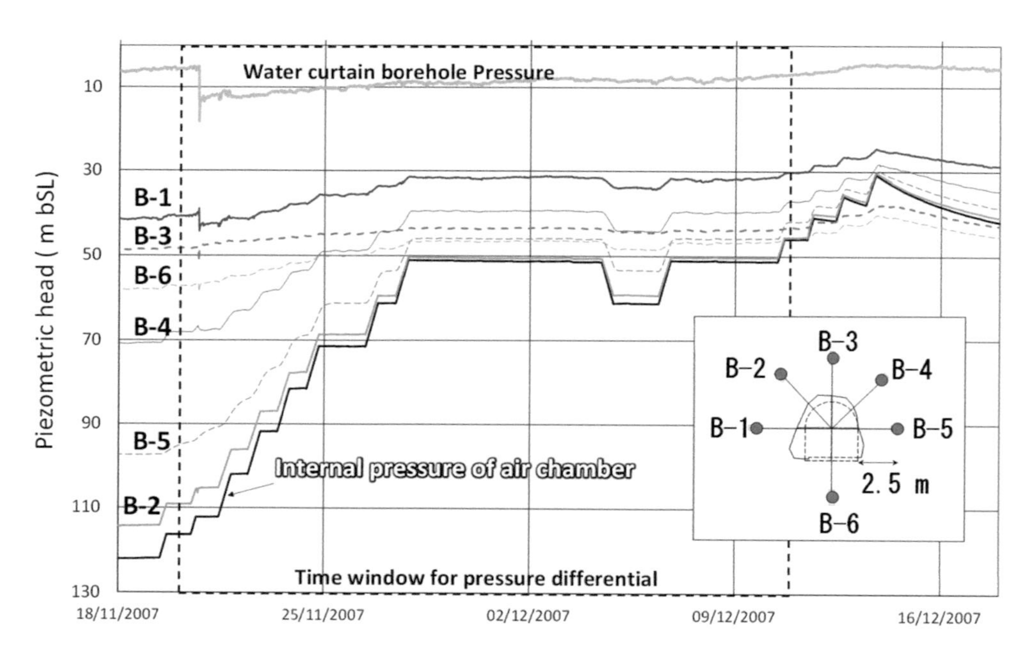

Figure 6.25 Temporal variation of the pore water pressure at injection into the cavern.

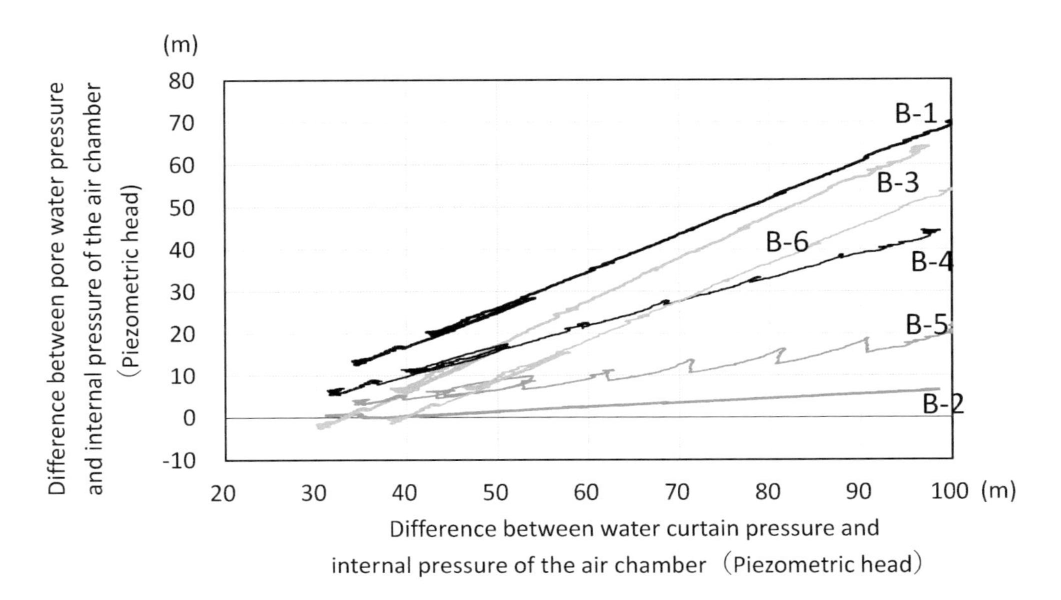

Figure 6.26 Correlation of pressure differences between the sealing water pressure and the pore water pressure based on the internal pressure of the cavern, at injection into the storage cavern.

6.2 Air-tightness Test of Underground Cavern

6.2.1 Air-tightness Test

The air-tightness of the underground hydraulic containment-type rock cavern storages at the Kurashiki and Namikata facilities was tested on completion of excavation. Its procedure is

i. Closing and sealing the cavern by installing a concrete plug between all the tunnels and caverns and inundating with water.
ii. Injecting pressurised air into the cavern to the specified design pressure (950 kPa for Kurashiki and 970 kPa for Namikata).
iii. After the internal pressure of the cavern has been stabilised, the change in the internal pressure ΔP for the standing period of 72 h is measured.

If the internal pressure maintains the design value, i.e., there is no loss in ΔP, then the air-tightness is deemed sufficient. At the same time, the reliability of the measurement is assured in comparison with the uncertainty $\varepsilon\Delta P$.

The water level of the water curtain system is set to 15 m bSL, the lowest operational limit. This is the test under the most stringent condition for air-tightness.

A conceptual diagram of the condition of the air-tightness test in the Kurashiki facility is shown in Figure 6.28.

Several metal pipes are installed in the shaft of the cavern to connect the storage cavern and the ground surface. These pipes are used for measuring the pressure in the cavern, draining seepage in the cavern and the injection and withdrawal of the LPG during operation. These pipes are also used for pressurised injection and for measuring the air-tightness for evaluation during the air-tightness test. Figure 6.29 shows its conceptual diagram.

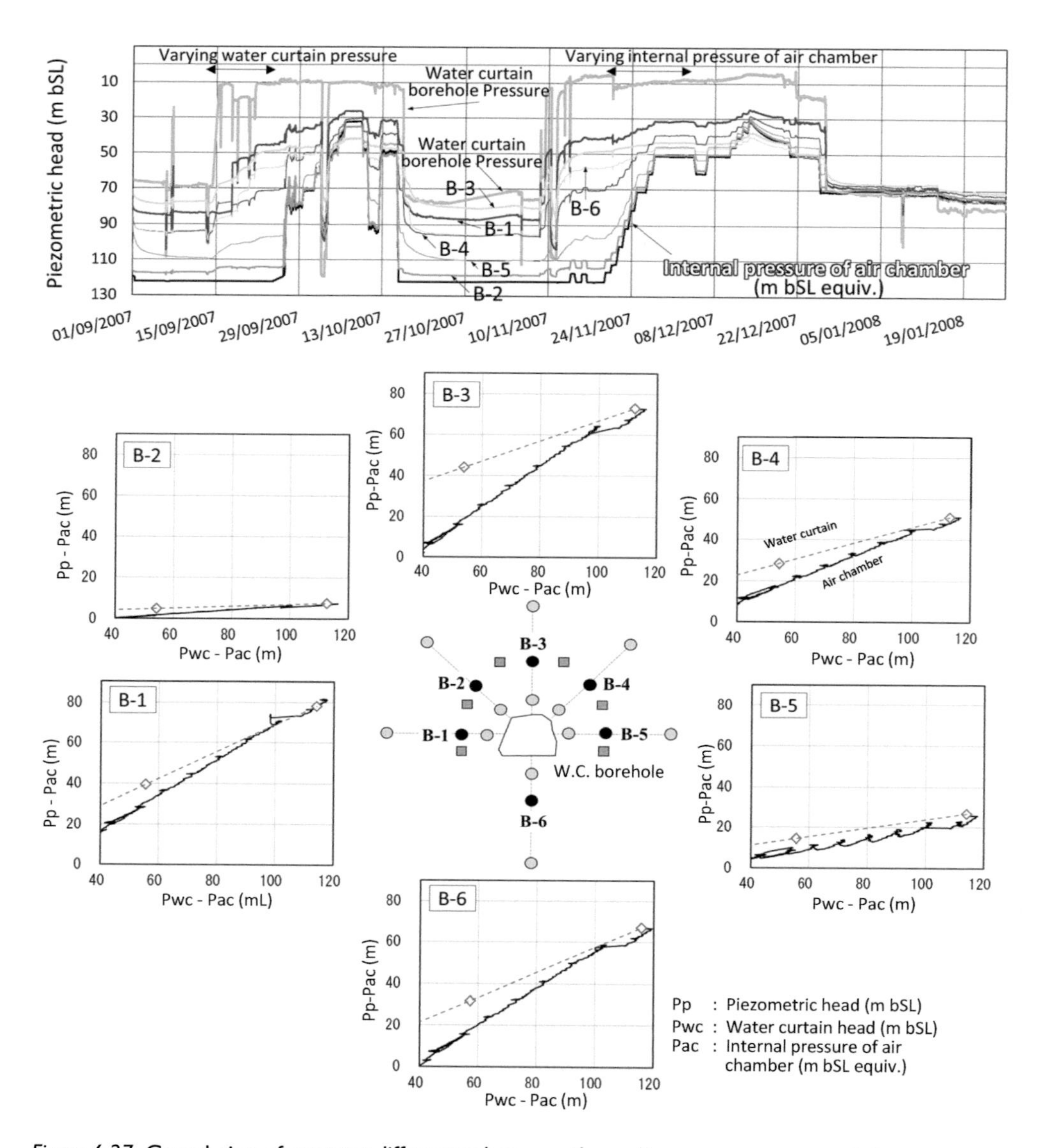

Figure 6.27 Correlation of pressure differences between the sealing water pressure and the pore water pressure based on the internal pressure of the cavern at the time of pressure injection and variation in the sealing water pressure.

The compressed air is injected into the storage cavern by a temporary compressor on the ground through a vent pipe. A pressure gauge, a thermometer, a flow meter and a hygrometer to monitor the water vapor content of the injected air are installed in the injection line.

The rate of pressurisation was set to 50 kPa/day as a base to ensure the functionality of the water curtains throughout the pressurisation process. During actual operation, pressurising was appropriately halted or the rate was varied in response to the measurement of the pore water pressure gauge around the caverns. This maintained the piezometric head difference

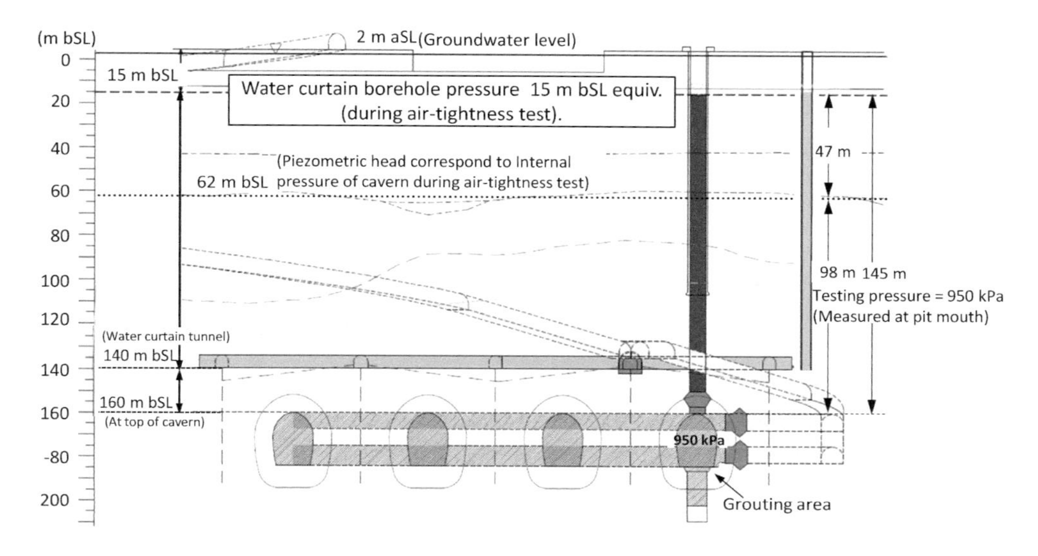

Figure 6.28 Conceptual diagram of the condition for an air-tightness test in the Kurashiki facility.

between the cavern and all the pore pressure gauges until the pressure reached the specified test value. This will be explained in Section 6.2.4.

6.2.2 Constructing Measurement System for Air-tightness Assessment

The internal pressure of the storage for air-tightness assessment is measured at the top of the pressure measurement tube connecting the storage and the ground surface through the piping shaft. The cable for the internal thermometers and water level gauges are connected to the data logger on the ground. These data are used for the appropriate assessment of correction of the internal pressure of the cavern ΔP. The seepage is drained from the bottom of the storage through the drainage pipes. A flow meter installed downstream from the drainage line monitors the volume of drainage. The drained water is also sampled for the measurement of dissolved air (Figure 6.29).

Thermometers of ±0.01°C precision were used as ascertained in the small tunnel experiment previously, considering their influence on the uncertainty of the internal pressure of the cavern, $\varepsilon\Delta P$. A 3D air flow analysis was used to examine the influence of the complex structure of the cavern and the seep in the cavern on the temperature. The arrangement of the thermometers was optimised according to this analysis.

6.2.2.1 Optimum Pattern of Thermometer Installation by 3D Air Flow Analysis

The variation in the internal pressure of the cavern is the indicator of air-tightness. It requires an accurate measurement of the average temperature, T_t, of the air in the cavern. The temperature is considered to vary with a considerable scatter. The capacity of the Kurashiki and Namikata facilities is as large as 600,000–800,000 m^3 with a complex layout. The temperature distribution in such caverns is affected by the boundary conditions of the rock walls and seeps. Data of this distribution are necessary for estimating the average temperature in the

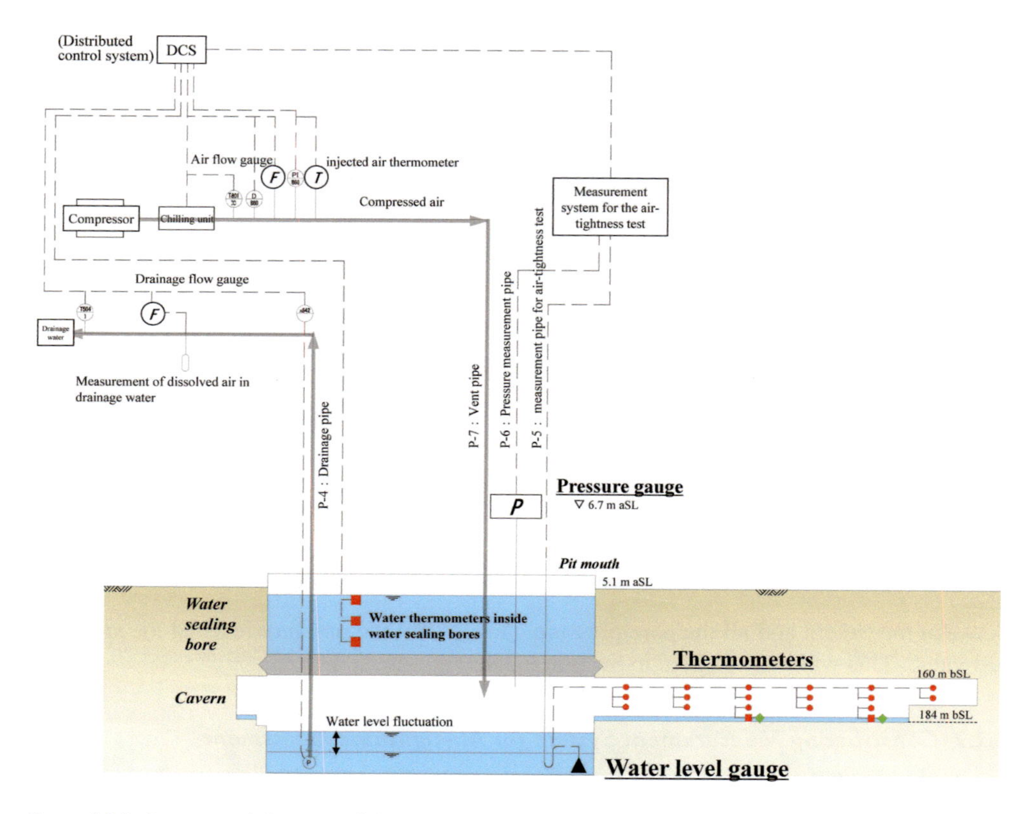

Figure 6.29 Conceptual diagram of the system of pressurised injection, draining seepage in the cavern and the measurement for air-tightness evaluation.

cavern, which was estimated by a 3D CFD analysis. According to the results of this analysis, the number of thermometers and their placements were optimised.

The Kurashiki facility has a complex structure with four storage caverns interconnected, and some cross sections were altered. The result of the air-tightness test in the small tunnel (Section 6.1) showed that the distribution of the temperature along the axes of the storage caverns was affected by the boundary condition, including the distribution of seepage on the walls of the storage cavern. The temperature distribution affected by the seepage was expected to be considerable in the Kurashiki facility, due to the large quantity of water coming from seeps unevenly spread in the complex hydrogeological structure.

The following sections detail these effects. The analysis software used was FLUENT 12 by ANSYS, its validity confirmed in an analysis of the temperature distribution in the pilot air-tightness test in the small tunnel (Section 6.1).

6.2.2.1.1 EFFECT OF THE COMPLEX STRUCTURE

An air flow analysis was run for a four-cavern model to examine the effect of the complex structure in the cavern on the temperature distribution.

The model shown in Figure 6.30 represents the complex morphology of the Kurashiki storage facility with the four connected storage caverns and leveled floor. The model parameters and the condition of analysis are shown in Table 6.4 and Figure 6.31.

Figure 6.32 is a contour map of the temperature distribution in the storage cavern two days after stopping pressurised injection of this analysis. The vertical section shows a uniform layered structure of the temperature in the storage cavern with little influence by the morphology of the cavern such as cross sections with different floor levels and branching tunnels. The vertical sections near the shaft where air is injected and the deep part of the caverns show similar temperature distribution. This suggests that the temperature distribution is dominated by the density flow after stopping pressurised injection.

These observations led to the following thermometer arrangement:

- Horizontal layout: Thermometers are vertically placed along the central axis of the storage cavern to confirm the stable density flow: three centre of gravity from the inflection points of the vertical temperature distribution curve (Figure 6.33).

6.2.2.1.2 EFFECT OF DISTRIBUTION OF SEEPS INSIDE THE CAVERN

An air flow analysis was carried out on a single storage cavern model to examine the effect of the irregularly distributed seeps in the storage cavern on the temperature. Figure 6.34 shows the predicted distribution of relative seepage at the time of completion of excavation. The boundary conditions include the distribution of seeps as well as the rock walls and water floor as considered in the previous paragraph.

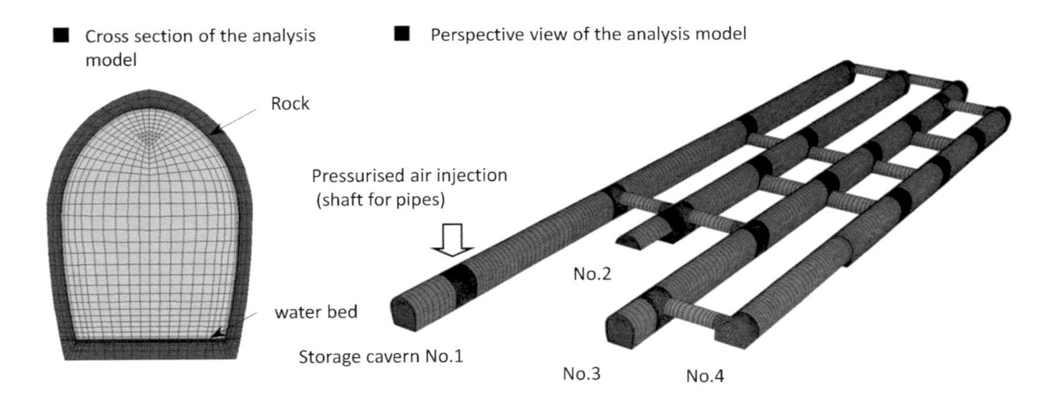

Figure 6.30 Model for storage system at the Kurashiki facility.

Table 6.4 Modelling methods for material in the analysis of the complex storage structure of the Kurashiki facility

Gas	Analysed with the standard k-ε turbulent current model, considering compressibility.
Rock	Mesh is created for the rocks 1 m around the cavity to estimate heat conduction; the outermost layer of the rock is set to a constant temperature.
Floor water	Mesh is created for the 20 cm thick floor water; it is modelled as a solid with consideration of the thermal conductivity.

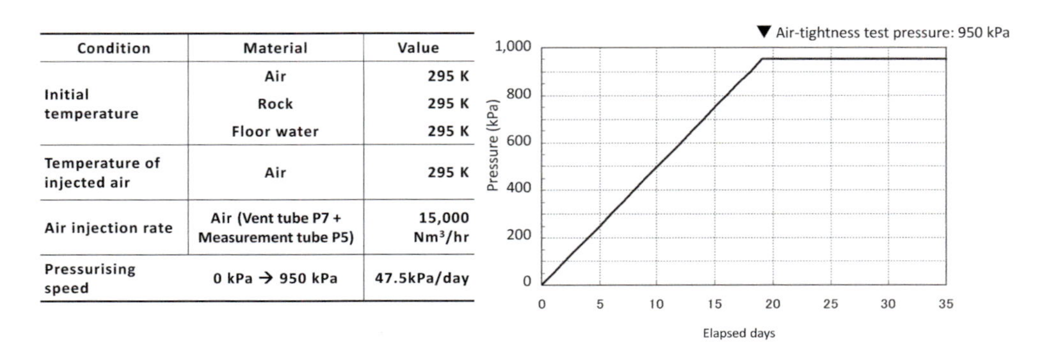

Condition	Material	Value
Initial temperature	Air	295 K
	Rock	295 K
	Floor water	295 K
Temperature of injected air	Air	295 K
Air injection rate	Air (Vent tube P7 + Measurement tube P5)	15,000 Nm^3/hr
Pressurising speed	0 kPa → 950 kPa	47.5kPa/day

Figure 6.31 Conditions for simulation in the analysis with the complex storage structure of the Kurashiki facility.

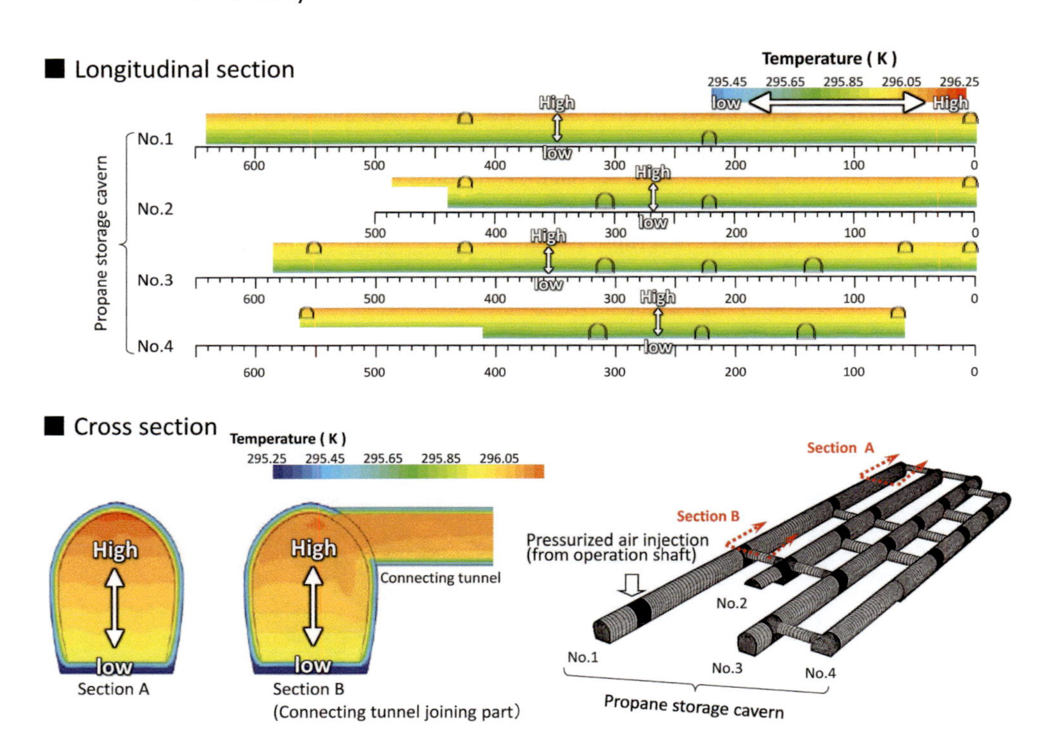

Figure 6.32 Result of air flow analysis for four connected storage caverns two days after stopping pressurised injection in the Kurashiki facility.

This test was a 3D unsteady constrictive fluid analysis for a single storage cavern such as cavern No. 1 of the Kurashiki facility with the distribution of seeps in the storage cavern as an additional boundary condition, by the same procedure and parameters as used in Section 6.2.2.1.1. The boundary conditions include the distribution of seeps as well as the rock walls and water floor as considered in Section 6.2.2.1.1. The boundary conditions were set up by a simplified model with a constant temperature of the wall based on the predicted

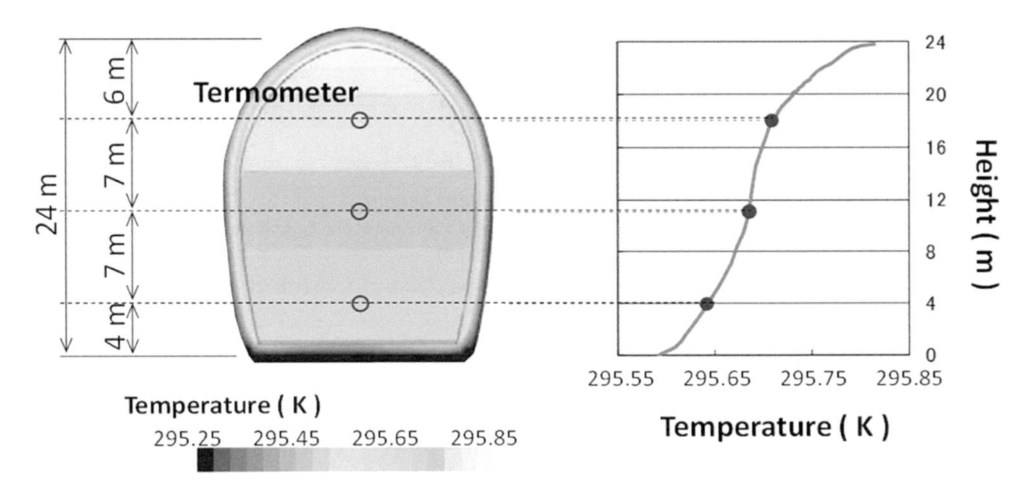

Figure 6.33 Arrangement of thermometers along the vertical plane of the storage cavern in the Kurashiki facility.

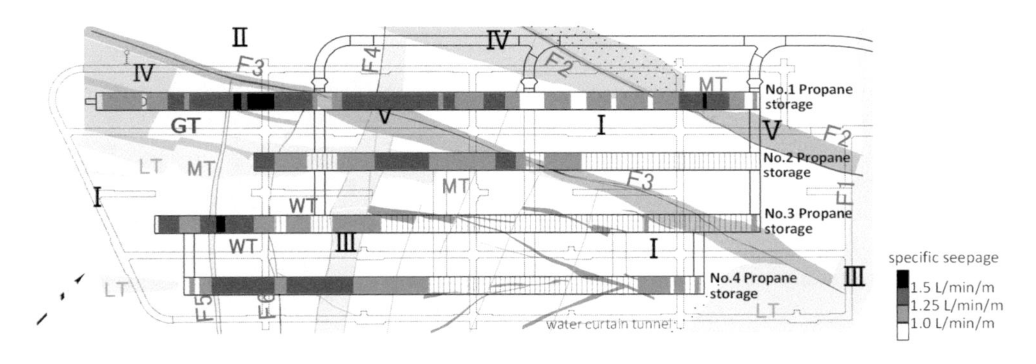

Figure 6.34 Distribution of specific seepage at the time of completion of tunnelling in the Kurashiki facility (predicted values from the 3D heterogeneous model).

distribution of water flow rate at completion of excavation (Figure 6.35). Figure 6.36 illustrates the analysis conditions.

Figure 6.37 shows the average temperature of the cross section seven days after stopping pressurised injection in this analysis. The figure shows that the temperature distribution largely corresponds to the distribution of seeps: lower in the areas with seeps than other parts. This is due to the high heat conductivity of the water compared with the rocks. From the flow velocity vectors, convection along the cavern axis is confirmed, which was not observed at the examination of the effect of the complex structure mentioned in Section 6.2.2.1. This suggested the necessity of measurement sections to capture the distribution of seeps.

The next examination was for the number of measurement sections necessary to achieve high accuracy in the average temperature in the storage cavern for this temperature

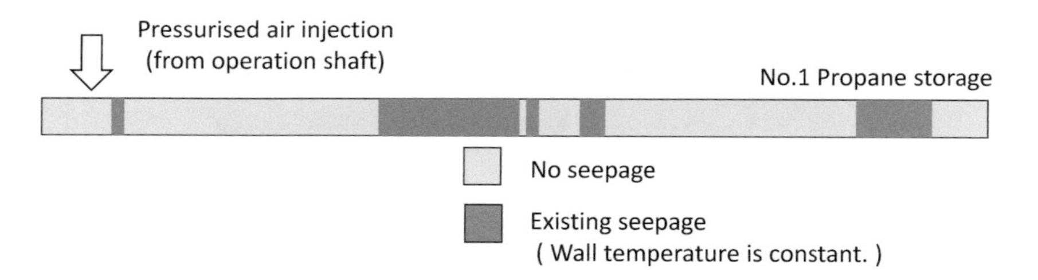

Figure 6.35 Concept of setting up the boundary condition with the distribution of internal seeps in the Kurashiki facility.

Condition	Material	Value
Initial temperature	Air	295 K
	Rock	295 K
	Floor water	295 K
	Seepage from cavern wall	295 K
Temperature of injected air	Air	295 K
Air injection rate	Air	15,000 Nm³/hr
Pressurising speed	0 kPa → 950 kPa	47.5kPa/day

Figure 6.36 Simulation condition of the analysis with the distribution of internal seeps in the Kurashiki facility.

distribution. Figure 6.38 shows the relationship between the number of measurement sections and the variance in the average temperature measured at the points. In the Kurashiki facility, the variance of the average temperature converged at the number of cross sections around 5–6. This information was incorporated in the arrangement of the thermometer.

- Layout on a vertical cross section: six cross sections (interval about 100 m) were used for No. 1 cavern for example, to capture the influence of the distribution of seeps.

From the results of Sections 6.2.2.1.1 and 6.2.2.1.2, the arrangement of the thermometers in the Kurashiki facility was determined as shown in Figure 6.39.

6.2.2.2 Precision Management of the Measurement System

In the air-tight test for the underground rock storage, a separate data transmission cable from each thermometer and water level gauge in the cavern for each sensor was necessary to cope with the risk of air leak. In particular, the thermometers with ±0.01°C precision used for the first time in the Kurashiki and Namikata facilities required careful monitoring of the variation of resistance of the cables, which could be affected by transport and installation processes, because the length of the cable was very long in the large caverns.

The precision management during transport and installation was in two stages: (1) the standard control values of a freezing point test, resistance between cables and an insulation resistance test were set up for each thermometer which had already been calibrated in the

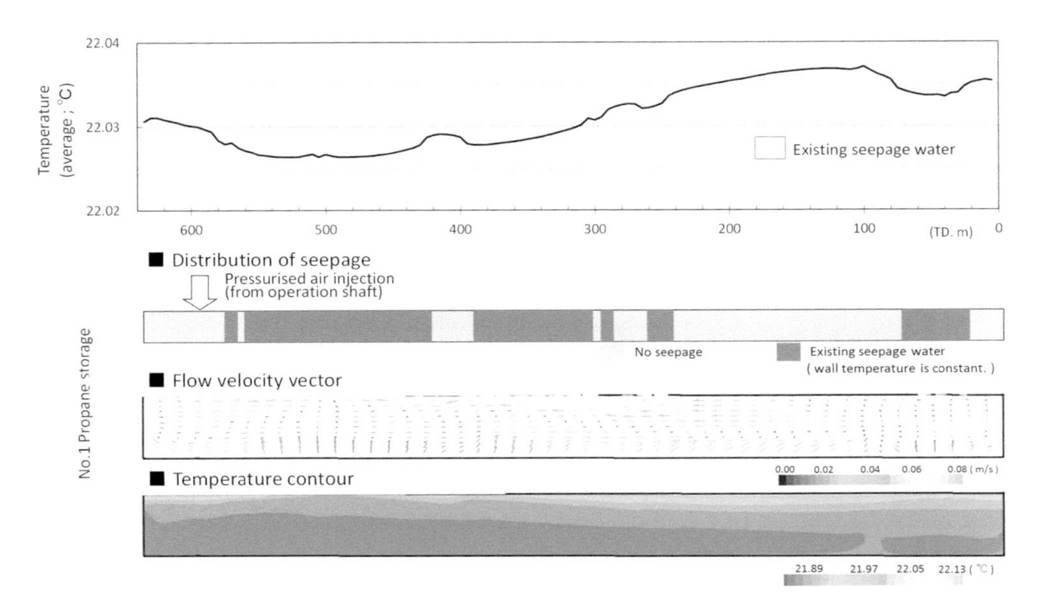

Figure 6.37 Distribution of the average temperature along the cavern section analysed with the locations of seeps in the Kurashiki facility (seven days after stopping pressurised injection).

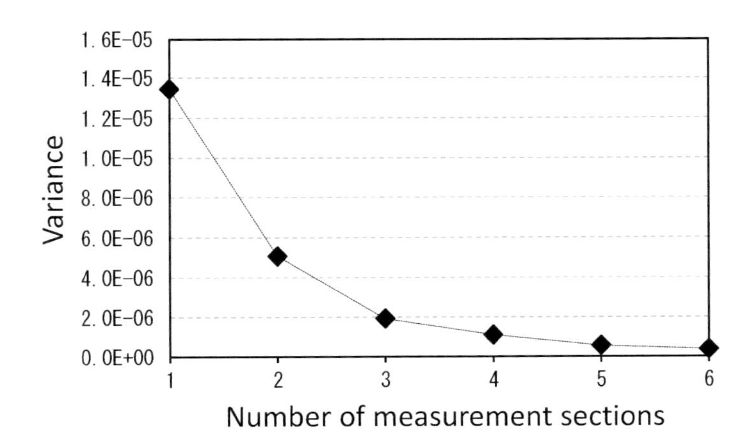

Figure 6.38 Relationship between the number of measurement sections and the variance in the average temperature at measurement points in No. 1 storage cavern of the Kurashiki facility.

factory; and (2) the required precision was assured by these standard control values after the installation of the instruments. Figure 6.40 shows the process of the precision assurance of the thermometers and management standard.

Figure 6.41 shows the result of the freezing point test at the time of final installation of the thermometer in the storage cavern of the Kurashiki facility, where the cable length reached 1500 m. The average deviation from the value at calibration in the factory was +0.0004°C with a maximum of +0.006°C. This assures the target precision of ±0.01°C.

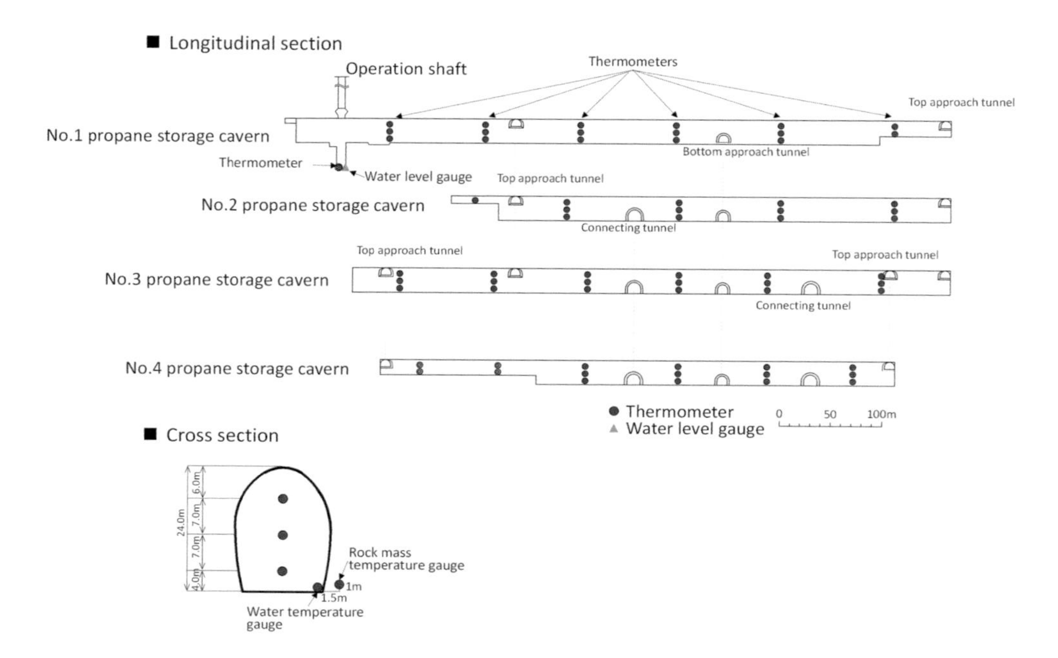

Figure 6.39 Arrangement of thermometers in the storage cavern of the Kurashiki facility.

6.2.3 Methodology and Judgement Criteria of the Air-tightness

The air-tightness is proven if the variation in the internal pressure of the cavern, ΔP, is less than its specified uncertainty, $\varepsilon\Delta P$, and the pressure is constant during the test period with the storage cavern closed after pressurised injection to a pre-specified level. The variation in the internal pressure of the storage cavern needs various corrections for this particular environment of rock storage.

An environment peculiar to rock cavern storage is that the internal pressure of the cavern is measured at the ground surface through the piping shaft. This condition is different from the experiment in the small tunnel described in Section 6.1 and necessitates some corrections. The necessary corrections include:

1. Correction for depth: The pressure inside the underground storage cavern is measured on the ground surface through a pipe along the shaft. From this, the measurement at the ground surface must be corrected for the air pressure at the depth of the cavern. This is different from the air-tightness test in the small tunnel described in Section 6.1.
2. Correction for temperature and volume: The measured pressure is then corrected for the internal temperature and volume of the cavern using the combined gas law (or Boyle–Charles' law).
3. Correction for the amount of air dissolved in the water: Groundwater constantly flows in the storage cavern from seeps as the pore water pressure is higher than the internal pressure in the cavern. This water, with dissolved air, is pumped out through a pipe along the shaft.

These corrections are illustrated in Figure 6.42 and are explained in detail below.

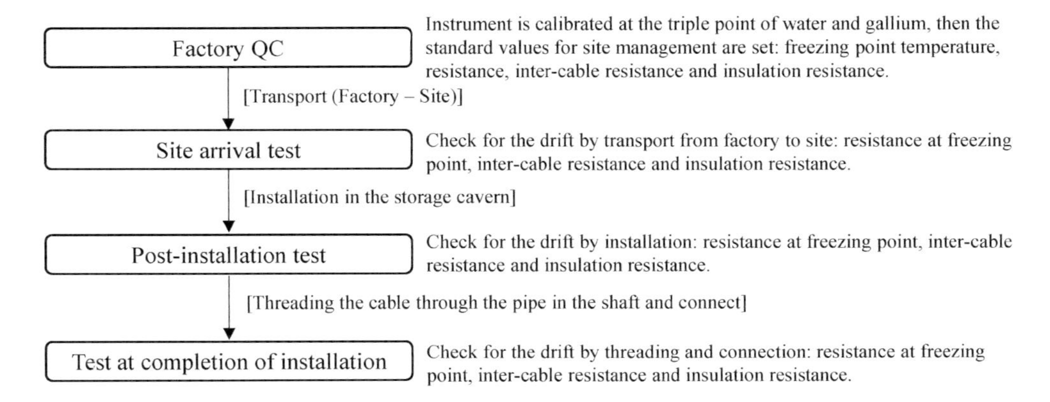

Test item	Evaluation	Result
Freezing point	Difference from QC test at factory (calculated from freezing point)	Test at arrival at site: ±0.003 °C
		Test at installation in the cavern: ±0.006 °C
		Final test after installation: ±0.009 °C
		Long-term drift allowance: ±0.001 °C
Cable resistance	Difference from QC test at factory	Lower than ±30Ω
Cable insulation resistance	Actual measurement	Higher than 100MΩ

Figure 6.40 Process of precision assurance of the thermometers and management standard.

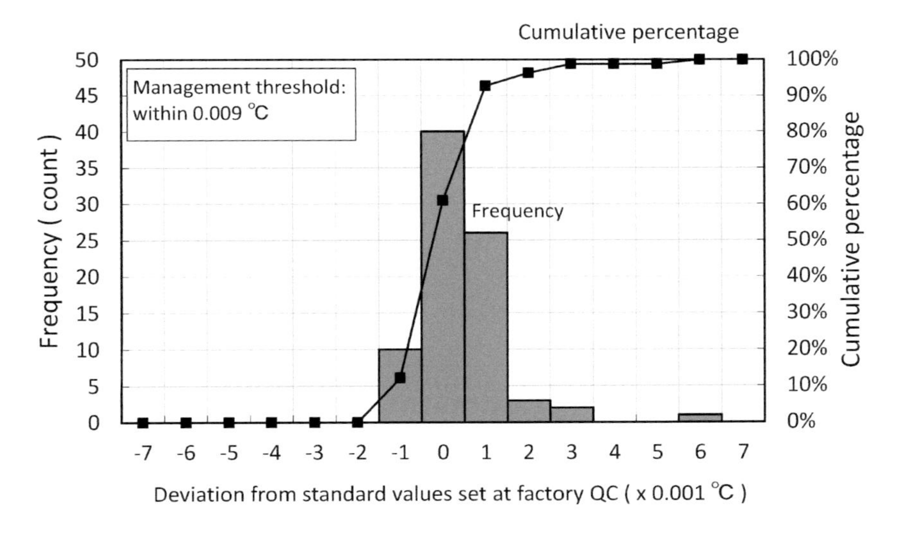

Figure 6.41 Result of the freezing point test at the time of final installation of the thermometer in the storage cavern of the Kurashiki facility.

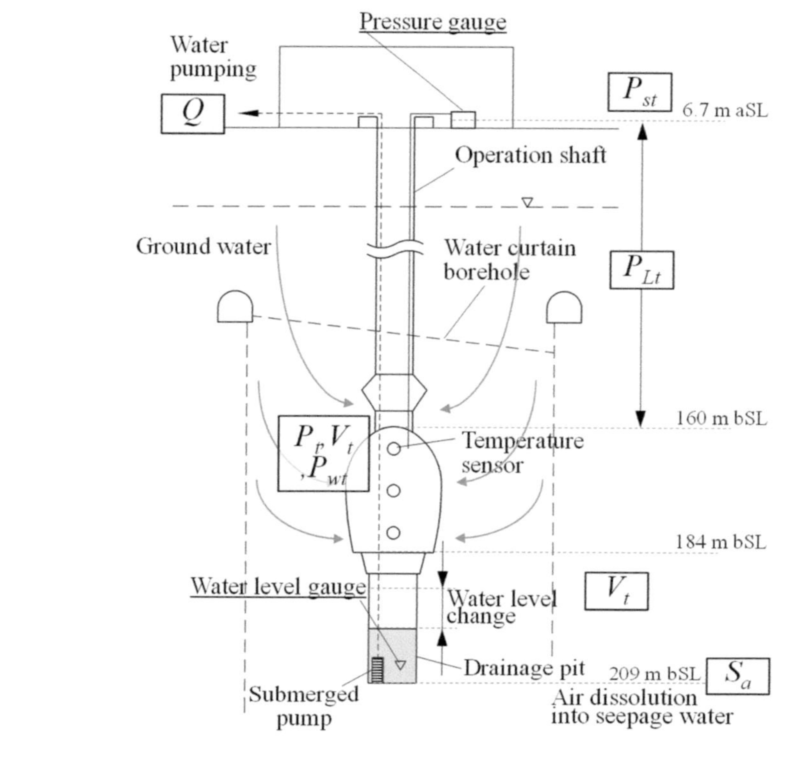

$$\Delta P = (P_1 - P_{W1}) - (P_2 - P_{W2}) \cdot \frac{T_1}{T_2} - (P_2 - P_{W2}) \cdot \frac{T_1}{T_2} \cdot \frac{A \cdot \Delta h}{V_1} - (P_2 - P_{W2}) \cdot \frac{T_1}{T_2} \cdot \frac{Q}{V_1} \cdot S_a$$

$P_{1\text{-}2}$: Pressure, $P_{W1\text{-}W2}$: Partial water vapor pressure, $T_{1\text{-}2}$: Temperature,
$V_{1\text{-}2}$: Gas phase volume, S_a : Air dissolution into seepage

Figure 6.42 Conceptual diagram of an air-tightness test in the storage cavern.

6.2.3.1 Depth Correction for Measured Air Pressure of the Storage Cavern

As the pressure inside the underground storage cavern is measured on the ground surface through a pipe along the shaft, the aerostatic pressure difference between the top of the pipe and the top of the storage cavern must be taken into consideration. There is always some seepage in the cavern causing water vapor pressure. This water vapor pressure component should be removed from the measured internal pressure of the storage cavern to obtain:

$$P_t = P_{st} + P_{Lt} - P_{wt} \quad (6.5)$$

where

P_t: corrected internal pressure of the storage cavern at time t (dry air pressure in the cavern) (Pa abs)

P_{st}: measured pressure from the top to the pipe in the shaft at time t (Pa abs)

P_{Lt}: aerostatic pressure inside the pipe in the shaft at time t (Pa)

P_{wt}: water vapor pressure in the storage cavern at time t (Pa)

P_{Lt} and P_{wt} are calculated as

$$P_{Lt} = \frac{\rho_t \times g \times h}{p_0} \times \left(P_{st} + \rho_t \times \frac{P_{st} \times g \times h}{p_0 \times 2} \right) \tag{6.6}$$

where

ρ_t: density of air in the pipe along the shaft at time t under standard atmospheric pressure (kg/m^3)

g: standard gravitational acceleration (constant 9.80665 m/s^2 is used)

h: length of the pipe along the shaft (level difference between the top of the pipe and the top of the storage cavern (m)

p_0: standard atmospheric pressure (101325 Pa)

The temperature of the air in the measurement pipe was considered almost the same as the temperature of the water in the water pipe along the shaft at the time of the test, as the water pipe was filled to near ground level. Therefore,

$$\rho_t = \frac{W_g}{V_g} \times \frac{273.15}{T_{Lt}} \tag{6.7}$$

where

T_{Lt}: average temperature of the water in the water pipe along the shaft at the time of the test (K)

W_g: molecular weight of 1 kmol air (28.9645 kg/kmol)

V_g: volume of 1 kmol ideal gas under the standard condition (at 0°C at 1 atm) (22.41399 m^3/kmol)

The seeps constantly supplied water into the cavern providing water vapor pressure. The air-tightness was tested after a stabilising period following pressurised injection, and the cavern was assumed to be saturated with water vapor. There are several approximation formulas for calculating the water vapor pressure, one of which by Tetens (1930) states:

$$P_{wt} = \left(1 \times 10^2 \times 6.11 \times 10^{\frac{7.5 \cdot (T_t - 273.15)}{T_t - 273.15 + 237.3}} \right) \cdot \frac{H_t}{100} \tag{6.8}$$

where

T_t: average internal temperature of the storage cavern at time t (K)

H_t: relative humidity inside the storage cavern at time (=100) (%)

6.2.3.2 Correction for Internal Temperature of the Storage Cavern and Gas Phase Volume

Eq. (6.5) corrects the measured value of the pressure (dry air pressure in the storage cavern). The following formula is used to correct the pressure for the internal temperature of the storage cavern and gas phase volume:

The formula of Eq. (6.1) is used to correct the pressure for the internal temperature of the storage cavern and gas phase volume.

$$\Delta P = P_0 - P_t \cdot \frac{T_0}{T_t} \cdot \frac{V_t}{V_0} \qquad \text{(6.1) Repeat}$$

As the seepage stored in the bottom drainage tank is periodically pumped to outside of the storage system (Figure 6.43), the gas phase volume of the storage cavern, V_t, is calculated by

$$V_t = V + (v - v_t) \tag{6.9}$$

where

V: gas phase volume above the water surface of the drainage tank (m^3)

v: cumulative volume of water from the bottom to the top of the drainage tank (m^3)

v_t: cumulative volume of water from the bottom to the water surface in the drainage tank at time t (m^3)

6.2.3.3 Correction for Variation of Air Volume due to Dissolution into Seepage

Some air in the cavern dissolves into the water bed and seepage, which are pumped out of the system. This affects the internal pressure of the cavern. Eq. (6.10) calculates the molar amount of the air lost by this dissolution from the start of the test to time t (in hours), Δn, and Eq. (6.11) calculates the change in pressure due to the loss, $\Delta P_t''$.

$$\Delta n = \frac{P_t \cdot \{\gamma \cdot (Q \cdot t)\}}{R \cdot T_t} \tag{6.10}$$

$$\Delta P_t'' = \frac{\Delta n \cdot R \cdot T_0}{V_0} \tag{6.11}$$

where

γ: solubility of gas in water under the condition of the storage cavern (m^3/m^3)

Q: seep flow rate (m^3/h)

t: duration since start of test (h)

R: gas constant

Substituting Δn of Eq. (6.11) into Eq. (6.10):

$$\Delta P_t'' = P_t \cdot \frac{\{\gamma \cdot (Q \cdot t)\}}{V_0} \cdot \frac{T_0}{T_t} \tag{6.12}$$

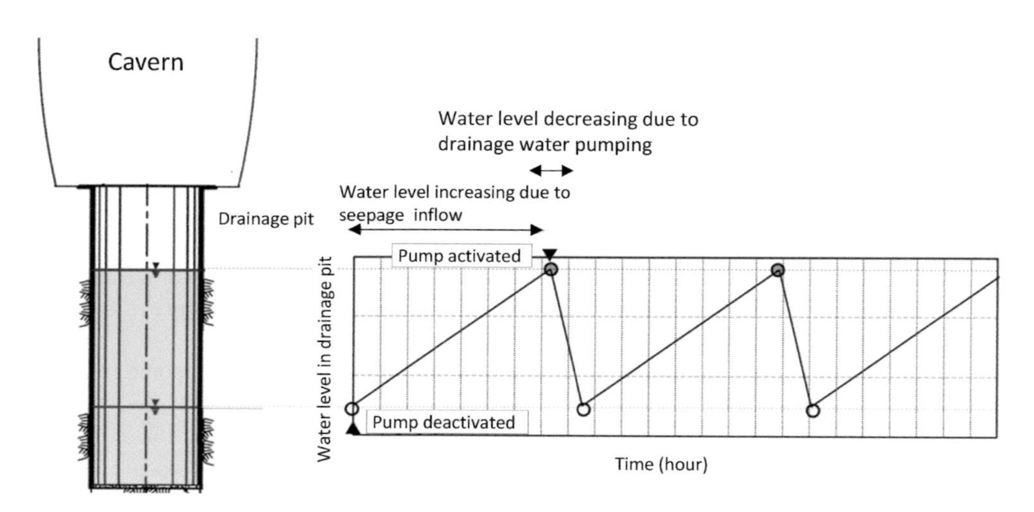

Figure 6.43 Conceptual diagram of the water level in the drainage tank.

Considering the variation in pressure due to air dissolution into seepage, $\Delta P_t''$, the pressure variation, ΔP, of Eq. (6.1) can be replaced by Eq. (6.13), and incorporating Eqs. (6.5)–(6.12) leads to Eq. (6.14):

$$\Delta P_t = \left(P_0 - P_t \cdot \frac{T_0}{T_t} \cdot \frac{V_t}{V_0} \right) - \Delta P_t'' \qquad (6.13)$$

$$\Delta P_t = \left(\left(P_{s0} + P_{L0} - P_{w0} \right) - \left(P_{st} + P_{Lt} - P_{wt} \right) \cdot \frac{T_i}{T_t} \cdot \frac{V_t}{V_i} \right) - \left(P_{st} + P_{Lt} - P_{wt} \right) \cdot \frac{\{ \gamma \cdot (Q \cdot t) \}}{V_0} \cdot \frac{T_0}{T_t} \qquad (6.14)$$

6.2.3.4 Criterion for Air-tightness

The criterion to judge the air-tightness was set, as stated in Section 6.1.1, as the internal pressure does not drop and the absolute value of the variation of pressure, ΔP, is smaller than its uncertainty throughout the test period by Eq. (6.2) and Eq. (6.3).:

$$|\Delta P| \leq \varepsilon \Delta P \qquad (6.2)\ \text{Repeat}$$

$$\varepsilon \Delta P = P_t \cdot \left\{ \left(\frac{\sigma P_t}{P_t} \right)^2 + \left(\frac{\sigma P_0}{P_0} \right)^2 + \left(\frac{\sigma T_t}{T_t} \right)^2 + \left(\frac{\sigma T_0}{T_0} \right)^2 + \left(\frac{\sigma V_t}{V_t} \right)^2 + \left(\frac{\sigma V_0}{V_0} \right)^2 \right\}^{1/2} \qquad (6.3)\ \text{Repeat}$$

The result of the air-tightness test in the small tunnel confirmed the applicability of thermometers with a precision of ±0.01°C (Section 6.1). The tests in the Kurashiki and Namikata sites used thermometers of this type. The precision of other instruments including pressure gauges and water level gauges was set to be equal to or better than those used by the predecessors for air-tightness tests of rock caverns.

Using these instruments allowed the uncertainty of the variation in the internal pressure of the cavern, $\varepsilon\Delta P$, less than 0.1 kPa, to be maintained. Therefore, the passing criterion for the tests at the Kurashiki and Namikata facilities was to satisfy the condition that "the internal pressure does not drop and the absolute value of the variation of pressure, ΔP, is smaller than its uncertainty throughout the test period".

6.2.4 Evaluation of Water-tightness at Pressurised Injection

6.2.4.1 Pressurised Injection Plan

The air-tightness test in the small tunnel (Section 6.1) found that the pore water pressure around the storage cavern varies with some scatter at the time of pressure injection before the air-tightness test. The differential hydraulic head between the pore water pressure and the internal pressure of the storage cavern must be kept at more than 0 m to ensure the air-tightness of the storage cavern.

On the other hand, the risk of losing water-tightness was expected from the decrease in the hydraulic head against the cavern due to the delayed response by the unsteady pore water pressure around the cavern at the time of pressurised injection.

The pressurising rate of 10 kPa/day was generally used by the predecessors and was adjusted in response to the unsteady behaviour. The Kurashiki and Namikata localities are known for a complex heterogeneous hydrological structure and a large unsteady behaviour was expected. This unsteady behaviour at the time of pressurised injection was predicted by analysis using a 3D heterogeneous model. Figure 6.44 shows an example of the behaviour prediction at Section TD 55 m of No. 1 propane storage cavern of the Namikata facility. In the unsteady seepage flow analysis in this book, the transient characteristics of groundwater are described by the hydration rate (K/Ss): a smaller K/Ss value represents a longer head change time. The hydraulic diffusivity in the 3D heterogeneous model, K/Ss, was set based on the response of the pore water pressure to the water curtain pressure during the construction of the storage cavern. The graphs in the bottom row of Figure 6.44 are the prediction of the pore water pressure for two assumptions of a pressure injection rate at P1A-3 situated on the hanging wall of the F-6 fault where the K/Ss is small. If the pressurised injection rate is a constant at 1000 kPa/day, the pore water pressure is predicted to drop to less than the internal pressure of the cavern before completion of injection due to unsteadiness (bottom left of Figure 6.44). On the other hand, it is predicted that the pore water pressure can be maintained higher than the internal pressure of the cavern until completion of injection if the pressure injection rate is 50 kPa/day (bottom right of Figure 6.44).

From the above observation, the speed of the pressurised injection in the Kurashiki and Namikata facilities was set to 50 kPa/day to maintain water-tightness during pressurised

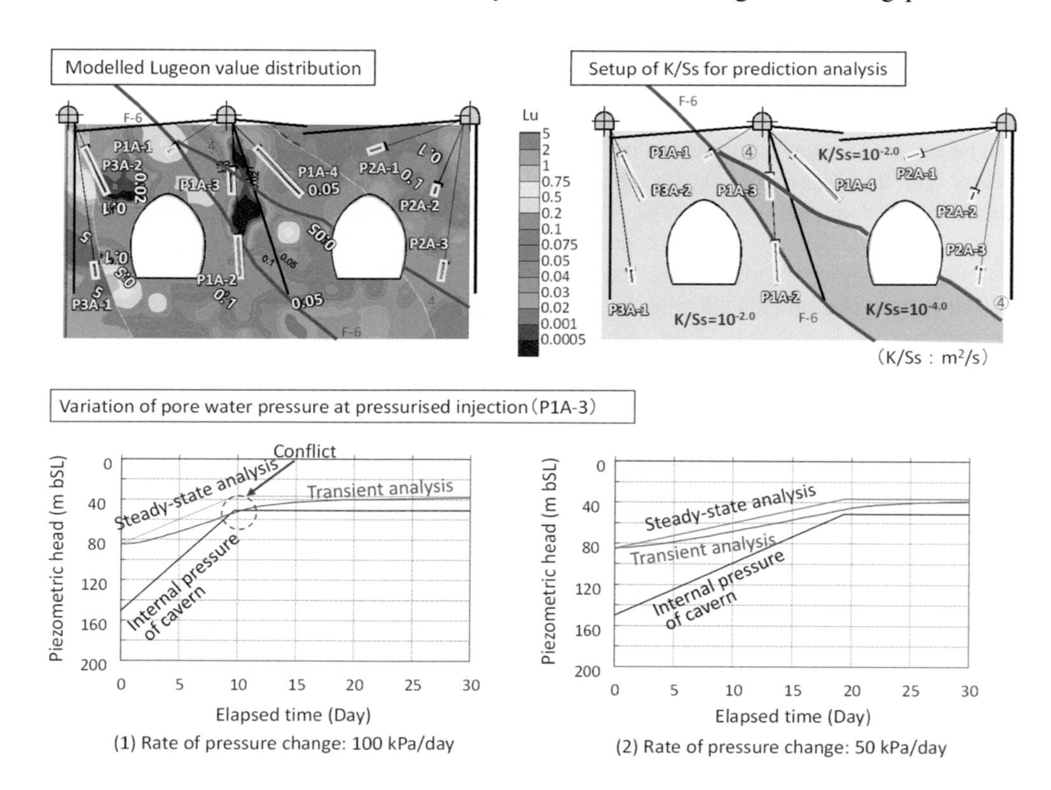

Figure 6.44 Behaviour prediction of pore water pressure on pressurised injection at the Namikata facility.

injection. The pore water pressure was closely monitored during pressurised injection using the difference relationship confirmed above. The pressurised injection was continued until the pressure reached the air-tightness test value maintaining the differential head between the pore water pressure and the internal pressure of the cavern by changing the injection speed and sometimes interrupting the injection (Figure 6.45).

6.2.4.2 Behaviour of Groundwater at Pressurising of the Cavern

This section explains the behaviour of the water level during pressurised injection and the pore water pressure near the cavern by a change in the pore water pressure over time from completion of excavation to the air-tightness test (excavation – filling the water seal tunnels and pressure boost – immersion of access tunnel – pressurised injection – air-tightness test). Then, the characteristic behaviour of the pore water pressure and the hydrogeological properties (geology and permeability) of the location are explained.

6.2.4.2.1 BEHAVIOUR OF GROUNDWATER AT PRESSURISED INJECTION IN THE KURASHIKI FACILITY

The air-tightness test at the Kurashiki facility was carried out with the internal pressure of the storage cavern at 950 kPa (measured at the tube at the top of the shaft); the level of the water curtain was 15 m below sea level (water seal bore and water tunnel level). The piezometric head on the storage cavern was 47 m.

6.2.4.2.1.1 Temporal Variation of Groundwater Level and Pore Water Pressure To illustrate the behaviour of groundwater shallower than the water curtain at the Kurashiki facility, Figure 6.46 shows data from the groundwater level gauges and the pore water pressure gauges from completion of excavation to the air-tightness test. On completion of excavation, the pore water pressure showed some scatter, but it largely recovered with the immersing and

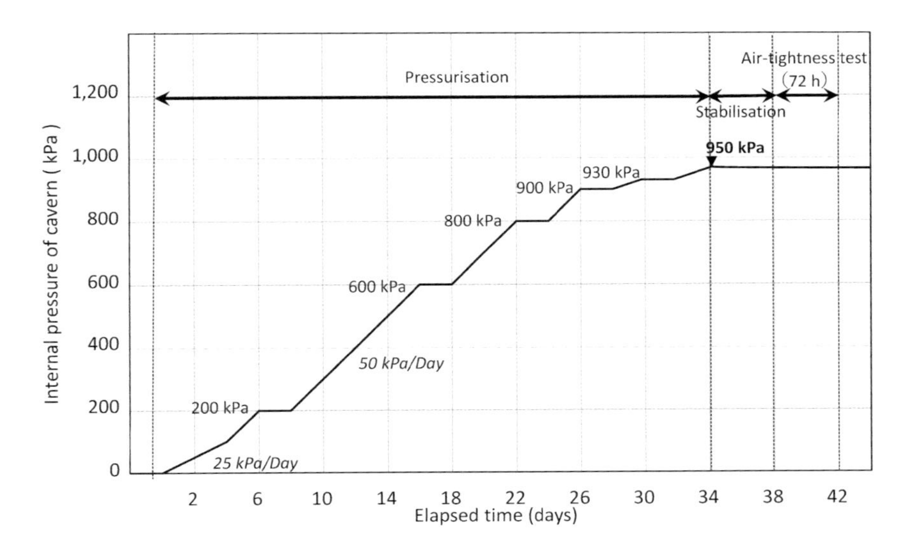

Figure 6.45 Schedule of pressurised injection.

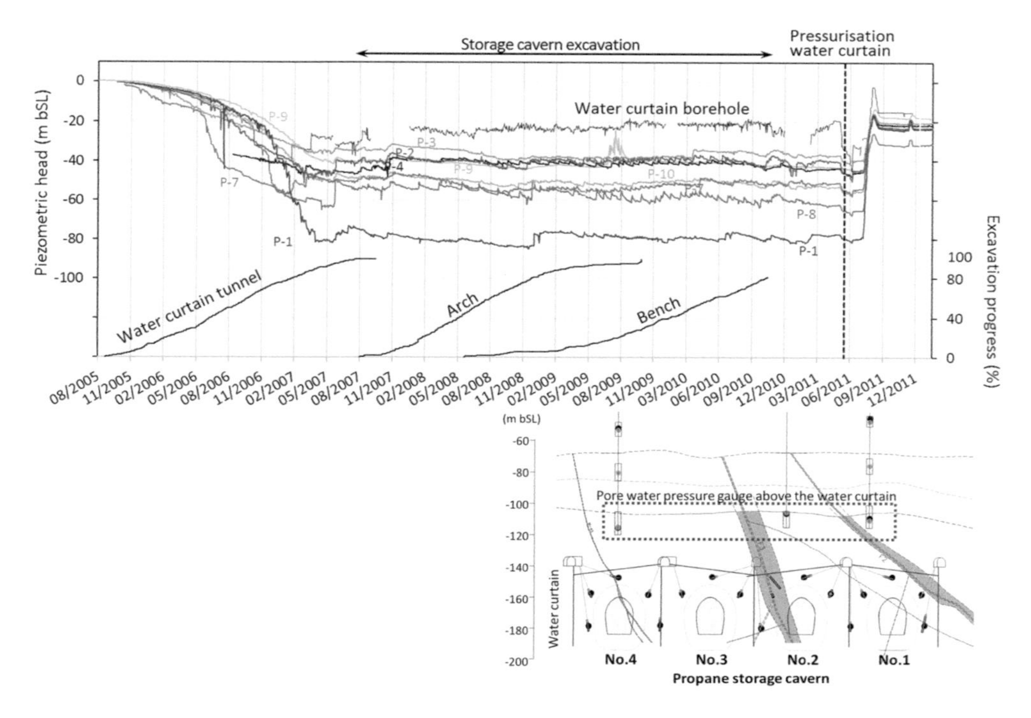

Figure 6.46 Behaviour of the pore water pressure from completion of excavation to the air-tightness test in the Kurashiki facility.

pressurising of the water seal tunnels and access tunnel. These values did not vary with the pressurised injection, indicating that an effective water curtain was built as a hydrological boundary.

The behaviour of the pore water pressure deeper than the water curtain (cavern side) is shown in Figure 6.47 at each depth level of the measurement instruments. At all depths of the instruments, the pore water pressure maintained a piezometric head higher than the level of the instruments, and an unsaturated zone was not formed. Then, the pore water pressure increased to a level close to the pre-excavation saturation by immersing and pressuring the water curtain tunnels and access tunnels, which were open to atmospheric pressure at the time of excavation.

On the other hand, some instruments were recognised as showing lower pore water pressure values than other instruments before pressurised injection. The pore water pressure at gauge A-10 situated at the shoulder (160 m below sea level) at the start of pressurised injection was about 130 m below sea level, which was lower than others by about 50 m. This instrument is located on the hanging wall of fault F3, one of the faults characteristic of the Kurashiki facility site. The measurement at this gauge recovered well with pressurised injection and exceeded the internal pressure of the cavern at the air-tightness test.

6.2.4.2.1.2 Temporal Variation of Flow of Seepage at the Pressurised Injection The variation in the flow of seepage in the storage cavern at the time of pressurised injection is shown in

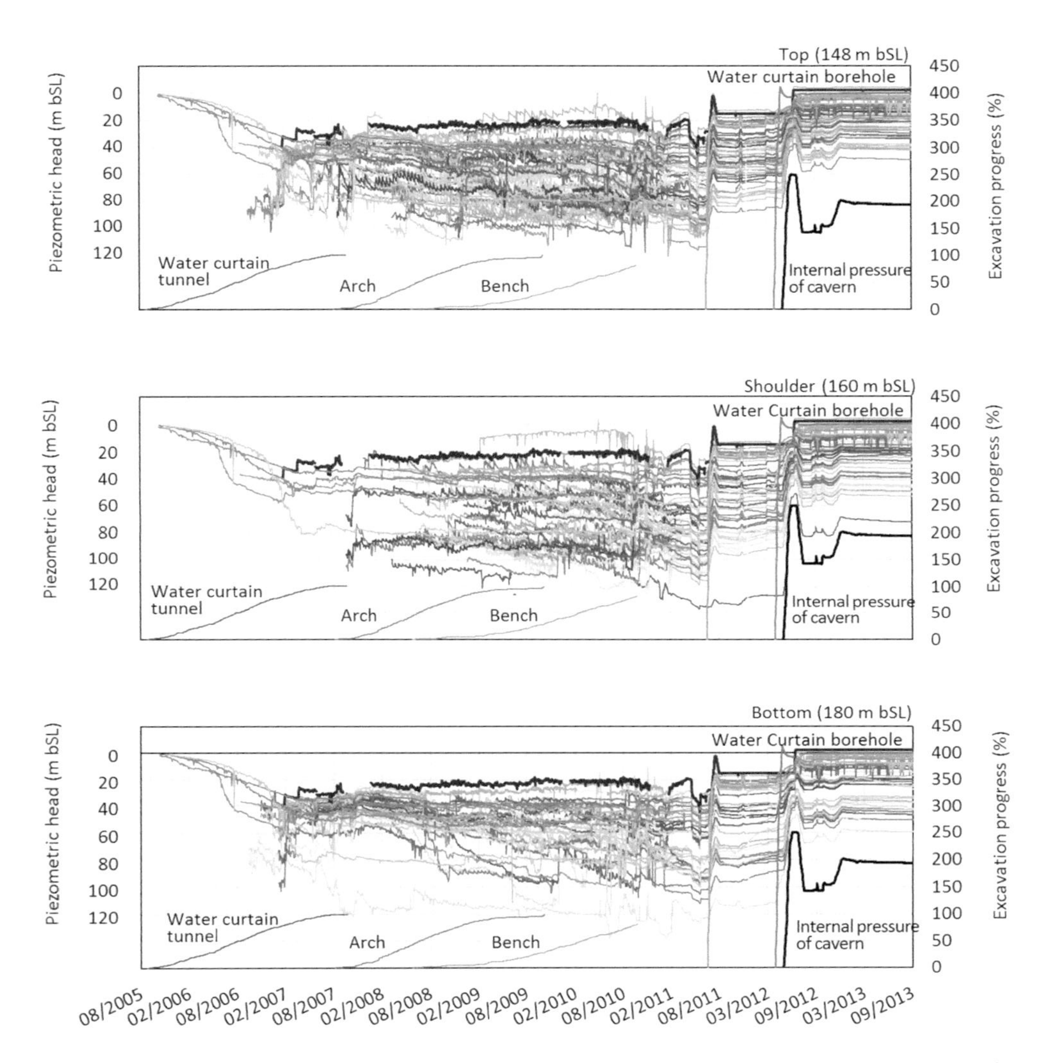

Figure 6.47 Behaviour of the pore water pressure inside the water curtain from excavation to the air-tightness test.

Figure 6.48. The volume of seepage in the cavern was around 150 m^3/h before the pressurised injection and decreased as the pressurised injection progressed, reaching around 70 m^3/h, as expected from the prediction.

6.2.4.2.1.3 Distribution of Pore Water Pressure at Air-tightness Test Next, the distribution of the pore water pressure at the time of the air-tightness test is discussed. Figure 6.49 shows the distribution of the actual measurement values at the pore water pressure gauges installed 148 m below sea level in contour. This map shows that a relatively low water pressure zone is locally formed near the faults, although a piezometric head higher than the internal pressure of the cavern was seen in some places, such as A-10, which were close to fault F3 and had

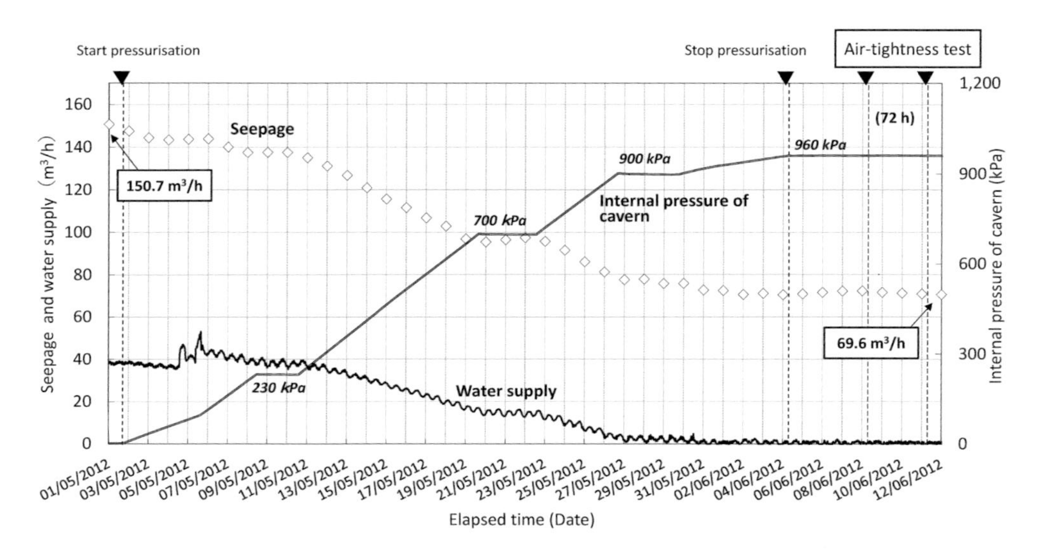

Figure 6.48 Temporal variation of the volume of seepage at pressurised injection in the Kurashiki facility.

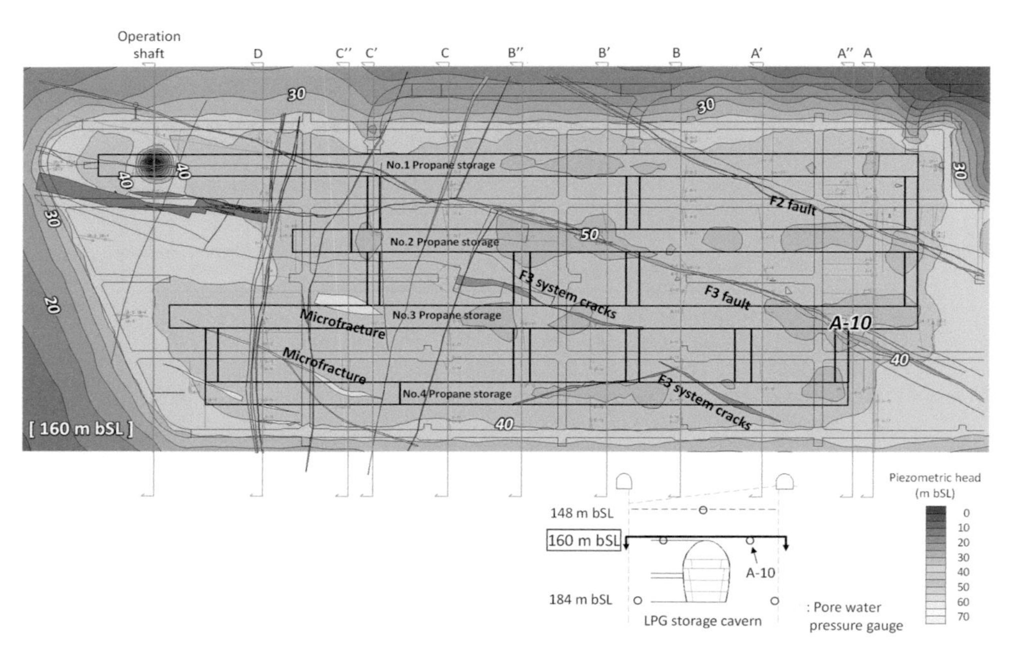

Figure 6.49 Distribution of the pore water pressure at the time of the air-tightness test (plan view at 148 m below sea level) of the Kurashiki facility.

lower pressure before pressurised injection. The behaviour of the pore water pressure gauge at A-10 is described in the following section.

On the other hand, the pore water pressure maintained relatively high values in cross section D at the southern end of No. 2 cavern, which is representative of the microfracture zone characteristic of the Kurashiki site.

Figure 6.50 shows the pore water pressure contour along section B as an example of the distribution of the pore water pressure across a vertical section. It illustrates that the pore water pressure recovered with the immersion of the water curtain tunnel and the pressurised sealing water, and this occurred not only in the vicinity of the water curtain system (water curtain tunnels and water curtain bores) but also in all the surroundings. This suggested that the water curtain was formed surrounding the storage cavern. At the time of the air-tightness test after pressurising, the pore water pressure between the water curtain system and the cavern recognisably increased while the pore water pressure outside the water curtain system did not vary much. This confirmed that the water curtain was effectively functioning as a hydrological boundary.

6.2.4.2.1.4 Behaviour of Pore Water Pressure Gauge Showing Low Values at the Test Two cross sections are discussed as the characteristic behaviour of the pore water pressure in the Kurashiki facility: (i) cross section A of No. 3 storage cavern which includes A-10 near the F3 fault that showed low pore water pressure value before pressurised injection; and (ii)

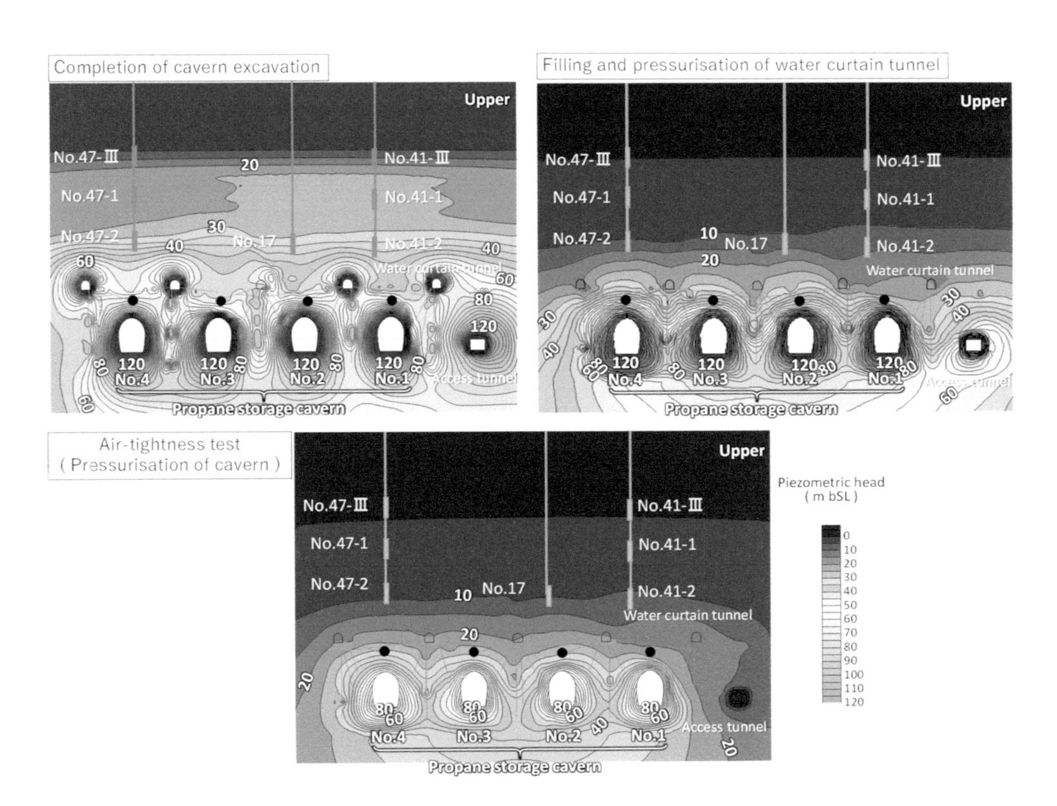

Figure 6.50 Distribution of the pore water pressure across vertical section B of the Kurashiki facility.

cross section D of No. 2 storage cavern which is within the microfractured zone with low groutability.

6.2.4.2.1.4.1 Section A of No. 3 Storage Cavern (Under F3 Fault) Figure 6.51 shows the temporal variation in the water pressure of section A of No. 3 storage cavern from completion of excavation to the air-tightness test. Instrument A-10 is located in the low permeability zone on the hanging wall of fault F3 (Figure 6.53). At this location, the pressure severely dropped during excavation and the shut-in test failed to satisfy the criterion. Therefore, many additional water sealing boreholes were drilled but the pore water pressure remained low at 140 m below sea level at the time of completion of excavation. On the other hand, this location responded well to pressurised injection and the pressure greatly increased, maintaining the differential piezometric head.

As discussed in Section 6.1, the water curtain pressure and the pore water pressure are represented in terms of deviations from the internal pressure. A correlation between these two parameters is used to evaluate the effectiveness of the water curtain. This correlation was calculated twice at the time of change in the water-tight pressure and the time of pressurised injection. Figure 6.52 shows a comparison of these correlations. At the time of variation in the water pressure, there is large unsteady behaviour and the correlation of the differences forms an arc. On the other hand, the pressure difference after stopping the pressurising agrees with the differential correlation at the variation of the sealing water pressure. This led to the belief that the water curtain around the location of this instrument was well intact.

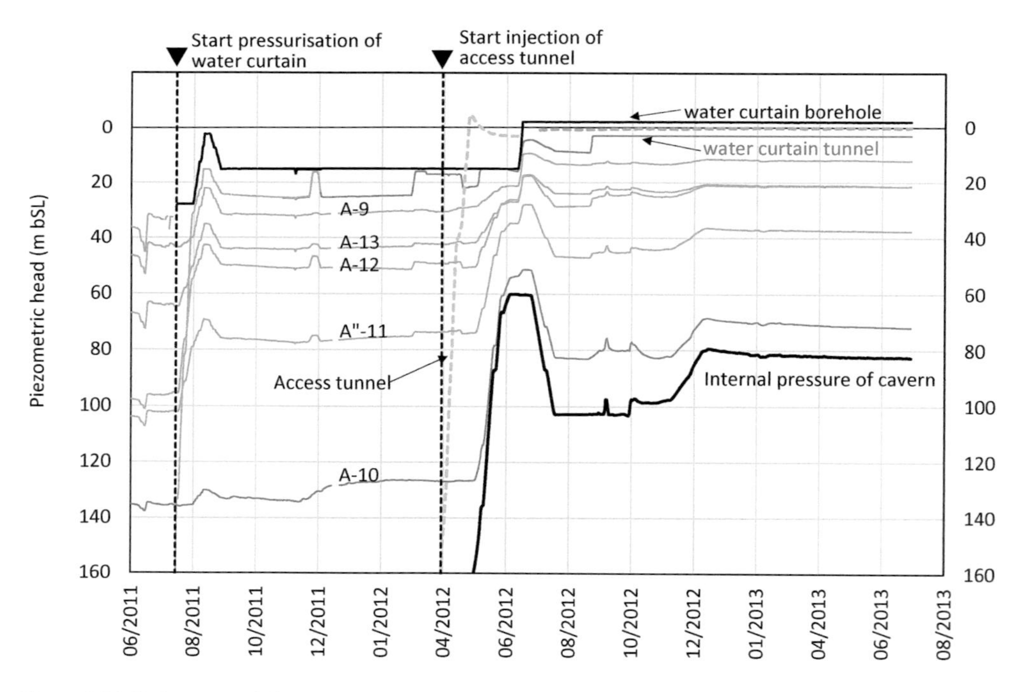

Figure 6.51 Behaviour of the pore water pressure from completion of excavation to the air-tightness test in the Kurashiki facility (areas deeper than water curtain toward the storage cavern: Section A of No. 3 storage cavern).

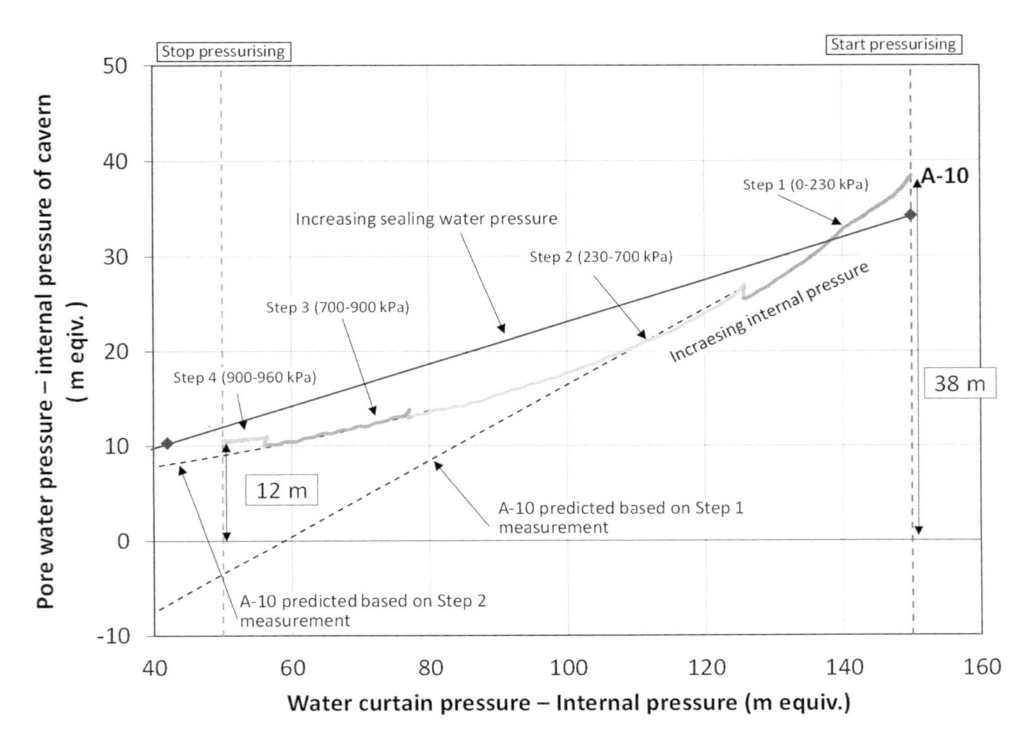

Figure 6.52 Correlation of the pressure difference measured at A-10 at water pressure increase and pressurised injection in the Kurashiki storage cavern.

Instrument A-10 is located in the zone with very low permeability (Figure 6.53). Therefore, the hydrological continuity between the sealing water bores and the instrument is poor. (i) The pore response to the variation in the water curtain pressure measured at A-10 was smaller compared with other instruments, while it recorded low values before pressurising. (ii) The pressure increase was greatest in response to pressurising of the storage cavern. The latter suggests that the sealing capacity of the water curtains against the pressurisation of the cavern is functioning sufficiently. It is considered that this is because an appropriate hydrological barrier was provided by the water curtain formed by the additional sealing water bores.

6.2.4.2.1.4.2 Section D of Storage Cavern No. 2 (Zone of Dense Microfractures) Figure 6.54 shows the temporal variation between the pore water pressure and the air-tightness test in Section D of No. 2 storage cavern, which had undergone modification of the shape of its cross section because of the difficulty of improving by means of grouting due to the dense microfractures characteristic of the Kurashiki site. The pore water pressure value at the lowest instrument was about 100 m below sea level with little scatter, unlike that observed near the fault. At immersion of the water curtain tunnel and water pressurising, by the time of pressurised injection ~~it~~ the pore water pressure had recovered and reached values higher than the internal pressure potential of the storage cavern at the air-tightness test.

During pressurised injection, a high piezometric head differential with little scatter was maintained as the pressure distributed through the air-tightness test. From this observation,

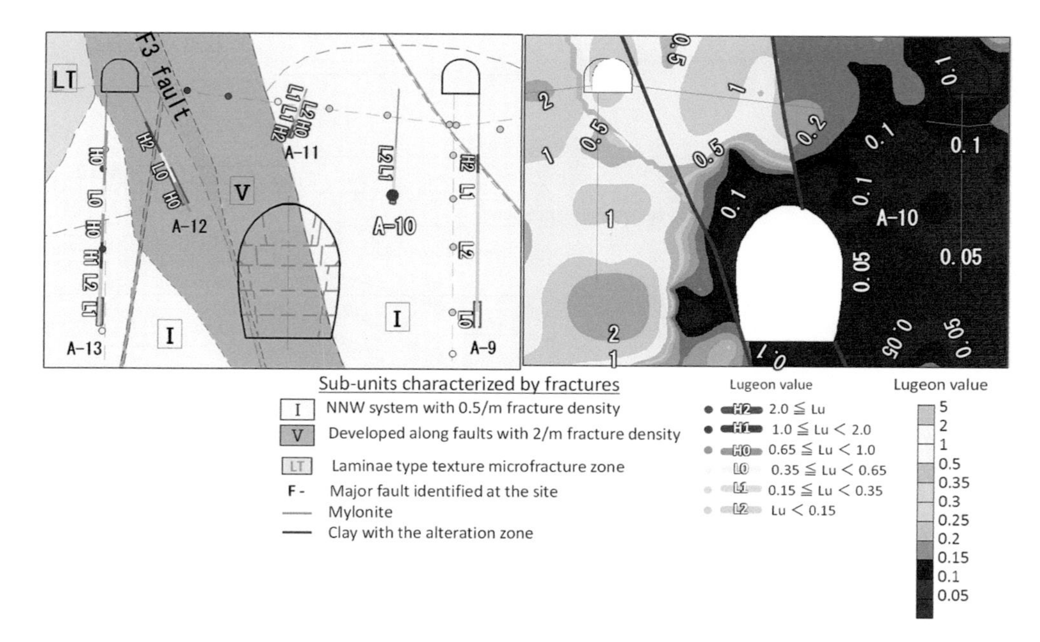

Figure 6.53 Geological cross section and distribution of the permeability at section A of storage cavern No. 3 of the Kurashiki facility.

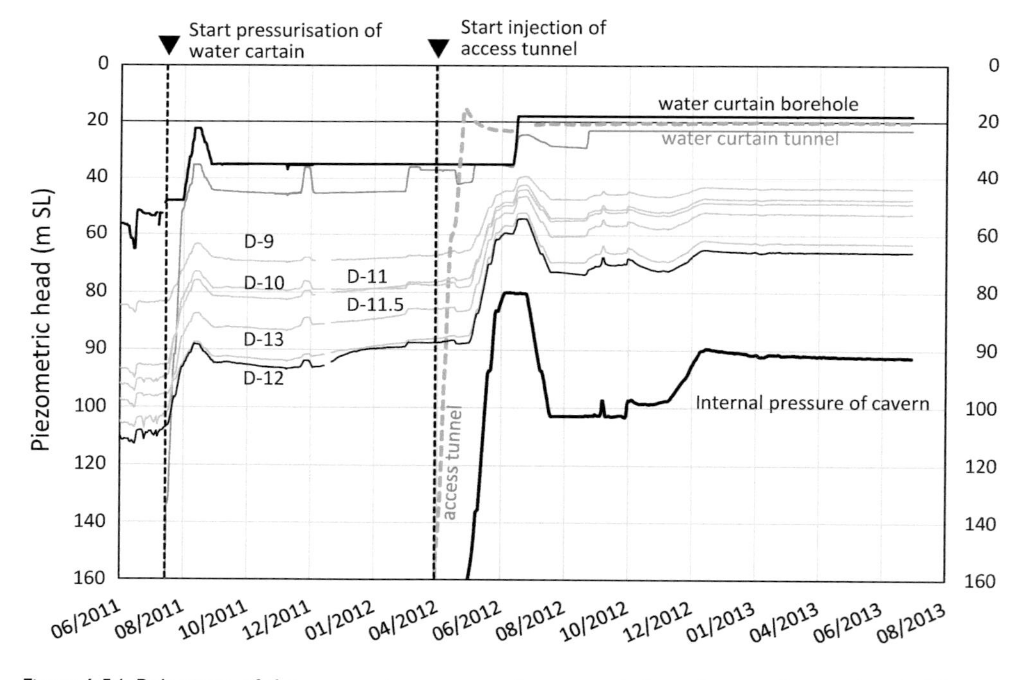

Figure 6.54 Behaviour of the pore water pressure from completion of excavation to the air-tightness test (area deeper than the water curtain side of the storage cavern: section D of No. 2 storage) in the Kurashiki facility.

the microfracture zone presented little issue of water-tightness, although it has lower groutability and permeability after grouting than other places. The reason for this is thought to be that grouting improved the water-tightness as a uniform zone, although high permeability remained after the grouting due to the abundance of microfractures.

Figure 6.55 shows the correlation of piezometric pressures between the water curtain and pore as a deviation from the internal pressure at pore water pressure gauge D-11.5 in the cross section in the microfracture zone. Unlike A-10 in the foot wall of fault F3 where permeability was low, unsteady behaviour in the correlation at pressurised injection and the behaviour of the pore water pressure were not observed but the behaviour of the pore water pressure was largely in agreement with the prediction by analysis.

6.2.4.2.1.4.3 BEHAVIOUR OF GROUNDWATER AT PRESSURISING OF THE CAVERN IN THE NAMIKATA FACILITY

The air-tightness of the Namikata storage cavern was tested at an internal pressure of 970 kPa and a water curtain level (water levels of water curtain bore level and water curtain level) 15 m below sea level. Under this condition, the piezometric head difference between the water level of the water curtains around the cavern and the internal pressure at the top of the storage cavern was 36 m.

6.2.4.2.1.4.3.1 Variation of Underground Water Level and Pore Water Pressure Figure 6.56 shows the measured data of the behaviour of groundwater shallower than the water curtain at the Namikata facility from excavation to air-tightness test. All the data show a potential higher than the internal pressure of the cavern open to the atmosphere on completion of excavation. Then, it shows values higher than the potential of the internal pressure at the air-tightness test, 47 m below sea level, in response to the filling of the water-tight tunnel, boosting the water seal pressure and immersing the access tunnels. ~~It~~ The measured pore water pressure maintained a constant value at pressurised injection.

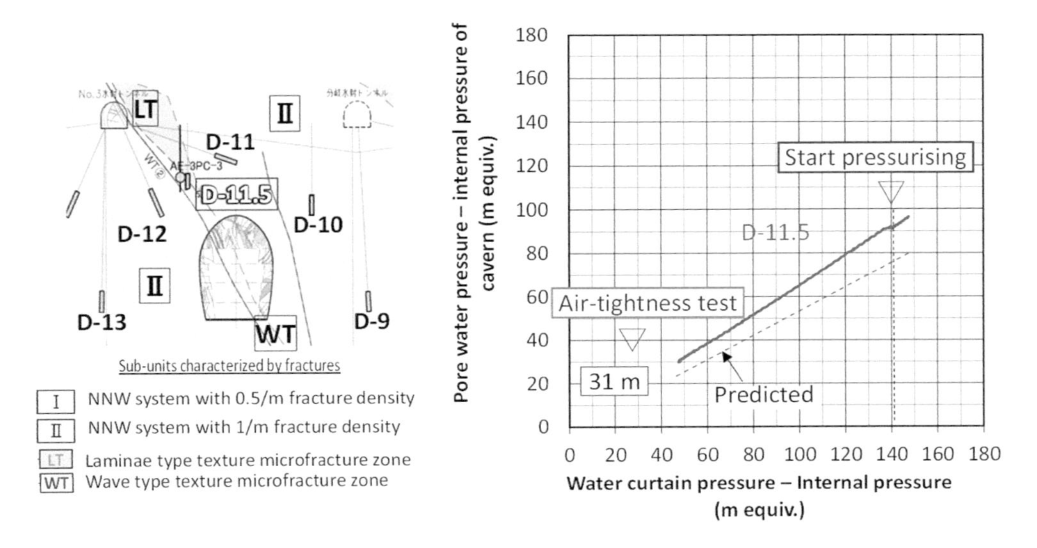

Figure 6.55 Differential correlation of section D of No. 2 storage cavern of the Kurashiki facility.

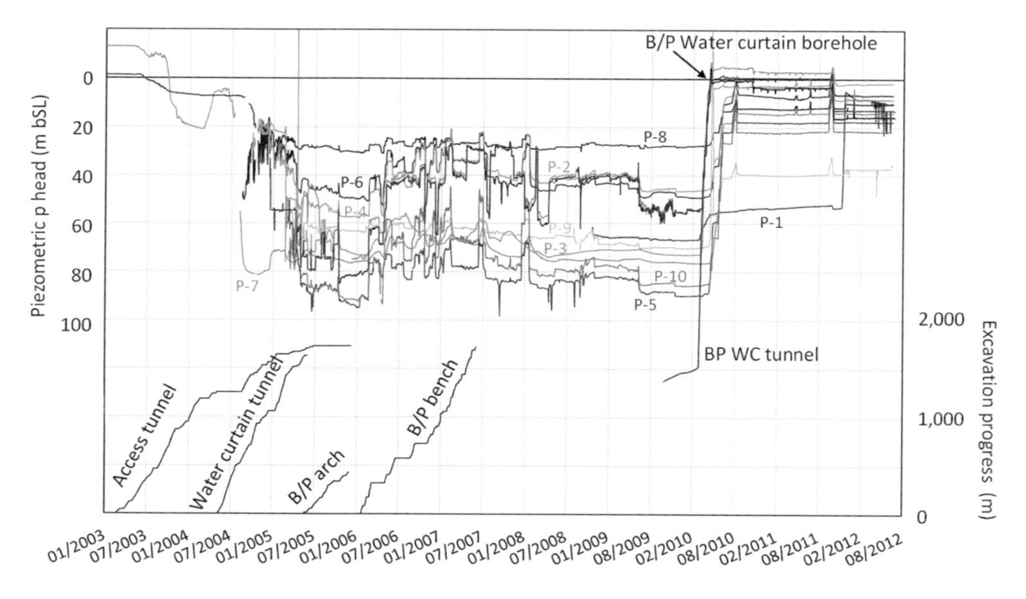

Figure 6.56 Behaviour of groundwater shallower than the water curtain at the Namikata facility from excavation to the air-tightness test.

This observation of the behaviour shows that the water pressure field shallower than the water curtains decreased in response to the water-tight tunnels and access tunnels opening to the atmosphere, but it recovered at the time of immersion, and that the water curtain worked well as a close hydraulic boundary at pressurised injection

Next, Figures 6.57 and 6.58 show the behaviour of the pore water pressure under the water curtain (toward the cavern) for storage caverns of butane and propane, respectively, at three depths. On completion of the pressure injection, all the records from the pore water pressure gauge showed that the piezometric head was maintained higher than the level of the installation of the instruments, not forming an unsaturated zone; the pore water pressure largely recovered along with immersing and pressurising the water-tight tunnels and access tunnels after they were open to atmospheric pressure.

On the other hand, some measurements of the pore water pressure, such as the one at B2C-1 at the bottom of the storage cavern (180 m below sea level), did not significantly recover and presented a low value before pressurised injection into the storage cavern. However, the values at these places recovered with pressurised injection and showed values higher than the internal pressure at the air-tightness test.

6.2.4.2.1.4.3.2 Temporal Variation of Volume of Seepage Figures 6.59 and 6.60 show the temporal variation in the flow rate of seepage during pressurised injection at the combined butane/propane storage and propane storage, respectively.

The volume of seepage in the combined butane/propane storage decreases from 10.8 m^3/h before pressurised injection to 5.7 m^3/h at the air-tightness test, as expected.

Similarly, the volume of seepage in the propane storage decreases from 43.6 m^3/h before pressurised injection to 18.4 m^3/h at the air-tightness test, also as expected.

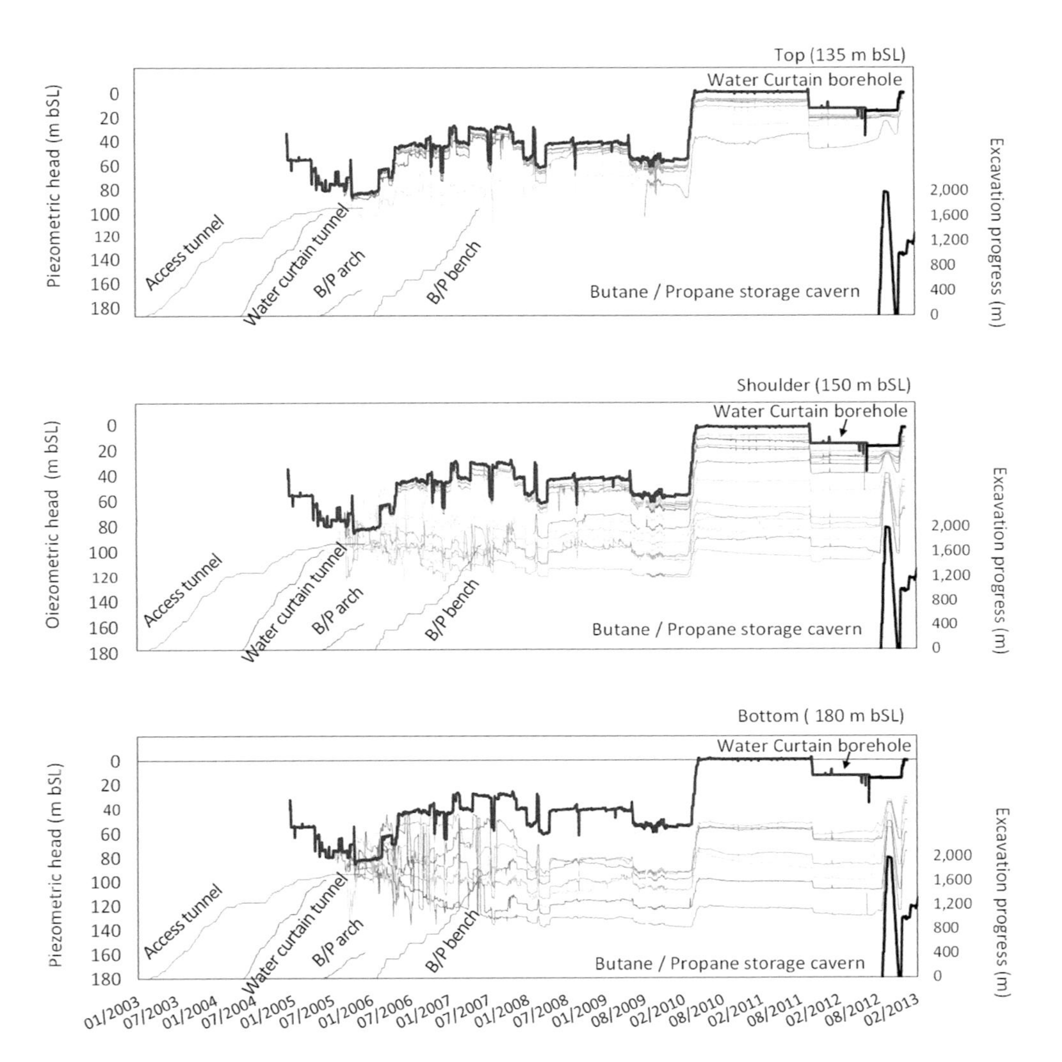

Figure 6.57 Behaviour of the pore water pressure under the water curtain of the combined butane/propane storage cavern in the Namikata facility.

6.2.4.2.1.4.3.3 Distribution of Pore Water Pressure at the Air-tightness Test This section discusses the distribution of the pore water pressure at the time of pressurised injection. Figure 6.61 shows a comparison between the analytical results with the 3D heterogeneous model (contours) and the actual measured values on distribution of the pore water pressure of the instrument situated on the shoulder of the storage cavern at 135 m below sea level. This figure illustrates that the pore water pressure maintains values higher than the internal pressure of the storage cavern, albeit there are local low pressure areas.

Figure 6.62 shows the pore water pressure contours on cross section C of the propane storage and cross section BC of the butane storage. Although the distribution of the pore water pressure inside the water curtain borehole shows some scatter due to heterogeneity in

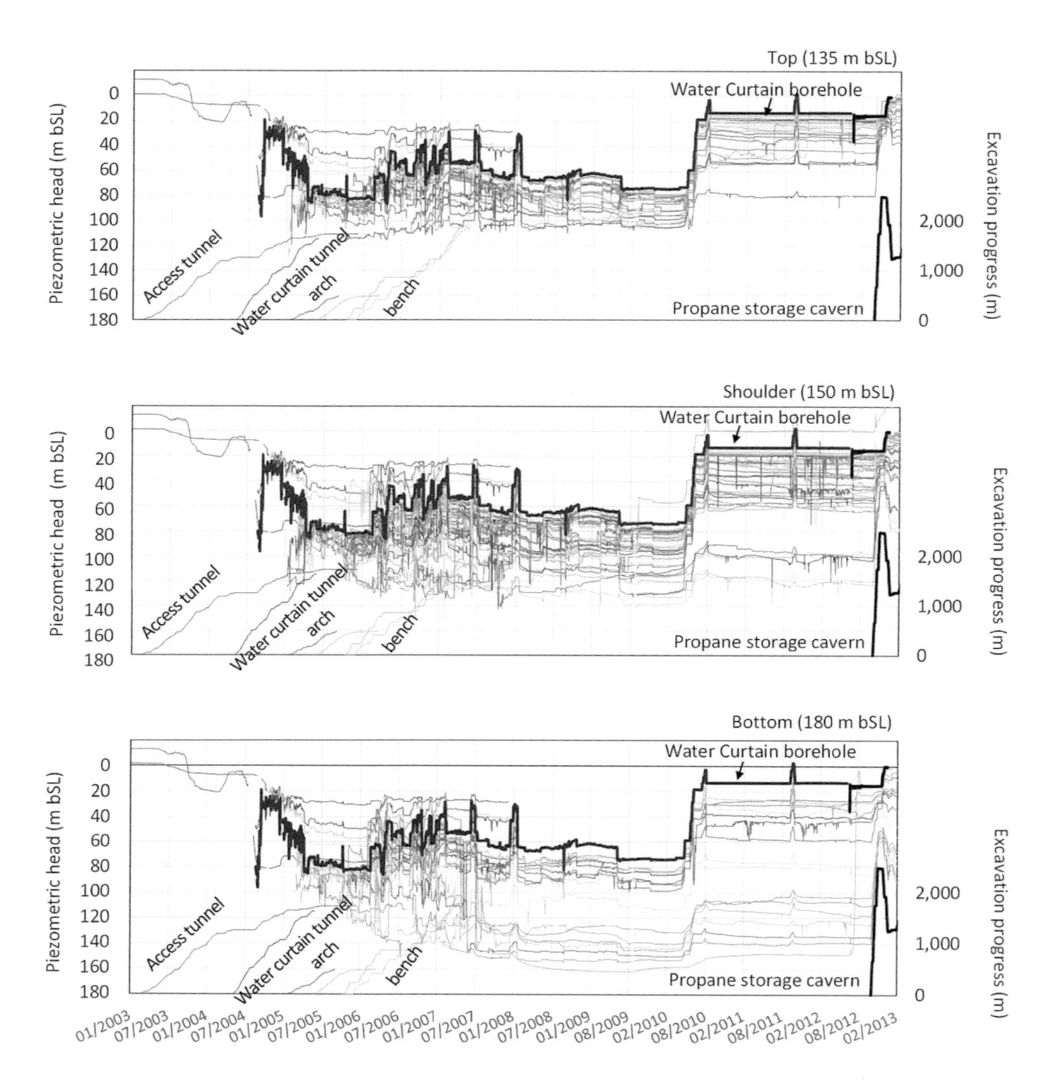

Figure 6.58 Behaviour of the pore water pressure under the water curtain of the propane storage cavern in the Namikata facility.

the permeability of the rock, the water pressure field shows a clear division across the water curtain borehole. This implies the presence of a water curtain.

6.2.4.2.1.4.3.4 Behaviour of Pore Water Pressure Gauge Showing Low Values at the Test Figure 6.63 shows the measured pore water pressure values at cross section CD of the butane storage situated at the bottom of the storage cavern (180 m below sea level) from completion of excavation to air-tightness test. From completion to just before the air-tightness test, the pore water pressure at B2C-1 is about 40 m lower than B1C-1 situated at a similar depth. This is the instrument that did not record an increase in the water pressure while it

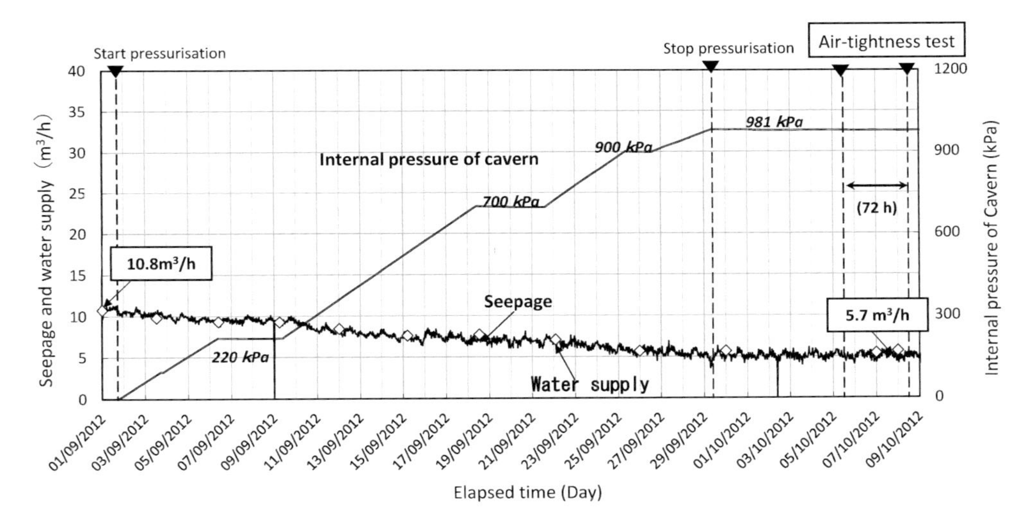

Figure 6.59 Temporal variation of the rate of seep at pressurisation in the combined butane/propane storage in the Namikata facility.

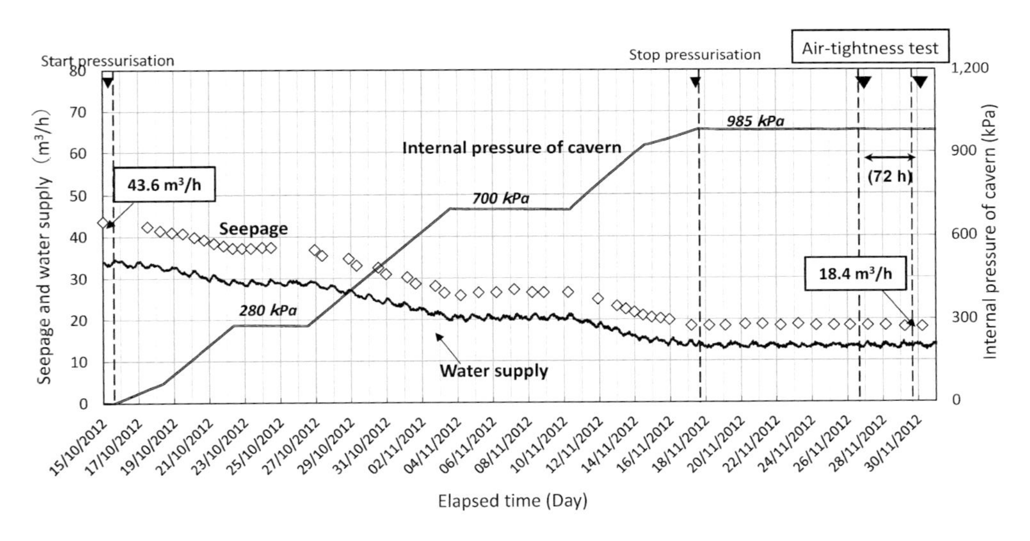

Figure 6.60 Temporal variation of the rate of seep at pressurisation in the propane storage in the Namikata facility.

satisfied the standard threshold of the shut-in test by an additional vertical water sealing borehole (see Section 4.1.3.2). It is located in the V sub-zone where high-angle joints of the N60W system are dominant in high density, and the permeability around the storage cavern is low.

On the other hand, it responded very well to pressurised injection, and the value surged at pressurised injection maintaining the piezometric head difference. As discussed in

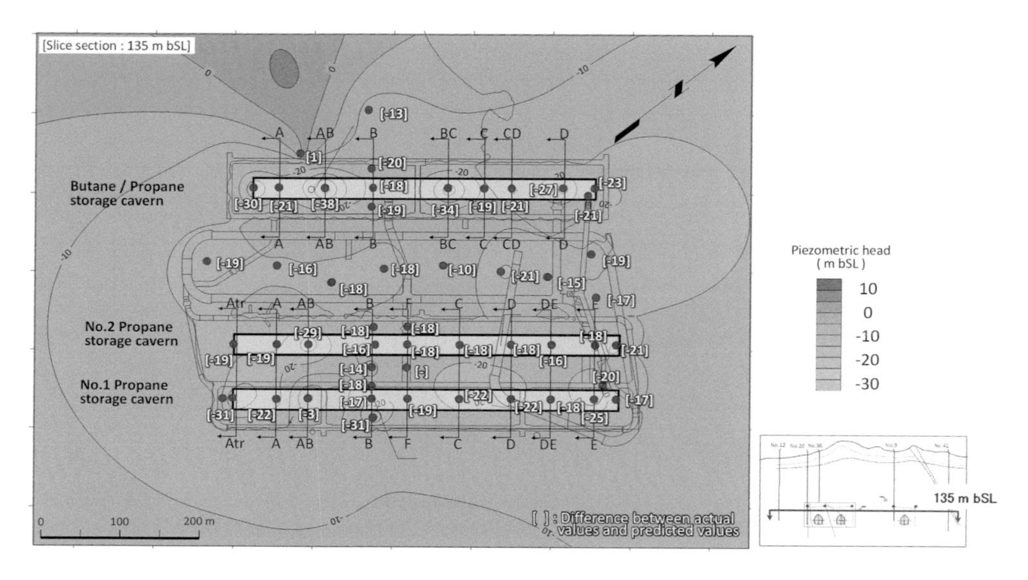

Figure 6.61 Distribution of the pore water pressure at the air-tightness test (plan view at 135 m below sea level) in the Namikata facility.

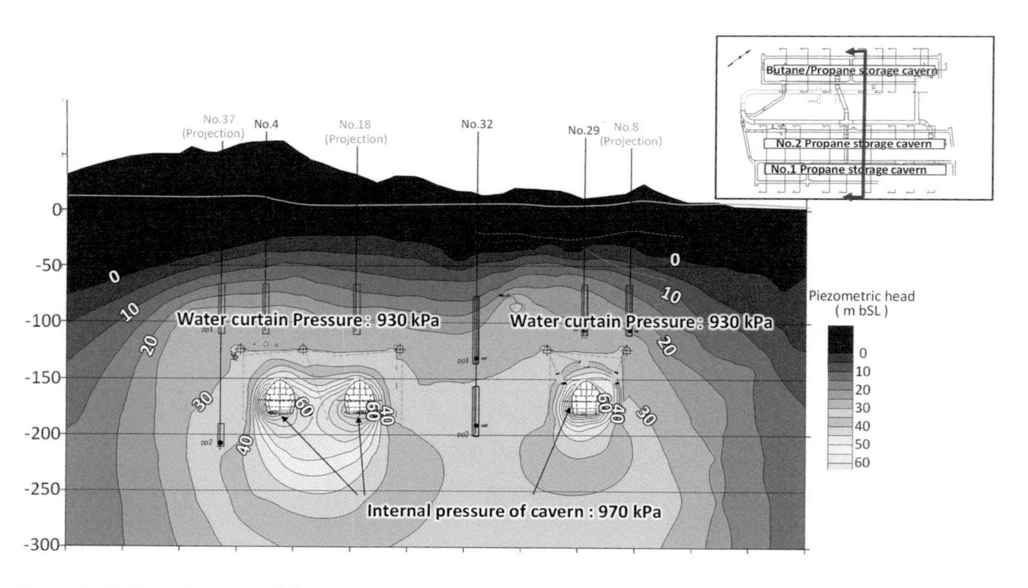

Figure 6.62 Distribution of the pore water pressure at the air-tightness test (cross sections at C of propane storage and BC of butane storage) in the Namikata facility.

Section 6.1, the water-tight pressure and the pore water pressure are represented in terms of deviations from the internal pressure. A correlation between these two parameters is used to evaluate the effectiveness of the water curtain. This correlation was calculated twice at the time of change in the water-tight pressure and the time of pressurised injection. Figure 6.64

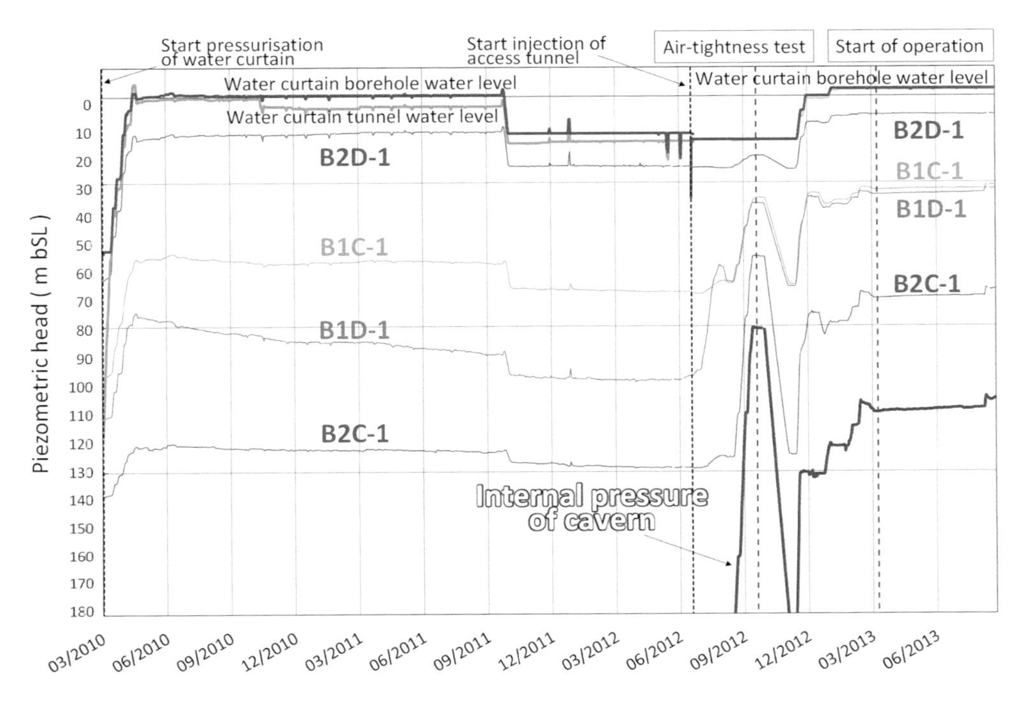

Figure 6.63 Behaviour of the pore water pressure at CD cross section of the butane storage from completion of excavation to the air-tightness test in the Namikata facility (areas deeper than water curtain toward the storage cavern: instrument at 180 m below sea level).

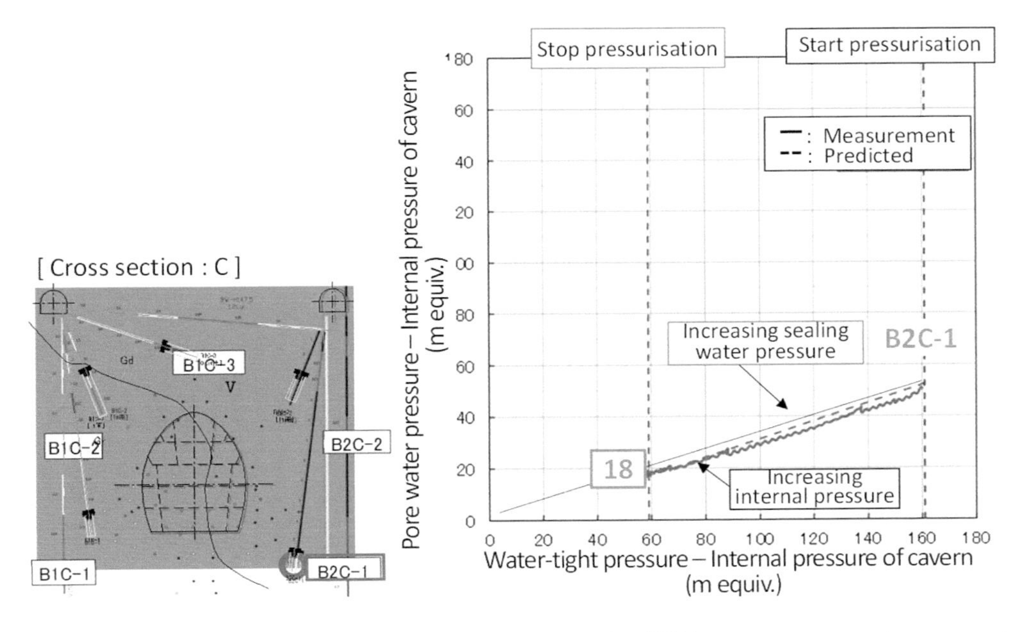

Figure 6.64 Correlation of the pressure difference measured at B2C1 at water pressure increase and pressurised injection in the Namikata butane storage cavern.

shows a comparison of these correlations. A very similar correlation of the differences at the time of variation in the water pressure and pressurised injection indicates the intactness of the water curtain around the location of this instrument.

The location of B2C-1, as seen in Figure 6.65, has a heterogeneous permeability distribution between the storage cavern and the water curtain borehole. The area between the instrument and the water curtain borehole is low in hydrological continuity due to the presence of the high-angle joints. On the other hand, the area between the instrument and the storage cavern is considered to have relatively high hydrological continuity. Therefore, the measurements near these areas showed lower readings than others before pressurised injection into the storage cavern, and ~~it~~ the measurements increased along with the increase in the internal pressure, because an appropriate water curtain acting as a hydrological boundary was built by the additional water curtain boreholes.

6.2.4.3 Evaluation of Air-tightness by Analysis by 3D Hydrogeological Model

The distribution of the pore water pressure over the entire storage cavern at the air-tightness test was evaluated by analysis with 3D hydrogeological modelling, which can reproduce the behaviour from excavation to pressurised injection.

This section discusses the distribution of the difference in the hydrometric pressure between the pore water pressure and the internal pressure of the cavern in the regions deeper than the water curtains at the air-tightness test by actual measurement and model prediction in the Kurashiki and Namikata facilities. This distribution diagram expresses the pressure differential at a constant distance around each storage cavern: 12 m in the Kurashiki facility and 15 m in the Namikata facility.

The distribution of the hydrometric pressure difference between the pore water pressure by actual measurement in the Kurashiki facility is shown in Figure 6.66. This shows that differential values for regions of low hydrometric pressure were found in No. 3 and No. 4 storages, and that the minimum pressure differential by measurement was 12 m at A-10.

The model prediction in Figure 6.67 shows general agreement with the actual measurements. The model prediction shows that the anomalous areas of low hydrometric pressure difference were limited to the close vicinity of A-10, and all the rocks surrounding the storage

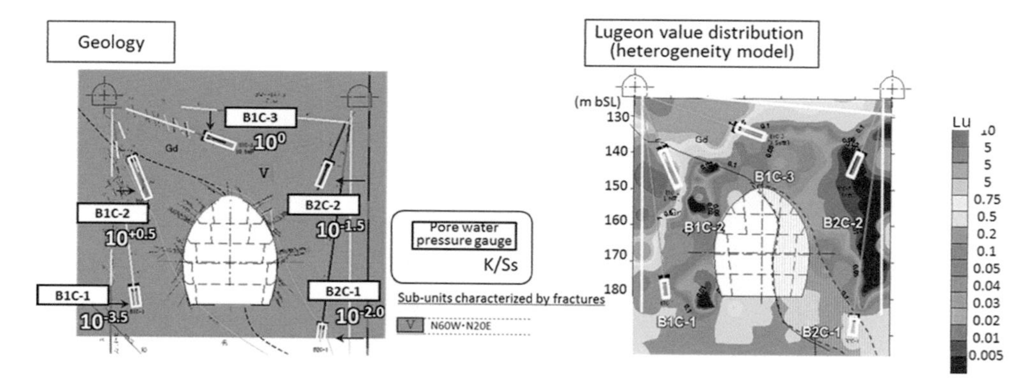

Figure 6.65 Geology and Lugeon value distribution along cross section C of the butane storage at the Namikata facility.

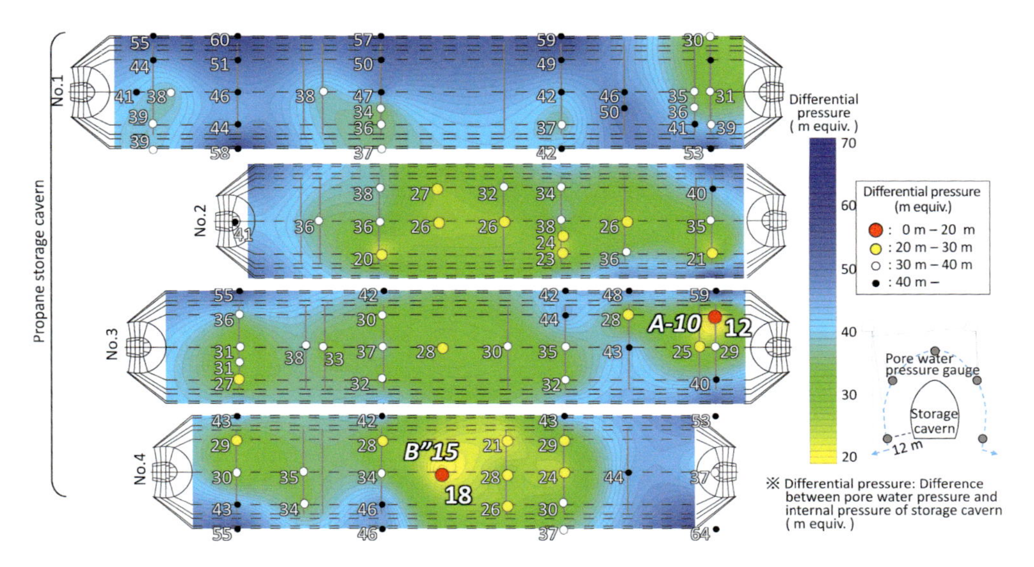

Figure 6.66 Contour map of the measured hydrometric pressure difference between the internal pressure and the pore water pressure at the air-tightness test in the Kurashiki facility.

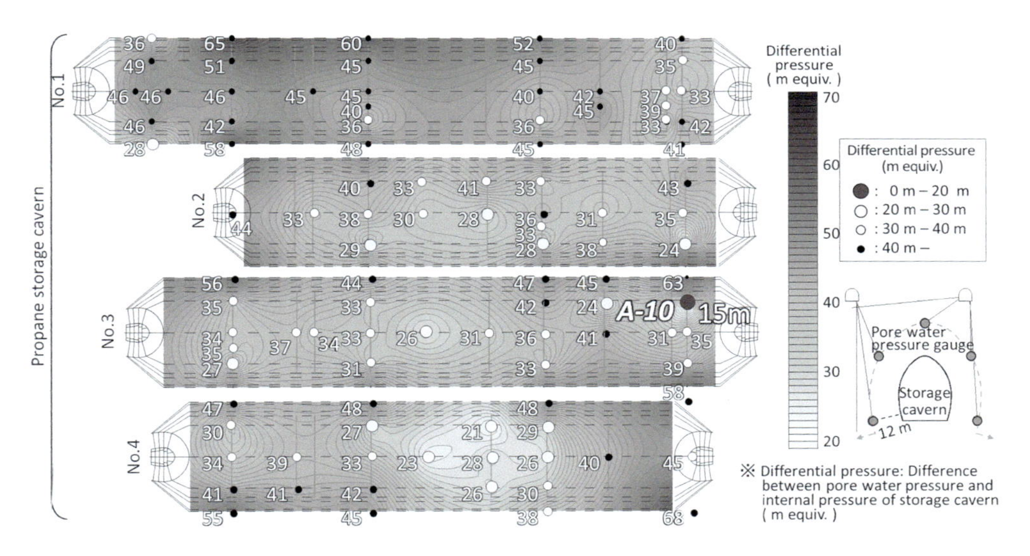

Figure 6.67 Contour map of modelled hydrometric pressure difference between the internal pressure and the pore water pressure at the air-tightness test in the Kurashiki facility (predicted values for the 3D heterogeneous model).

cavern were largely considered to maintain the pore water pressure higher than the internal pressure of the storage cavern under the condition of the air-tightness test.

The distribution of the hydrometric pressure difference by actual measurement in the butane storage cavern of the Namikata facility is shown in Figure 6.68 and the model prediction is shown in Figure 6.69. Both figures show the lowest hydrometric pressure differential

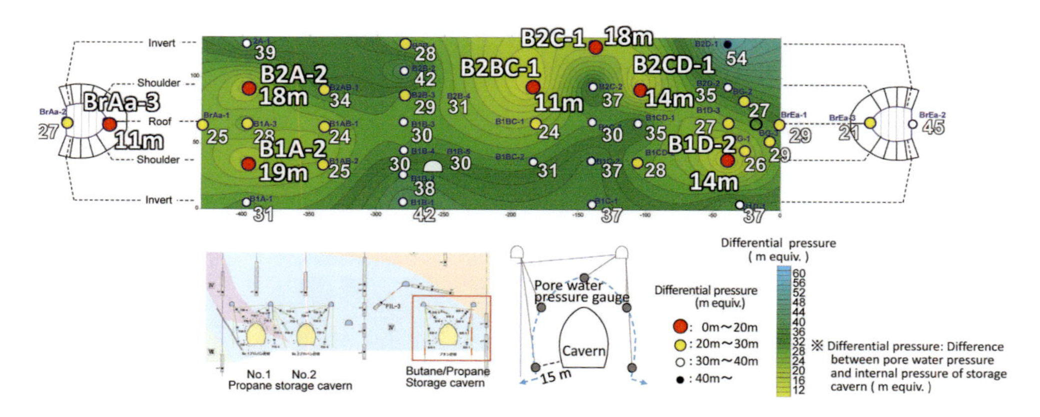

Figure 6.68 Contour map of the measured hydrometric pressure differential between the internal pressure and the pore water pressure at the air-tightness test at the butane/propane storage cavern of the Namikata facility.

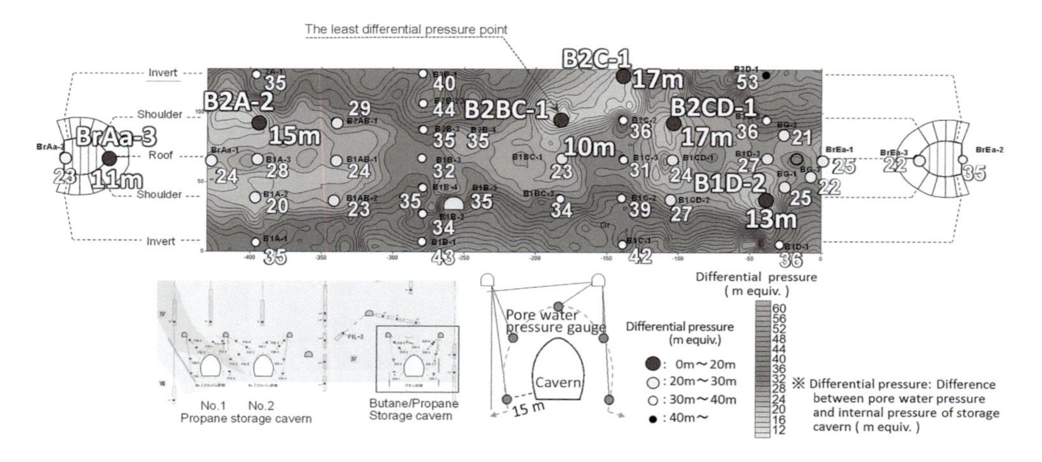

Figure 6.69 Contour map of modelled hydrometric pressure differential between the internal pressure and the pore water pressure at the air-tightness test at the butane/propane storage cavern of the Namikata facility (predicted values for the 3D heterogeneous model).

around 10 m near the B2BC-1 instrument. The distribution of the hydrometric pressure differential by actual measurement in the Kurashiki facility is shown in Figure 6.70 and the prediction using a 3D hydrogeological model is shown in Figure 6.71. The low hydrometric pressure differential measured near the 1AB-3 instrument was generally reproduced by the analysis. The pore water pressure of all the rocks surrounding the butane and propane storage caverns was largely considered to maintain the pore water pressure higher than the internal pressure of the storage cavern under the condition of the air-tightness test.

6.2.5 Results of the Air-tightness Test of Storage Caverns

This section describes the internal pressure of the storage cavern to calculate its variation, ΔP, an indicator of the air-tightness of the cavern measurement factors for data correction

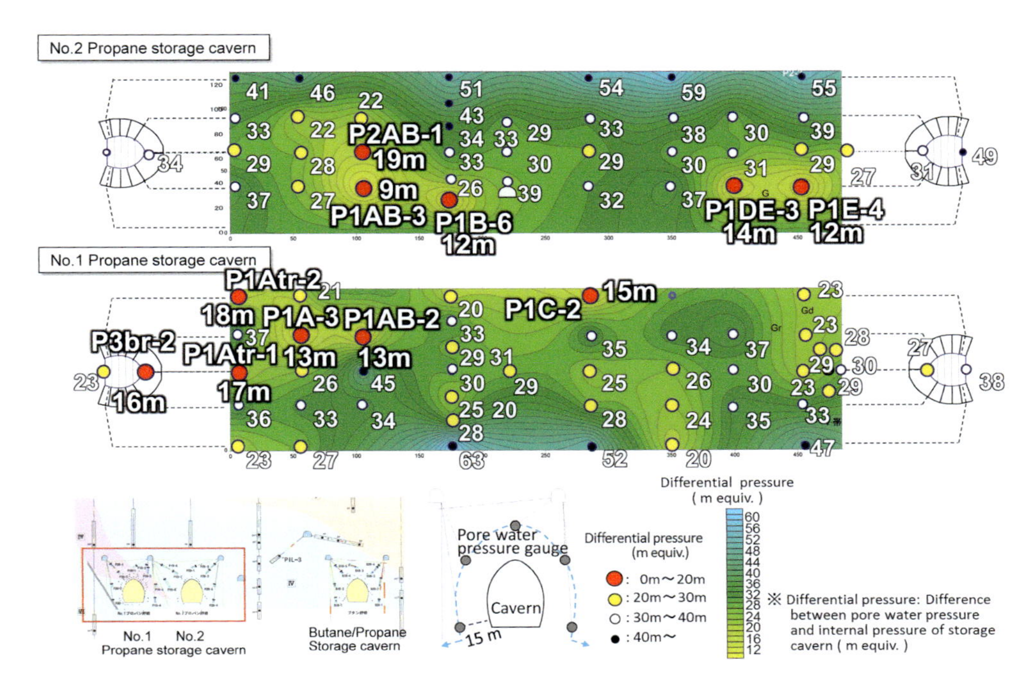

Figure 6.70 Contour map of measured hydrometric pressure differential between the internal pressure and the pore water pressure at the air-tightness test at the propane storage cavern of the Namikata facility.

and the results of a subsequent evaluation of air-tightness, taking the Kurashiki facility as an example.

Figure 6.72 shows graphs of the storage pressure, the internal temperature at 64 thermometer stations, the water level of the bottom drainage tank and the air flow and its temperature over the time from pressurised injection to the end of the air-tightness test. The pressurised injection was carried out step by step controlling the rate of pressure from its base value of 50 kPa/day according to the actual behaviour of the pore water pressure at pressurising. The base value of 50 kPa/day was set considering the delayed response of the surrounding pore water pressure to pressurising (see Section 6.2.4). The 72-h air-tightness test took place after about four days of the stabilising period following the end of pressurising.

Temperature is the greatest factor influencing the internal pressure of the storage cavern. Therefore, a stabilising period is necessary to settle the distribution of the internal pressure of the storage caverns so that the uncertainty in the calculation of ΔP can be reduced. The storage cavern was left undisturbed until the rate of change of all the measured temperatures fell below 0.1°C/day.

6.2.5.1 Evaluation of Temperature Stability at Pressurised Injection

Figure 6.73 shows graphs of the internal temperature and its rate of change, the internal pressure and the water level of the bottom drainage tank from the actual measurement at the air-tightness test in the Kurashiki facility.

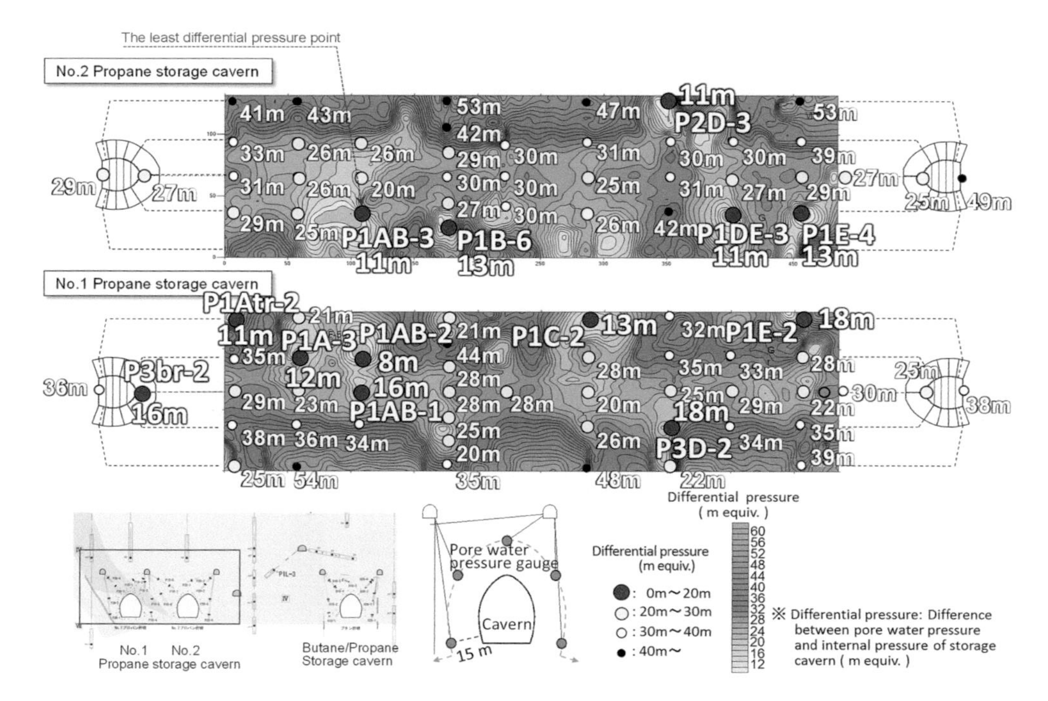

Figure 6.71 Contour map of modelled hydrometric pressure differential between the internal pressure and the pore water pressure at the air-tightness test at the propane storage cavern of the Namikata facility (predicted values for the 3D heterogeneous model).

It shows the downward tendency of the internal pressure of the cavern as the temperature starts to drop after pressurised injection. This is due to the effect on the internal pressure by the drop in the internal temperature due to the escape of heat to the surrounding rock and water as input of heat, which was produced by compression of the air, and stops at the end of injection.

On the other hand, some disturbance of the internal temperature was noted in some parts of the cavern. This disturbance was limited to the vicinity of the bottom water drain of No. 1 storage, and the splash of the water injected to maintain the bottom water level was considered to be the cause. This disturbance in the temperature was settled by injecting water from the top of the LPG reception line pot pipe and heightening the bottom water drain level. At the start of the air-tightness test, the spatial scatter of the temperature among the 64 measurement points was maintained around 0.14°C, and its rate of change was small at about 0.03°C/day at all the instruments. Hence, it is considered that the distribution of the internal temperature converged to a steady state and remained stable.

6.2.5.2 Evaluation of Air-tightness of Storage Cavern

Figure 6.74 shows graphs of the internal pressure of the storage cavern, the internal temperature at the 64 thermometer stations and the water level of the bottom drainage tank from the actual measurement at the air-tightness test in the Kurashiki facility.

The thermometers at the 64 locations could measure very small variation in the internal temperature throughout the air-tightness test. All 64 thermometers even detected the minute

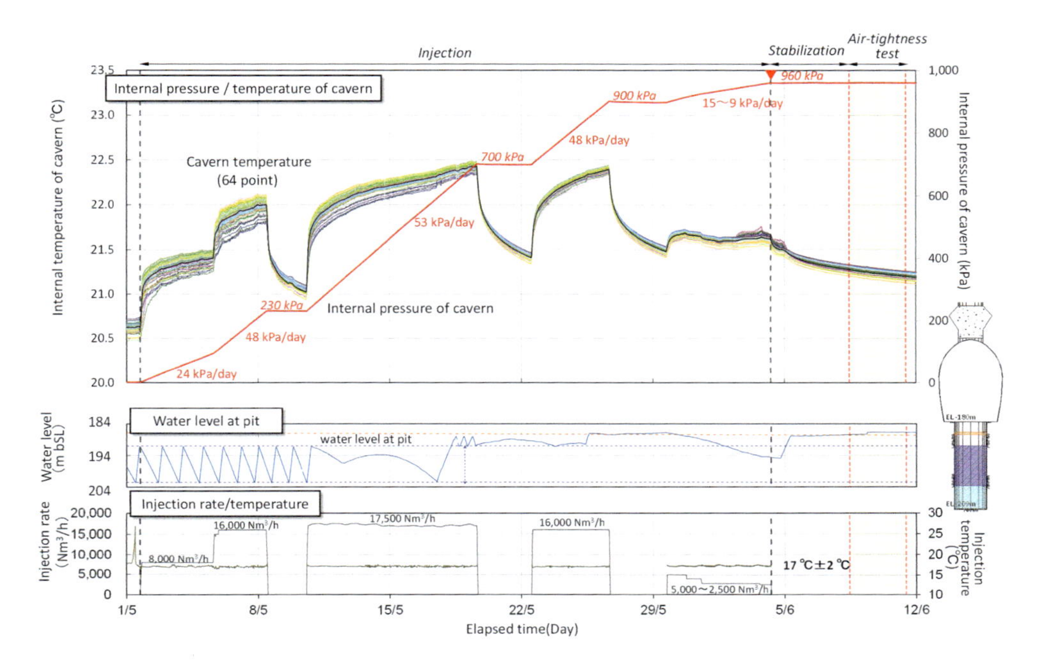

Figure 6.72 Temporal variation in the pressure in the storage cavern, the internal temperature, the water level of the bottom drainage tank and air flow and its temperature from pressurised injection to the end of the air-tightness test in the Kurashiki facility.

temperature variation between 0.003°C and 0.01°C due to compression in the gas phase caused by temporary trouble in the drainage pump causing a minor rise in the water level of the bottom drainage tank on June 9.

The internal pressure of the storage cavern measured at the top of the pipe along the shaft was processed in three steps as described in Section 6.2.1:

- conversion to the air pressure at the depth of the storage cavern;
- correction for the effect of the internal temperature of the storage cavern and volume change; and
- correction for the effect of the volume of air dissolved in the seepage in the cavern.

These corrections are shown in Figure 6.75.

The internal pressure of the storage cavern measured at the top of the pile along the shaft dropped by 0.286 kPa during the 72-h air-tightness test. This value was

- converted for the depth to 0.291 kPa;
- corrected by 0.178 kPa for temperature to 0.113 kPa; then
- corrected by 0.111 kPa for air dissolution,

finally reaching the variation of the internal pressure of the cavern $\Delta P = 0.002$ kPa.

Figure 6.76 shows the variation in pressure after all the corrections through the air-tightness test. At the end of the 72-h test, the variation in the internal pressure of the cavern ΔP

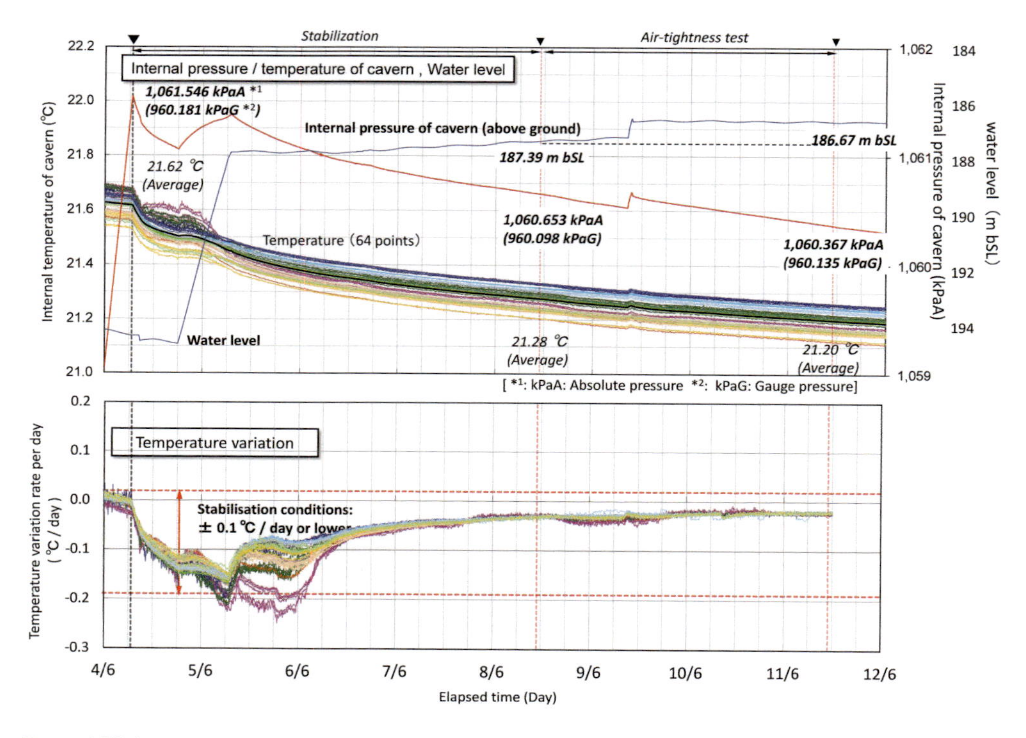

Figure 6.73 Internal temperature and its temporal variation after stopping pressurising the caverns in the Kurashiki facility.

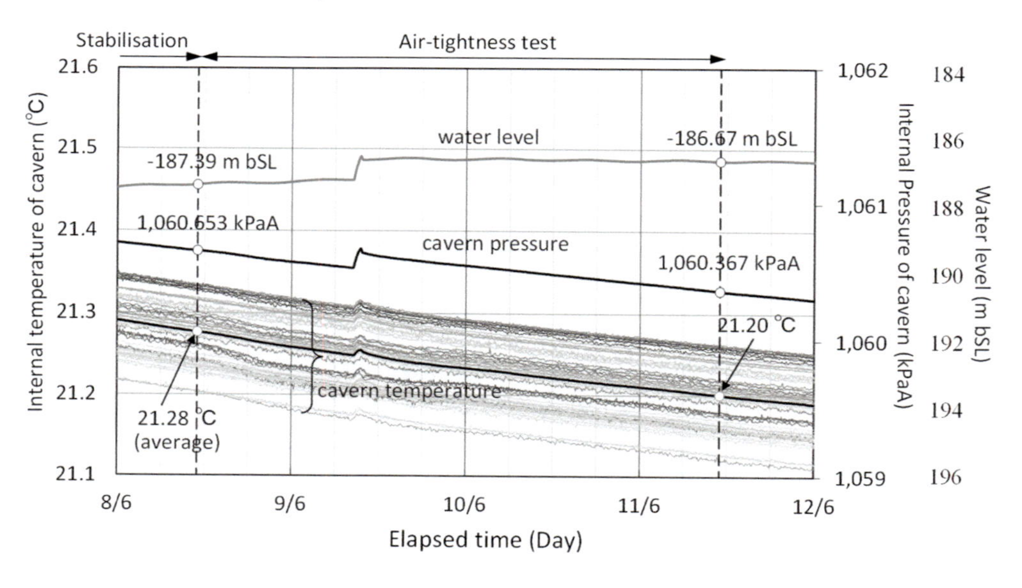

Figure 6.74 Temporal variation in the internal pressure of the storage cavern, the internal temperature and the water level of the bottom drainage tank from the actual measurement at the air-tightness test in the Kurashiki facility.

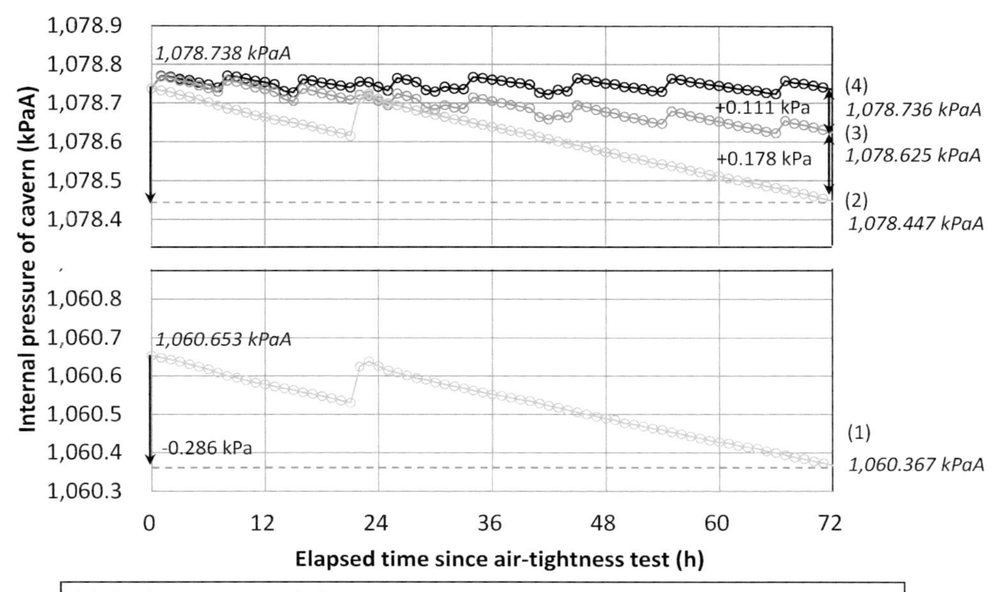

Figure 6.75 Measured and correction values of the internal pressure of the storage cavern during the air-tightness test in the Kurashiki facility.

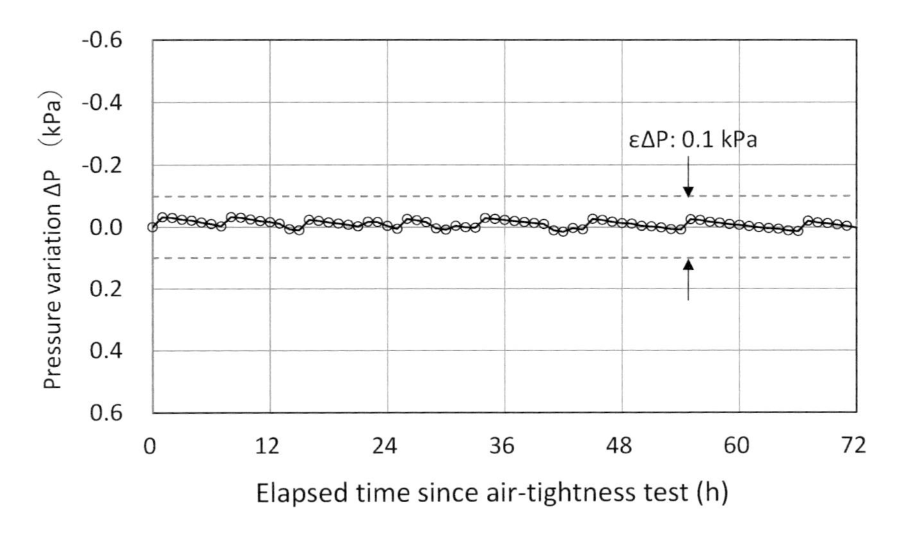

Figure 6.76 Variation of pressure ΔP at the air-tightness test in the Kurashiki facility.

was 0.002 kPa with a maximum of 0.033 kPa. Through the 72-h test period, the absolute value of the variation in the internal pressure of the cavern ΔP stayed within its uncertainty range of $\varepsilon\Delta P = 0.1$ kPa, and the value did not decline. This proved the effective air-tightness of this storage cavern at an internal pressure of 950 kPa.

Finally, the variation in the internal pressure of the cavern ΔP after all the corrections through the air-tightness test in the Namikata facility is shown in Figure 6.77 for the combined propane/butane storage and Figure 6.78 for the propane storage.

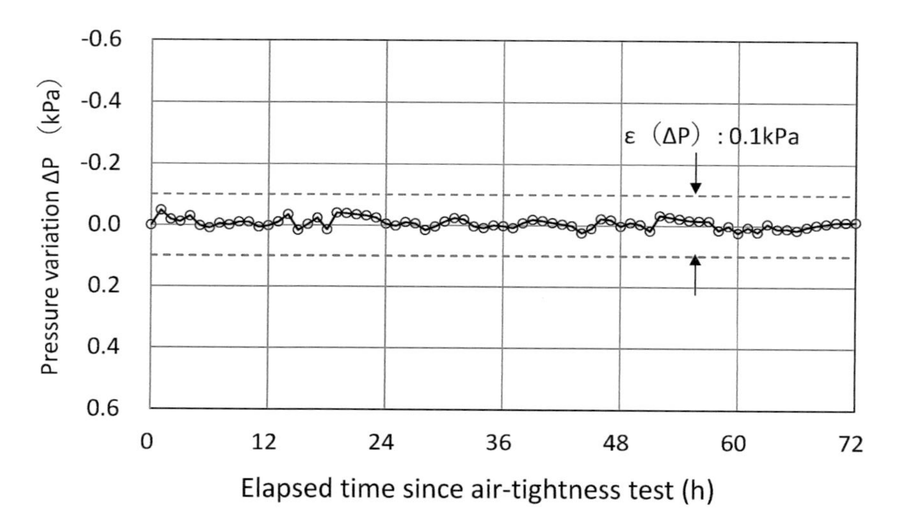

Figure 6.77 Variation of pressure ΔP at the air-tightness test of the combined propane/butane storage cavern in the Namikata facility.

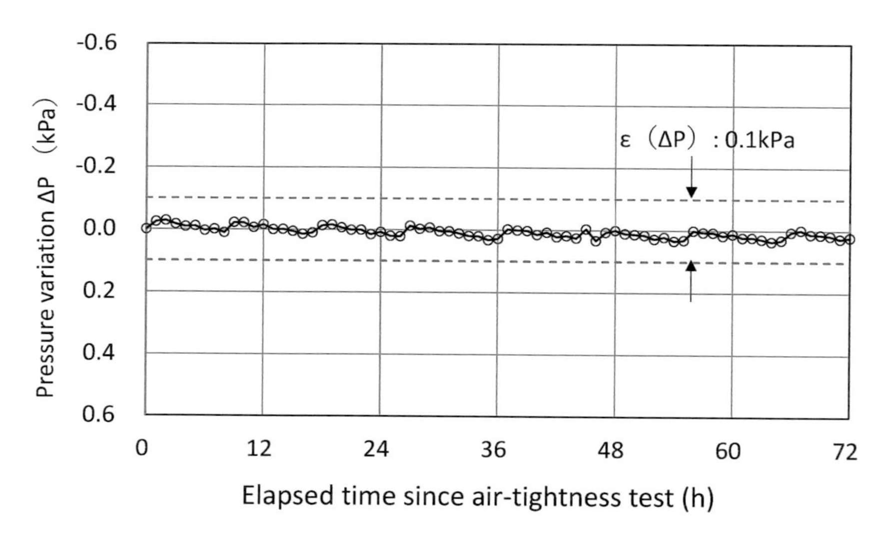

Figure 6.78 Variation of pressure ΔP at the air-tightness test of the propane storage cavern in the Namikata facility.

The variation in the internal pressure of the cavern ΔP at the end of the 72-h test was –0.012 kPa (maximum during the test was 0.048 kPa) for the combined propane/butane storage cavern and 0.022 kPa (maximum during the test was 0.035 kPa) for the propane storage cavern. From these results, the absolute value of the variation of the internal pressure of the cavern ΔP stayed within its uncertainty range of $\varepsilon\Delta P = 0.1$ kPa and no trend in reduction was recognised through the 72-h test. This proved the air-tightness of these storage caverns at the Namikata facility, too.

References and Further Readings

ANSYS, Inc (2009) ANSYS FLUENT Release 12.0 Theory Guide.

Bergman M. (1977) Storage in excavated rock caverns. *Proceedings of the First International Symposium*, Stockholm, 832.

Johansen P.M., Madsen O.K. (1989) The Rafnes propane storage cavern −12 years of successful operation. *Proceedings of the International Conference on Storage of Gases in Rock Caverns*, Trondheim, 303–306.

Kjorholt H., Goodall D.C., Johansen P.P., Stokkebo O. (1989) Water curtain performance at the Kvilldal air cushion. *Proceedings of the International Conference on Storage of Gases in Rock Caverns*, Trondheim, 87–94.

National Astronomical Observatory of Japan (2000) Chronological Scientific Tables, Maruzen.

Nilsen B., Olsen J. (1989) Storage of gases in rock caverns. *Proceedings of the International Conference on Storage of Gases in Rock Caverns*, Trondheim, 398.

Shimo M., Shimaya S., Kurose H., Okazaki Y., Maejima T. (2014) Borehole and chamber tests for evaluation of gas-tightness of a cavern in a fractured rock mass. *International Symposium −8th Asian Rock Mechanics Symposium*, Sapporo, 2122–2141.

Tetens, O. (1930) Uber einige meteorologische Begriffe. *Z. Geophys.*, Vol.6, 297–309.

Chapter 7

Cavern Storage and Surrounding Groundwater During Operation

7.1 Operation Facility of the Storage Caverns

7.1.1 Safety Management of Storage Cavern

The important issues in the safe operation of cavern storage are the long-term safety of the cavern, mitigation of the adverse effects on the environment and prevention of a deterioration in the quality of the stored material.

Regulations on the safety of liquefied petroleum gas (LPG) cavern storage include the High Pressure Gas Safety Act (1951 as amended) of Japan and recent safety rules by the EU such as the Seveso II Directive, EN 1918 Part 4 & 5 (Gas Supply Systems –Underground Gas Storage – Part 4: Functional Recommendations for Storage in Rock Caverns/Part 5: Functional Recommendations for Surface Facilities) ⇒ Seveso II & EN 1918, UNECE Convention on the Transboundary Effects of Industry Accidents, 2002 by the UN and Hydrocarbon Storage in Mined Caverns – A Guide for Safety Regulators, 2000 by the US Interstate Oil and Gas Compact Commission (IOGCC). Facilities with a high level of safety were designed to comply with these regulations.

To prevent an accident from escalating and the damage expanding, lessons from past serious accidents were studied and risks were analysed based on risk scenarios for each part of the facility. From this study, a failsafe system was devised to minimise the effect of an accident. Control measures were compiled and documented in a “serious accident prevention plan” and a “safety plan”.

According to this risk analysis, safety equipment (valve) called “failsafe valves” were installed as a means of preventing the spread of a serious accident. The failsafe valves were installed in the underground piping: at the pump inlets of the LPG discharge line and the plug positions of other LPG lines. These valves are usually open, and when a crack, opening or failure occurs in a pipe or shaft or on the ground, it closes to stop the release of the LPG from storage into the atmosphere.

The LPG level is monitored by simultaneous measurement of two physical properties: electrostatic capacitance and ultrasonic properties. This combination prevents measurement errors caused by the deterioration and malfunctioning of one of the instruments.

The pipes in the shaft are the reception line, discharge line, liquid level monitoring line, bed water drain line, vent line, pressure monitor line and purge line. They are fixed in the shaft with concrete plugs. The discharge line is installed with the LPG discharge pump for dispatch in the water drain pit. The bed water drain line is installed with the water draining pump to drain the seepage in the cavern and also in the water drain pit.

DOI: 10.1201/9780367822163-7

In operation, the shaft is filled with water. This water and the concrete plug play a sealing function to prevent LPG in the cavern from escaping into the atmosphere through the shaft (Figure 7.1).

7.1.2 Operational Equipment

Figure 7.2 shows the flow of the reception and discharge of LPG at the LPG storage facility. An outline of the operation equipment is listed here:

7.1.2.1 Reception Equipment

LPG is initially unloaded from an ocean vessel (very large gas carrier [VLGC]) to an above-ground low-temperature tank at the facility. A booster pump then moves the LPG into the cavern storage through a heater in the normal-temperature high-pressure LPG state.

7.1.2.2 Dispatch Equipment

LPG is discharged by a discharge pump into an above-ground high-pressure tank, and is dispatched using coastal vessels or tank lorries. The Namikata facility has a freezing plant to convert normal-temperature high-pressure LPG in the cavern storage to low-temperature normal-pressure LPG, and through an above-ground low-temperature tank it is capable of transporting low-temperature LPG to the customer.

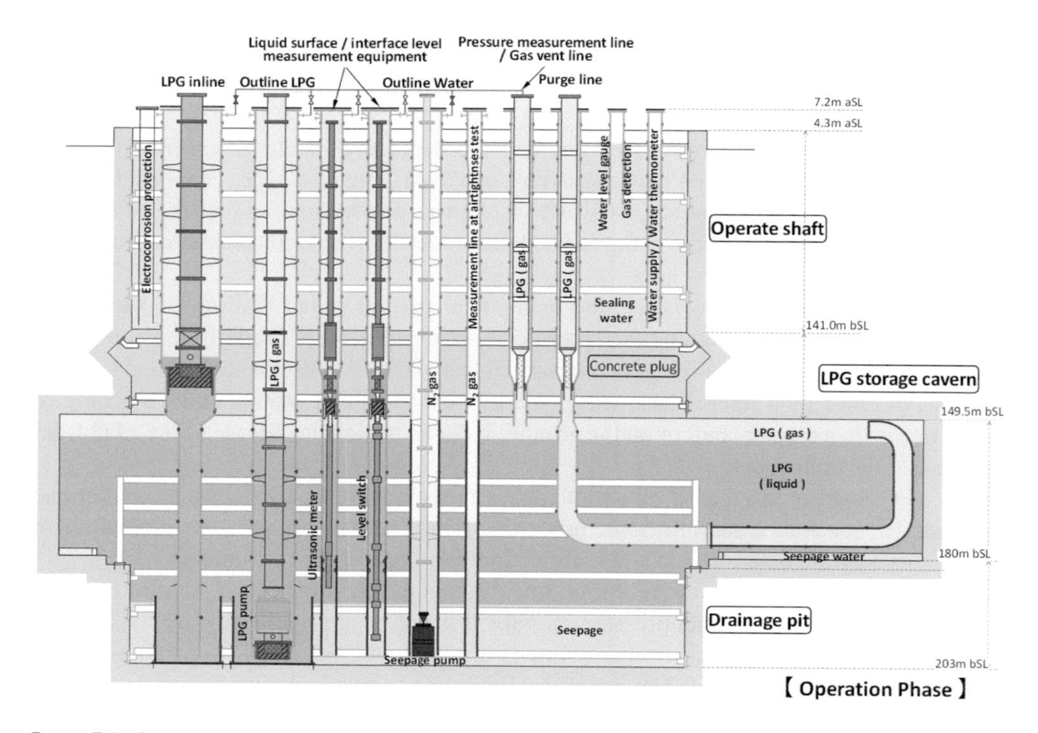

Figure 7.1 Concept of piping in the shaft in the Namikata facility.

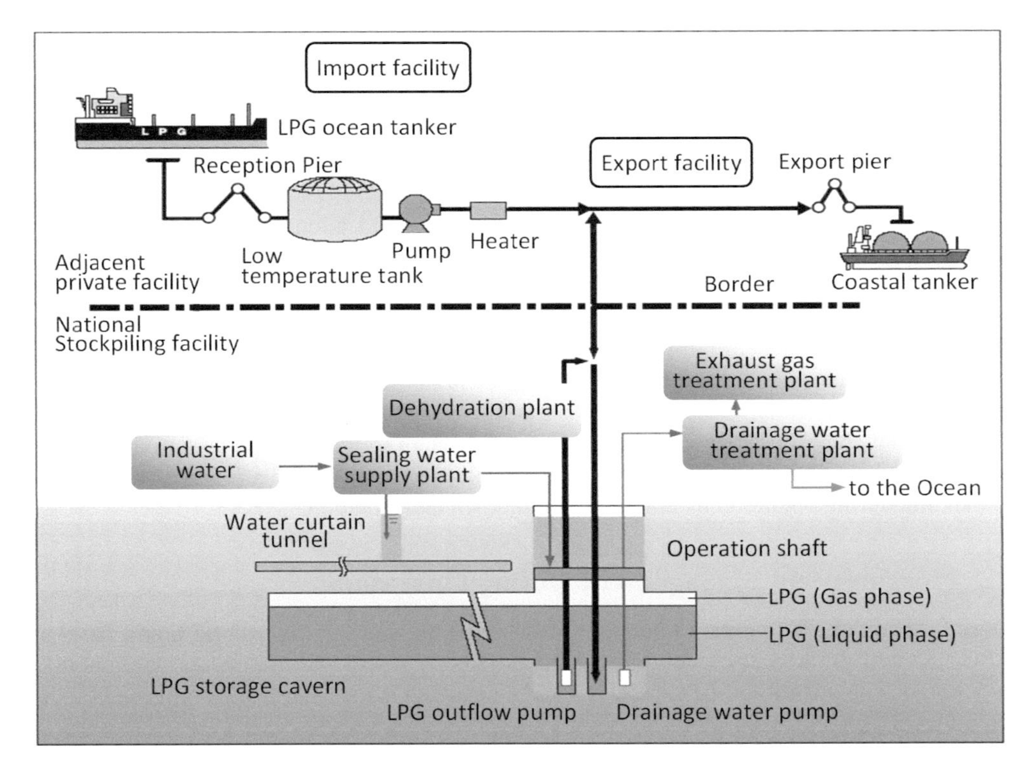

Figure 7.2 Reception–dispatch system diagram of the Namikata facility.

7.1.2.3 Bed Water Drainage Equipment

The seepage in the cavern is drained by a bed water draining pump in the draining pit. The discharged water is processed by de-gassing, pH adjustment, precipitation and filtering, and is subsequently released into the sea after the water quality has been checked.

7.1.2.4 Sealing Water System

A system to supply sealing water was installed to provide pressurised sealing water to the water curtain boreholes and tunnels. The sealing water in the Kurashiki facility is the industrial water from the nearby river treated by membrane filtration. In the Namikata facility where there is no river nearby, sea water desalinated by membrane filtration is used. These processes suppress clogging due to calcium scaling and slime caused by bacteria growth.

7.1.2.5 Safety and Disaster Prevention System

Fire prevention and extinguishing equipment, a disaster prevention monitoring system and an emergency shutdown system were installed for safety and disaster prevention.

The fire prevention and extinguishing systems comprise fire prevention equipment and fire extinguishing equipment. The fire prevention system includes sprinklers in the upper part of the storage cavern shaft, water curtains around the piping shaft and fire hydrants. The extinguishing system consists of outdoor hydrants and fixed water guns.

The disaster prevention monitoring system comprises communication monitoring, a leaked gas detection monitor, strong motion seismographs and maintenance electricity systems. The communication monitoring system includes an automatic fire alarm, a paging system and an closed circuit television (CCTV) system. The automatic fire alarm system complies with the Fire Service Act. It is installed in the management block and is centrally monitored in the instrument room. The paging system is installed at several locations in the facility including the management block, and is capable of emergency broadcasting to the entire site. The ITV system is installed at the entry and exit points of the facility, and is centrally monitored in the instrument room. The gas detectors are installed in the storage and production areas, and are centrally monitored in the instrument room. Strong motion seismographs are installed above- and underground of the facility, and are centrally monitored in the instrument room. An emergency generator provides electricity to the instruments.

The emergency shutdown system comprises a detector to identify any abnormality of the system and emergency shutdown valves. It is installed to prevent accidents and disasters from expanding during operation by emergency shutdown and isolation.

7.2 Measurement Management During Operation

During operation, the internal operating pressure of the storage cavern, the liquid level of LPG, the bed drain water level and the conditions of storage operation are continuously monitored. The quantity of LPG received and dispatched is measured, and confirmed by the liquid level.

To ensure the effectiveness of the water sealing function in the storage cavern and the safety of the void, a number of items are constantly monitored throughout the operation, including the groundwater level, pore water pressure, internal pressure of the storage, flow rate of bed drain water and behaviour of the groundwater by measuring the flow rate of sealing water intake. The mechanical stability of plugs and faults is also monitored by measuring ground motion during earthquakes. A change in the quality of the sealing water and drainage is also monitored.

7.2.1 Behaviour of Groundwater

At both the Kurashiki and Namikata facilities, groundwater observation boreholes and pore water pressure observation boreholes were installed to understand the behaviour of groundwater in operation (Figures 7.3 and 7.4)

Management criteria values are set up for the groundwater level and pore water pressure and they are monitored every day for any irregularity.

The difference in piezometric head from the internal pressure of the storage cavern is monitored by measuring the pore water pressure around the cavern using pore water pressure gauges installed around the cavern inside the water curtains (Figures 7.5 and 7.6).

7.2.2 Mechanical Stability of the Caverns

Dynamic measurement instruments including acoustic emission (AE) sensors, accelerometers and vibration sensors were installed in the rocks surrounding the cavern and around

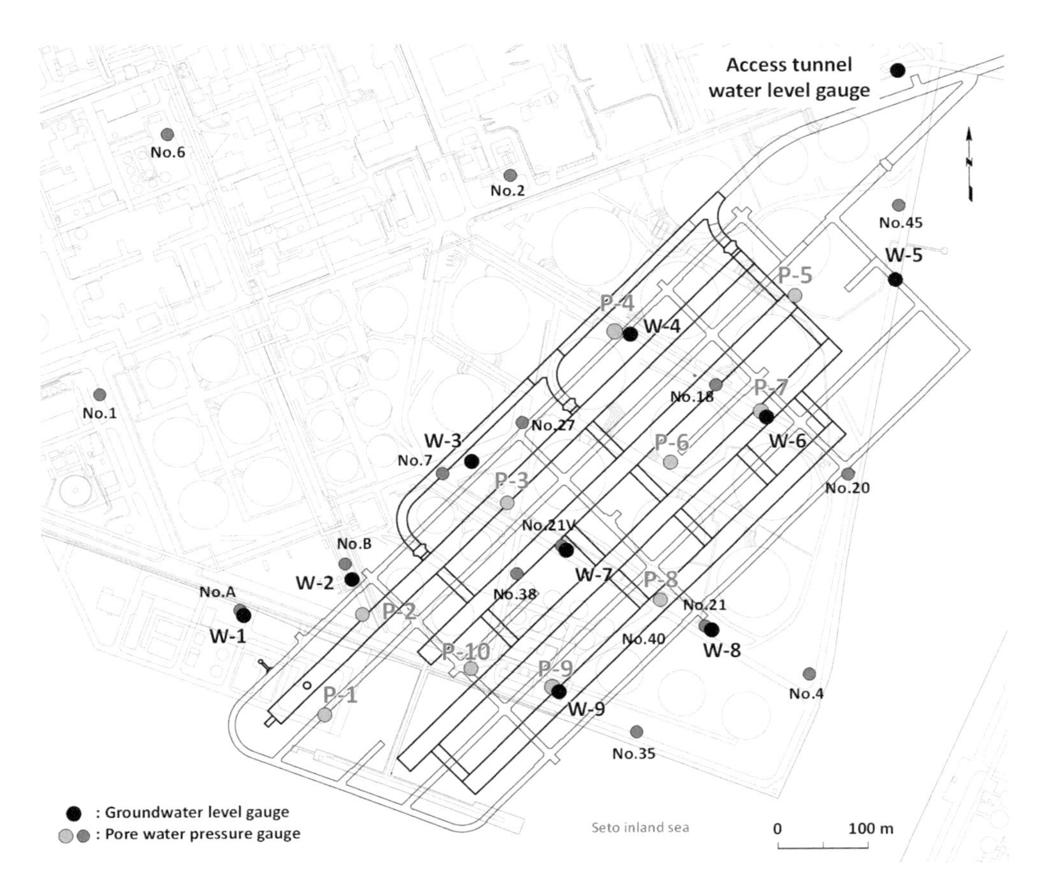

Figure 7.3 Arrangement of the groundwater observation boreholes and pore water pressure observation boreholes in the Kurashiki facility.

the concrete plugs to monitor the stability of the void and detect gas leaks during operation (Figures 7.7 and 7.8).

1. AE Measurements

AE sensors were installed at the plugs, faults and microfractures, to count the occurrence, frequency and amplitude of AEs. From these data, stress change in the rocks was analysed and the health of the rocks was confirmed. AE sensors use optical fibre, and occurrence per hour, amplitude, frequency and *m*-values were recorded every hour.

2. Measurements of Acceleration

Accelerometers were installed on the piping shaft near the widened part of the plugs to confirm the health of the plugs together with the AE measurements.

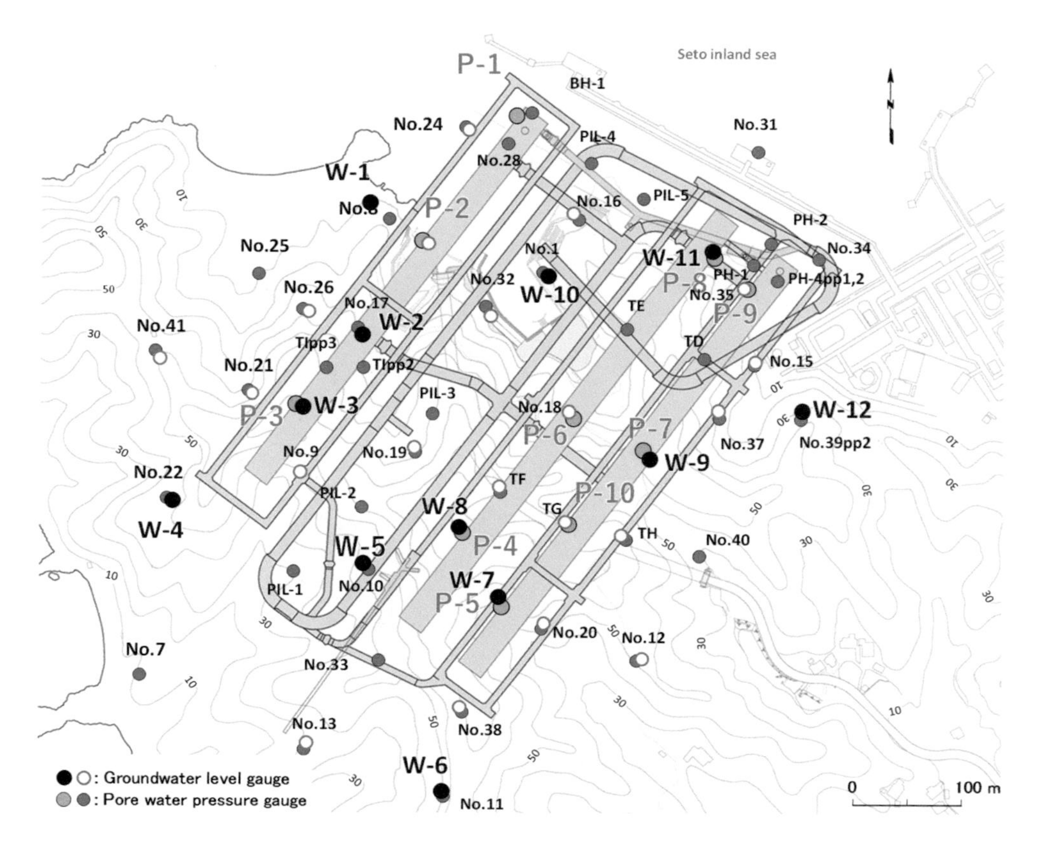

Figure 7.4 Arrangement of the groundwater observation boreholes and pore water pressure observation boreholes in the Namikata facility.

3. Vibration Measurements

To observe the unlikely event of a rockfall, several geophones were installed near the faults in the bed of the cavern, and the health of the rocks was confirmed.

4. Measurement of Earthquakes

Seismometers were installed both above- and underground to measure movement due to earthquakes at the facilities in operation. During an earthquake, the seismic acceleration is measured and the movement is analysed, and a timely response, patrol and inspection of the facility take place (Figures 7.9 and 7.10).

7.2.3 Water Quality Management

The quality of the sealing water, bed drain water and surrounding groundwater is monitored to manage corrosion of the pipes, leaks, influence on the environment and quality of the storage. An example from the Namikata facility is shown in Table 7.1.

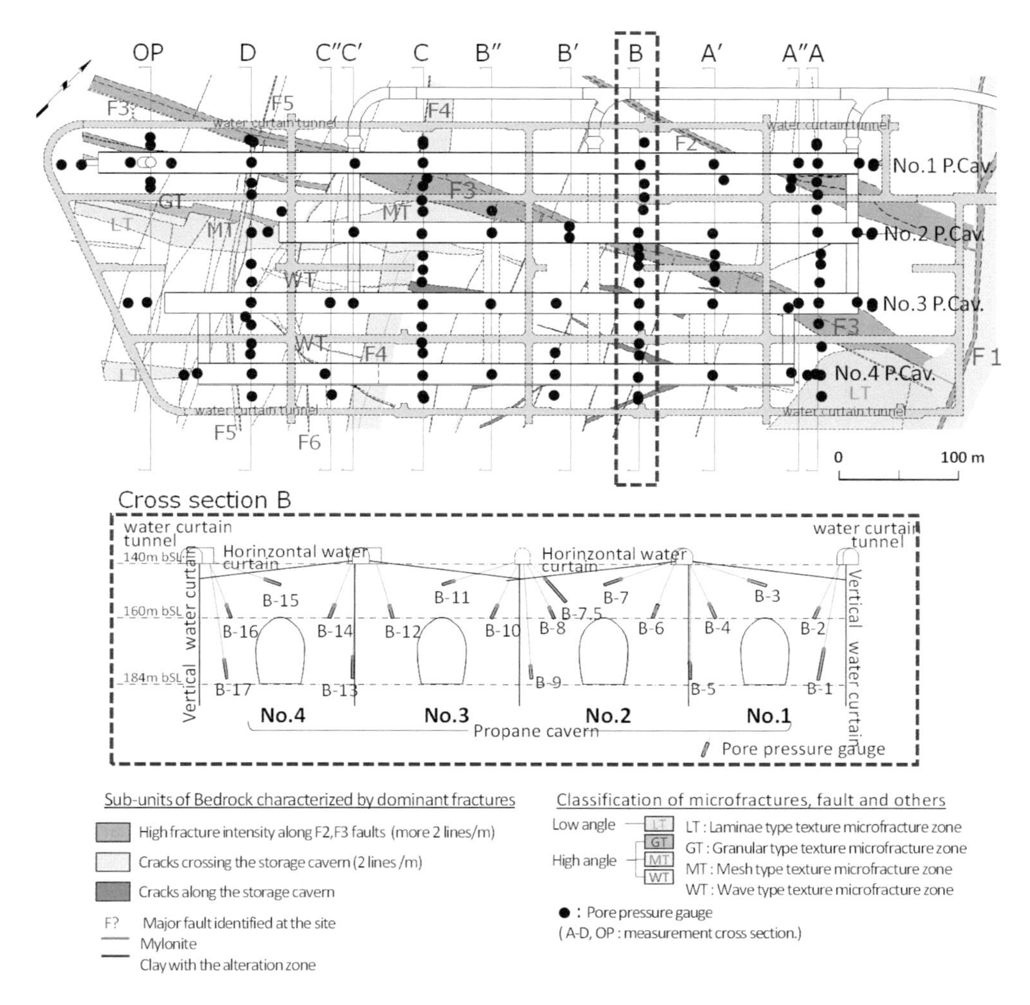

Figure 7.5 Plan view of the arrangement of the pore water pressure gauges inside water curtains in the Kurashiki facility.

7.3 Method of Test Run

7.3.1 Procedure of Test Run

After the air-tightness test, the operation from first reception to storage of LPG is a test run of the cavern storage. The procedure from air-tightness to reception of the first LPG is shown in Figure 7.11.

1. Depressurisation and Water Filling

To ensure the inertness of the storage cavern, the internal pressure of the cavern storage, which had been increased to the test pressure, was reduced and the cavern was filled with water so that the flow rate of gas was minimum.

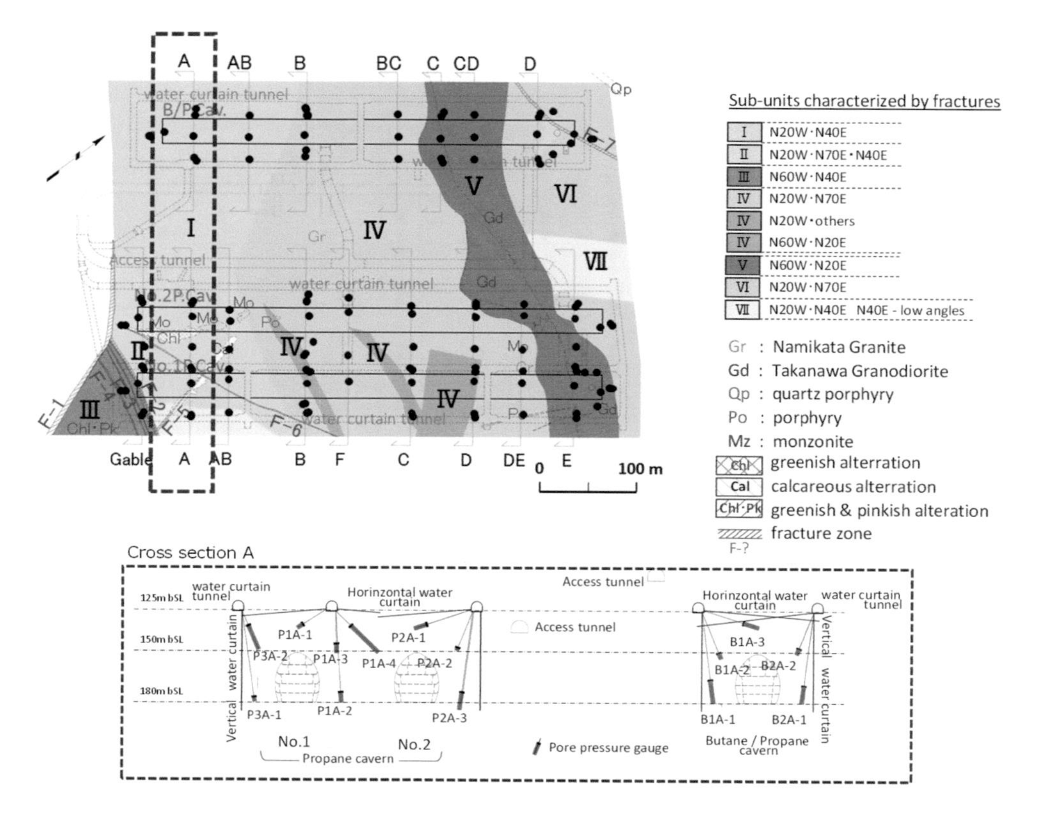

Figure 7.6 Plan view of the arrangement of the pore water pressure gauges inside water curtains in the Namikata facility.

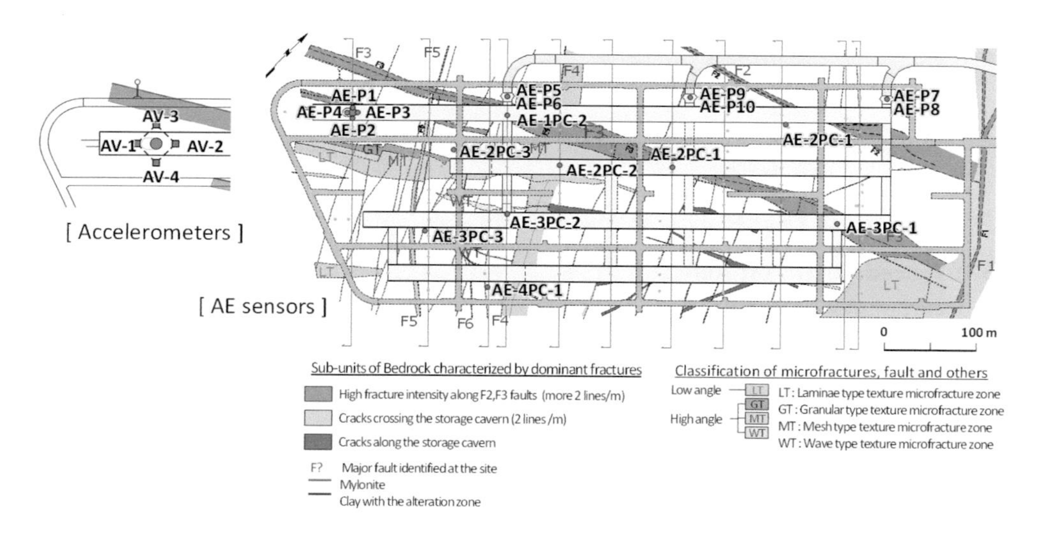

Figure 7.7 Arrangement of AE sensors and accelerometers in the Kurashiki facility.

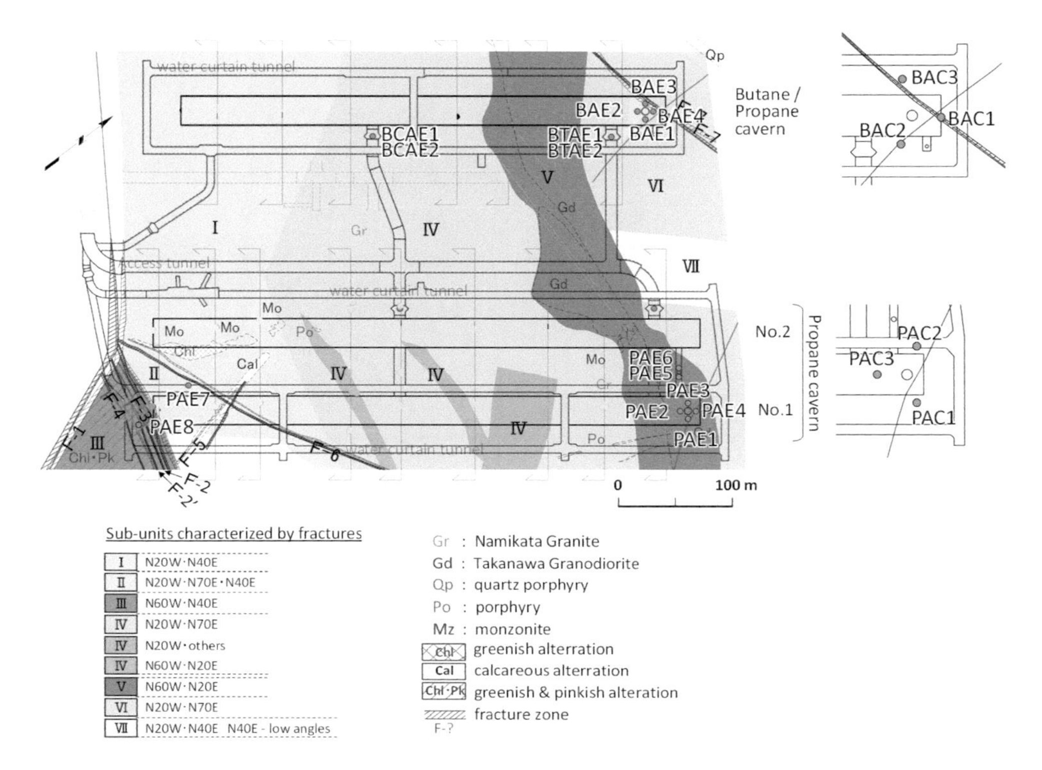

Figure 7.8 Arrangement of AE sensors and accelerometers in the Namikata facility.

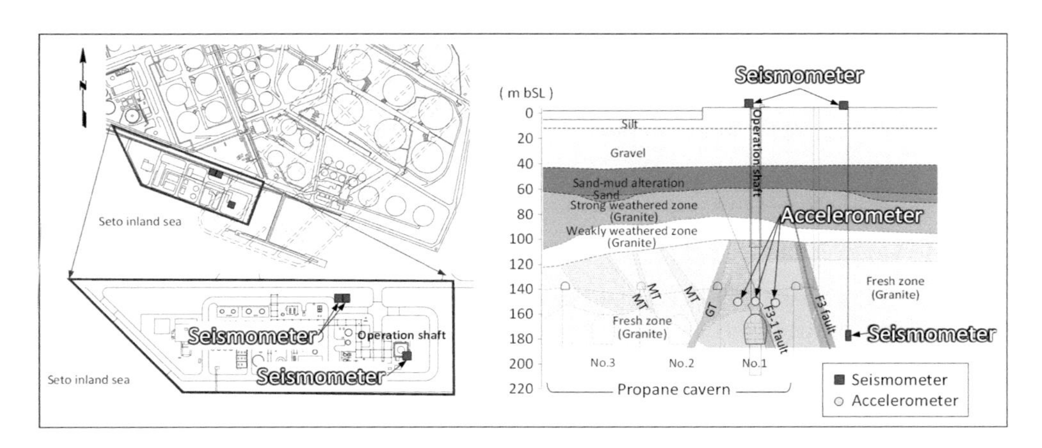

Figure 7.9 Arrangement of seismometers in the Kurashiki facility.

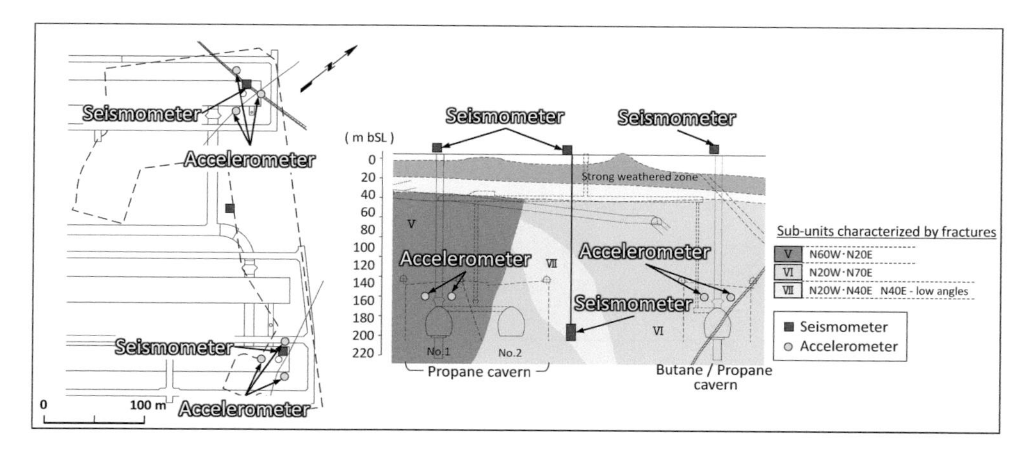

Figure 7.10 Arrangement of seismometers in the Namikata facility.

Table 7.1 Items for water quality measurement

Object	*Location*	*Measured items*
Sealing water	(a) Supply tunnel for water curtain borehole (supply water) (b) Operation shaft (stored water) (c) Return pipe (residual water inside return pipes)	• Measured on site (pH, EC, ORP, water temperature) • Principal ion analysis • (Na, K, Ca, Mg, Cl, SO_4, HCO_3, NO_3, T-Mn, T-Fe, SiO_2, F, B) • Organic analysis • (TOC, phosphorus, nitrogen, NO_2) • Scale (DO, Ca)
Floor drain water	(d) Bed water drainage pipe above ground	• Bacteria analysis • (aerobic bacteria, anaerobic bacteria, sulfite-reducing bacteria, iron-reducing bacteria, total microbial count)
Surrounding groundwater	(e) Groundwater observation boreholes	• Dissolved gas analysis • (C1, C2, C3, C4)

2. *Inertness*

To eliminate the danger of LPG exploding, the gas in the cavern storage was made inert by the "air purge" process whereby the air in the cavern was replaced by nitrogen.

3. *Gas Injection and Draining*

LPG in the gas phase was injected into the storage cavern purging the nitrogen. Injection continued while draining the water down to the operation level.

4. *Initial Reception of LPG*

The first batch of LPG in the liquid phase was injected. In this process, the gas capacity in the storage cavern considerably fluctuates depending on the water injection and draining. The

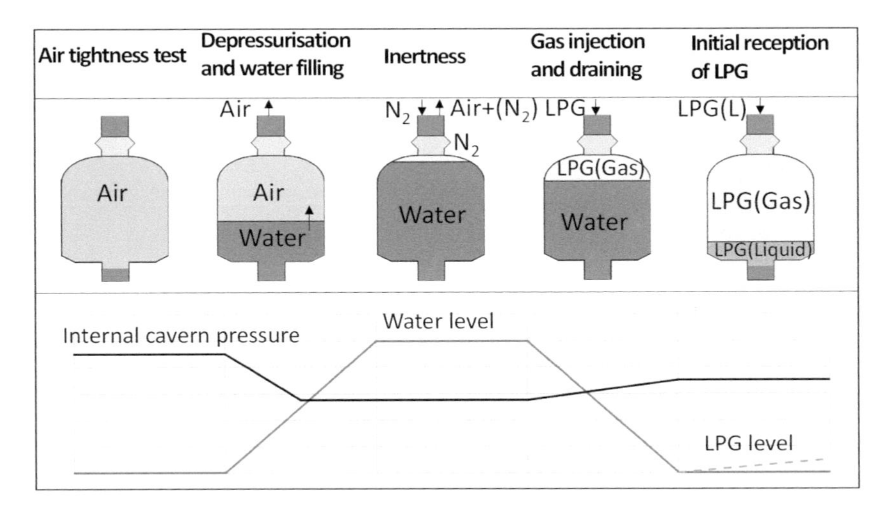

Figure 7.11 Variation of internal pressure and water level through the test operation procedure.

variation in the internal pressure of the cavern was controlled by injection and ventilation of the gas.

7.3.2 *Example of Test Operation*

7.3.2.1 Cavern Depressurisation and Water Filling

The flow rate of the gas phase was minimised by filling the storage cavern with water to ensure the inertness of the contents of the large cavern. Considering the uneven ceiling of the storage caverns, in the Kurashiki facility, the water filling level was set to 161 m below sea level for the lowest part of the ceiling (160.5 m below sea level). This resulted in a gas capacity of about 16,000 m^3 compared with the capacity of the storage at approximately 840,000 m^3. Due to the fault zones around the storage caverns, the groundwater pressure was expected to be unstable. Considering the stability of the void of the cavern, the rate of depressurisation of the storage was set to under 0.025 MPa/day, which is half the pressurisation rate of the air-tightness test.

7.3.2.2 Nitrogen Inerting

The cavern was made inert by replacing the air with nitrogen. This operation is called "air purge". The standard criterion for completion of this operation was set to an O_2 concentration of less than 4%; originally 21%.

Figure 7.12 shows the water level, pressure and O_2 concentration in the cavern during the inerting operation in the Kurashiki facility. The internal pressure of the cavern was reduced from 0.95 to 0.55 MPa at the end of the air-tightness test so that the gas capacity of the cavern reached about 16,000 m^3. Nitrogen was injected through the purge line to the deepest part of the gas phase, and the air was expelled through the vent line. Through this operation, the O_2 concentration remaining in the cavern was reduced to less than 4%.

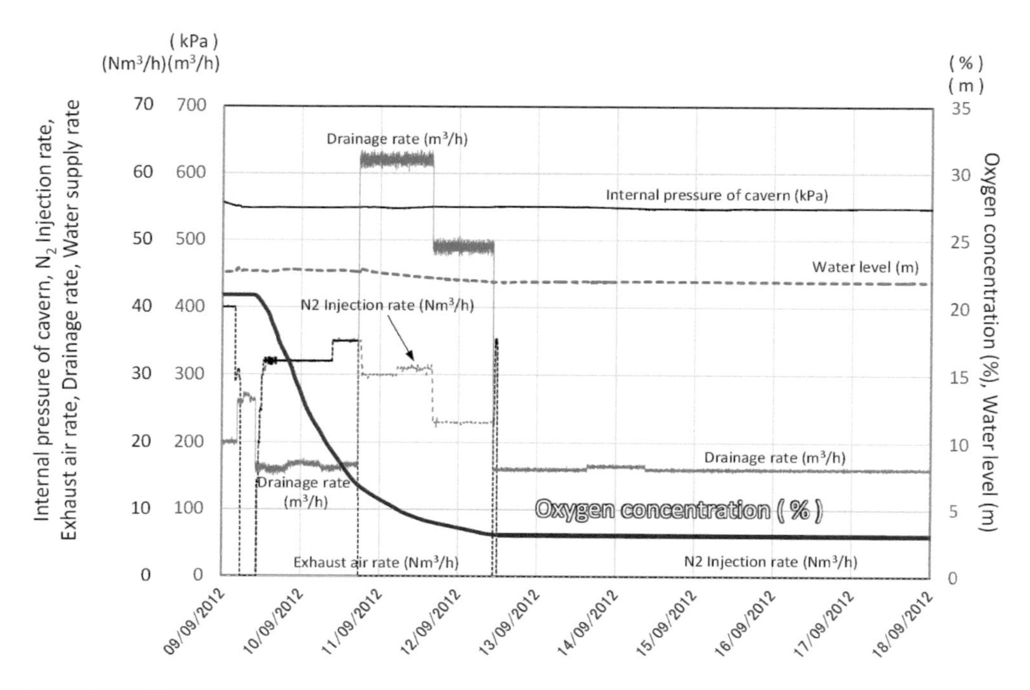

Figure 7.12 Water level, pressure and O_2 concentration in the cavern during the nitrogen inerting operation.

7.3.2.3 Gas Filling

After the air purge operation, propane gas was injected into the cavern, replacing the nitrogen. The propane gas was injected into the gas-phase part of the cavern through the purge line. At the same time, mixed gas coming out of the vent line was burned and exhausted into the atmosphere. This propane replacement continued until the nitrogen concentration in the cavern reached under 7% (Figure 7.13).

After the gas injection operation, the water filling the cavern was drained down to the operation level while LPG was injected in the gas phase and the storage was filled with the LPG in the gas phase. During this process, the pressure in the cavern was increased from 0.55 to 0.74 MPa, the operating pressure. This is equivalent to the saturated vapor pressure for the temperature of the rocks.

7.3.2.4 First Reception of LPG

The first batch of LPG in the liquid phase was received into the storage at normal temperature and high pressure. At the reception of the first liquid phase of LPG, the flow rate and temperature of the reception were carefully managed to avoid an increase in pressure by compression of the gas-phase LPG in the storage and the sudden vaporisation of the liquid-phase LPG (Figure 7.14).

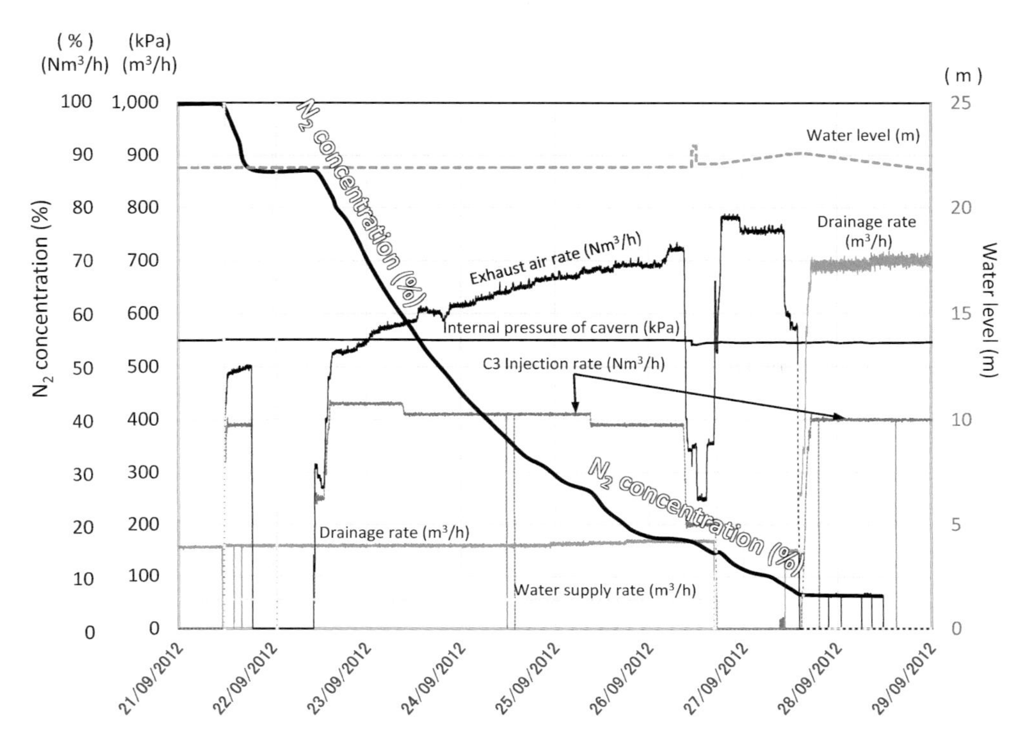

Figure 7.13 LPG injection of the Kurashiki facility.

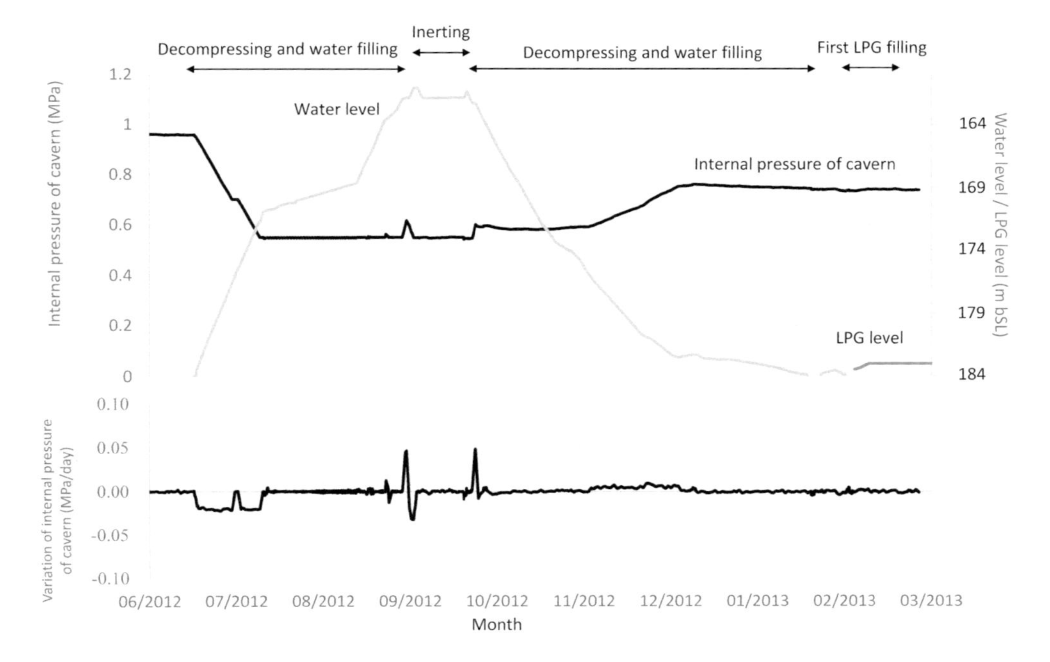

Figure 7.14 Pressure, water level and rate of pressure variation from air-tightness test to first reception of LPG in the Kurashiki facility.

7.3.2.5 Assessment of Water-tightness and Stability of Cavern at Test Operation

During the test operation, the internal pressure of the storage cavern changed from 0.95 MPa at air-tightness test, 0.55 MPa at air purge to 0.8 MPa at the first reception of LPG. This section explains the behaviour of groundwater and the stability of the cavern in response to this variation in the pressure.

In both the Kurashiki and Namikata facilities, a linear relationship between the seepage rate in the storage cavern, the sealing water supply and the internal pressure of the storage cavern during the test operation and its gradient is similar to one obtained from the air-tightness test, in which the internal pressure of the cavern was changed from atmospheric pressure to the design value. This confirmed that there was no change in the permeability characteristics of the surrounding rocks (Figures 7.15 and 7.16).

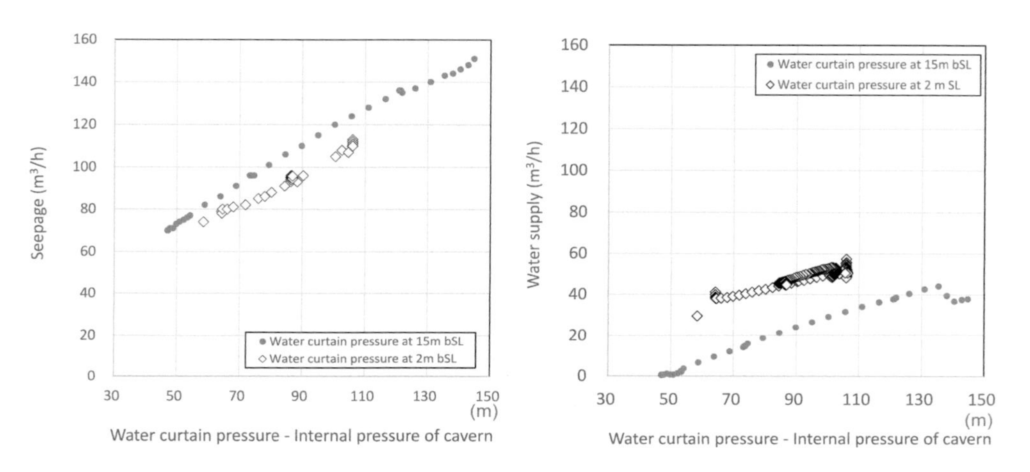

Figure 7.15 Relationship of potential difference with flow rate of seepage in the storage cavern and with sealing water supply in the Kurashiki facility.

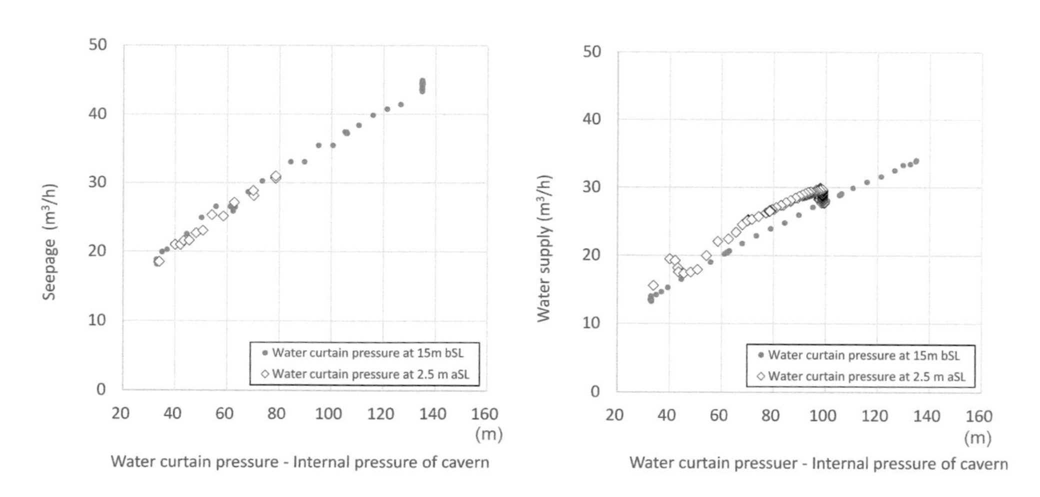

Figure 7.16 Relationship of potential difference with flow rate of seepage in the storage cavern and with sealing water supply in the Namikata facility.

The result of the AE measurement showed that the maximum frequency of AE events was 37 occurrences per hour. This is lower than AE occurrences significant to the calculation of the *m*-value (100/h), and no AE event indicated abnormality in the rocks (Figure 7.17).

7.4 Evaluation of the Groundwater Control System

7.4.1 Result of Measurements During Operation

7.4.1.1 Example from the Kurashiki Facility

7.4.1.1.1 BEHAVIOUR OF LPG IN STORAGE

The reception of LPG at the Kurashiki facility started on December 16, 2013 and concluded on November 23, 2017. The capacity of the Kurashiki facility is about 322,000 t.

The storage pressure was 737.7 kPaG at the start and 760.3 kPaG at the end of reception. Currently, it is about 770 kPaG. The temperature in the cavern is in the range of 20.8–21.3°C and the pressure is equivalent to the saturated vapor pressure of LPG.

The water level of the water curtain boreholes is constantly managed at 2 m above sea level. It stably maintains the difference in piezometric head of 80 m from the internal pressure at the ceiling of the cavern (160 m below sea level) converted to piezometric head equivalent (Figure 7.18).

7.4.1.1.2 FLOW RATE OF SEEPAGE AND SEALING WATER SUPPLY

The seepage rate in the cavern declined from 95 to 79 m^3/h. At the same time, the rate of supply of sealing water also decreased from 45 to 32 m^3/h. The reduction in these rates

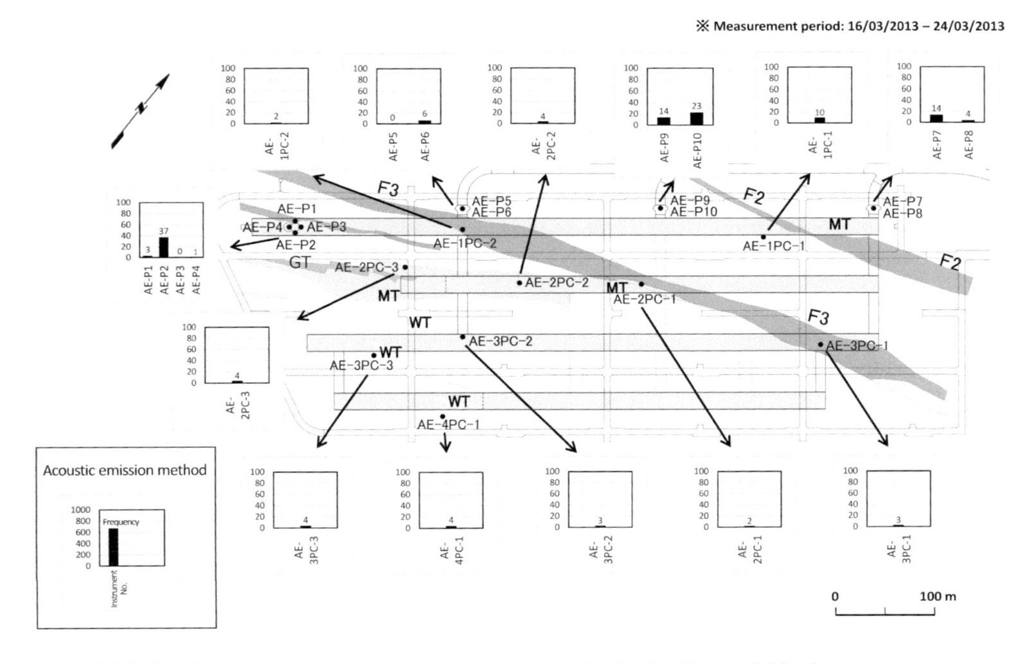

Figure 7.17 Result of AE measurement at the test operation in the Kurashiki facility.

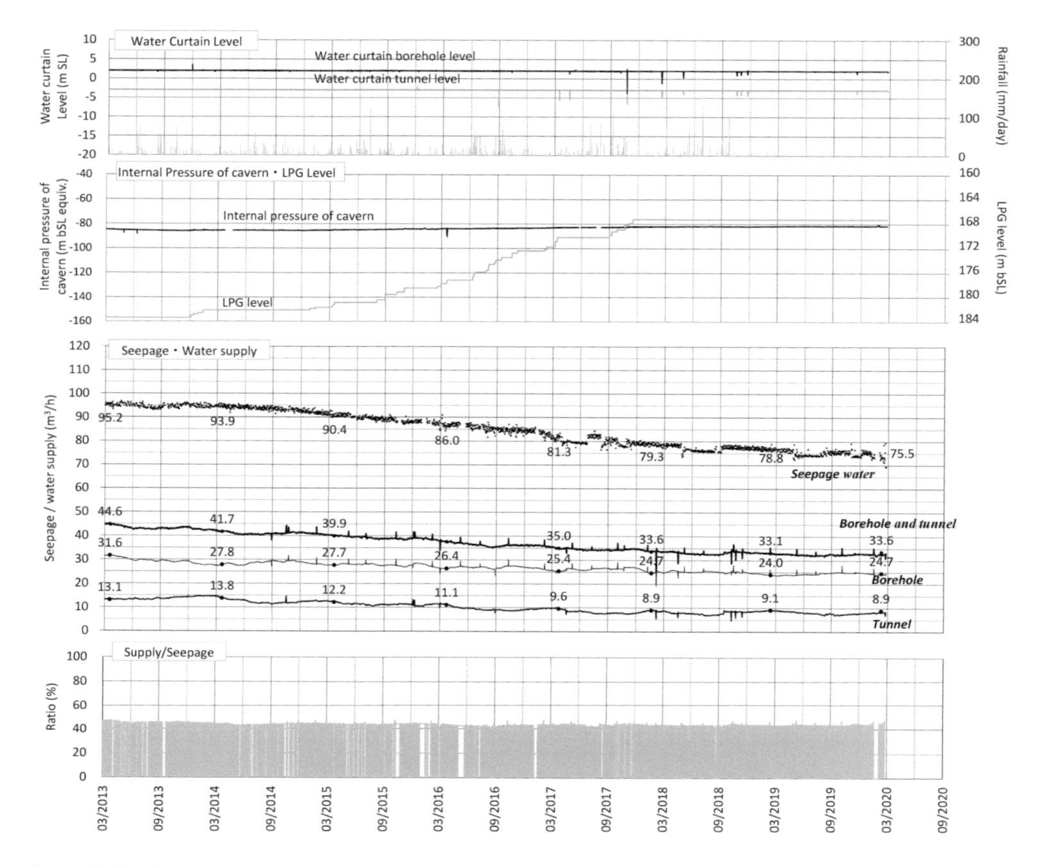

Figure 7.18 Temporal variation of internal pressure, flow rate of seepage and sealing water supply in the cavern in the Kurashiki facility.

was due to the rise in the liquid-phase LPG level and the internal pressure of the cavern at LPG reception. As the proportion of the sealing water supply in the seepage is largely constant, the water curtains are considered to constantly contribute to the water sealing function (Figure 7.18).

7.4.1.1.3 BEHAVIOUR OF GROUNDWATER IN THE STORAGE AREA

The water level in the groundwater observation hole in the upper part of the storage cavern is harmonious with the tide and its variation is stable within ±0.25 m (Figure 7.19).

The variation in the pore water pressure near the top of the water curtains (in fresh granite at 115 m below sea level) since the start of operation is also stable within ±1.5 m (Figure 7.20).

7.4.1.1.4 BEHAVIOUR OF PORE WATER PRESSURE AROUND THE STORAGE CAVERN

Figure 7.21 shows the result of the pore water pressure measurement on the ceiling of the storage cavern (148 m below sea level).

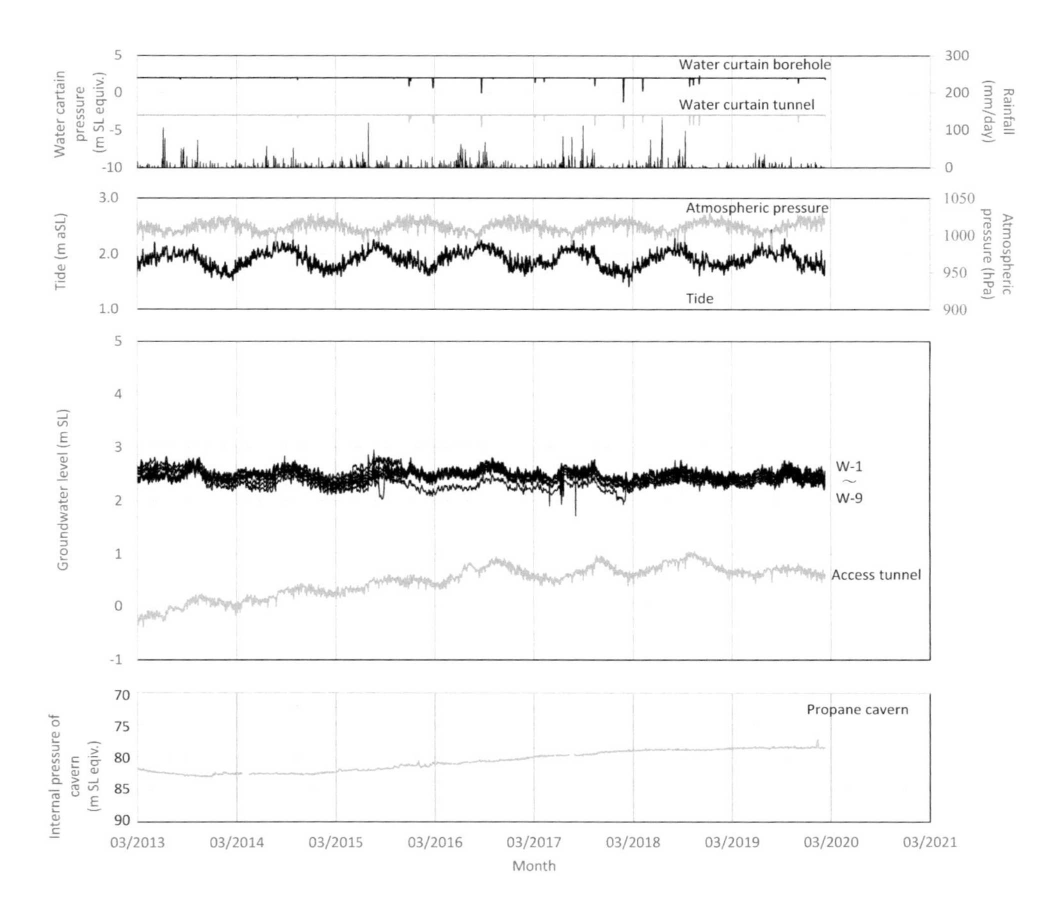

Figure 7.19 Temporal variation of the groundwater level.

Although some brief anomalies were recognised in some of the pore water gauges, the difference in piezometric head between the pore water pressure and the internal pressure of the cavern is adequately high and the health of the water sealing function and the air-tightness of the cavern has been maintained.

7.4.1.1.5 MECHANICAL STABILITY OF THE VOID

The result of the AE measurements in the entire storage cavern is shown in Figure 7.22. Both the event frequency and amplitude of signal are small, and no abnormality in the mechanical strength of the void was detected.

The largest vibration the Kurashiki facility ever experienced since its inauguration was the earthquake on March 13, 2014, with the epicentre in Iyo Nada between Shikoku and Kyushu islands. The maximum acceleration recorded by the seismometers in the Kurashiki facility was 51.2 gal on the surface and 16.4 gal underground (Figure 7.23).

No AE occurred at the time of the earthquake. It is known that vibration underground is smaller than on the surface, and, indeed, the maximum acceleration observed underground was only 13.6 gal. Hence, this earthquake did not influence the plugs or surrounding rocks.

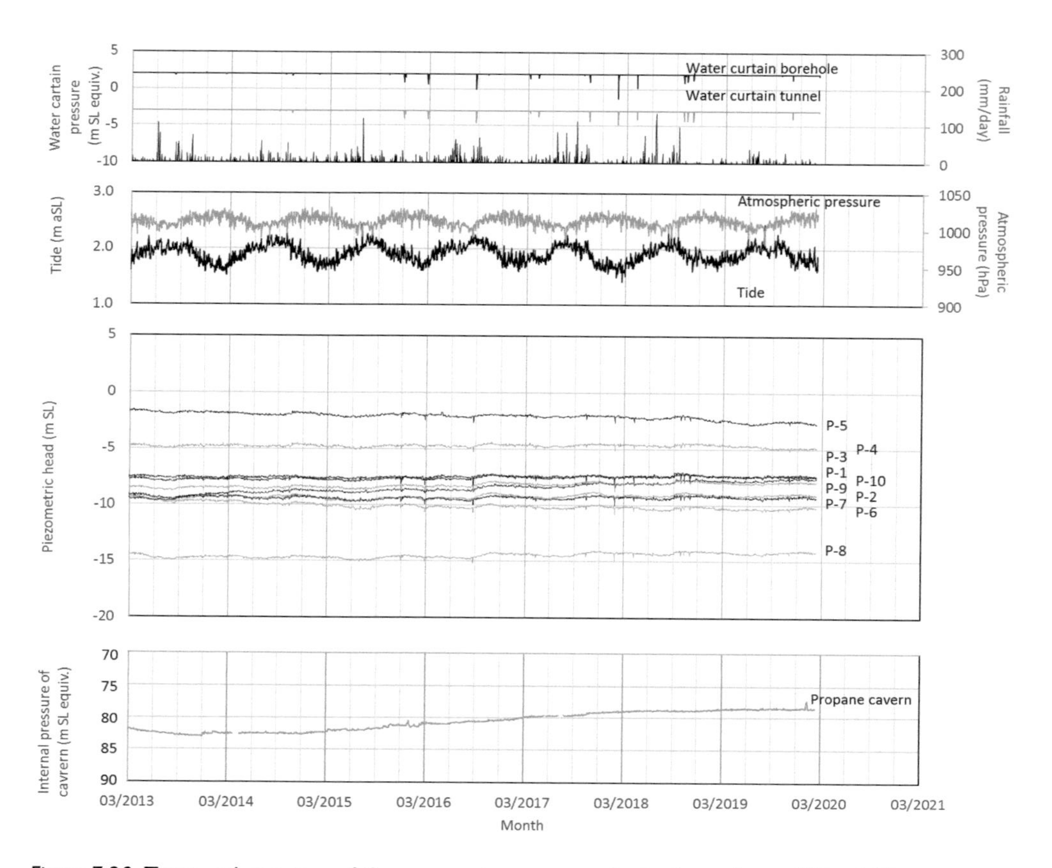

Figure 7.20 Temporal variation of the pore water pressure (in fresh granite at 115 m bSL).

7.4.1.1.6 WATER QUALITY

Figure 7.24 shows the change in seepage quality in the LPG storage cavern from the initial reception. The water quality of the bed drainage tank, which collects seepage in the cavern, is stable at an electric conductivity (Ec) of 2000 mS/m and a pH of about 7–8.

7.4.1.2 Example from the Namikata Facility

7.4.1.2.1 BEHAVIOUR OF LPG IN STORAGE

The reception of LPG at the Namikata facility started in August 2013 and concluded in December 2016. The capacity of the butane-propane storage at the Namikata facility is about 129,000 t and propane storage is 290,000 t.

The storage pressure during this period fluctuated at reception but stabilised at about 0.73 MPa at the start of operation and is currently at 0.735 MPa; currently about 770 kPaG. The temperature in the cavern is in the range of 19.8–20.3°C, and the pressure is equivalent to the saturated vapor pressure of LPG.

The water sealing pressure is constantly managed at 2 m above sea level. It stably maintains the difference in piezometric head of 77 m from the internal pressure of the cavern (75 m below sea level) (Figure 7.25).

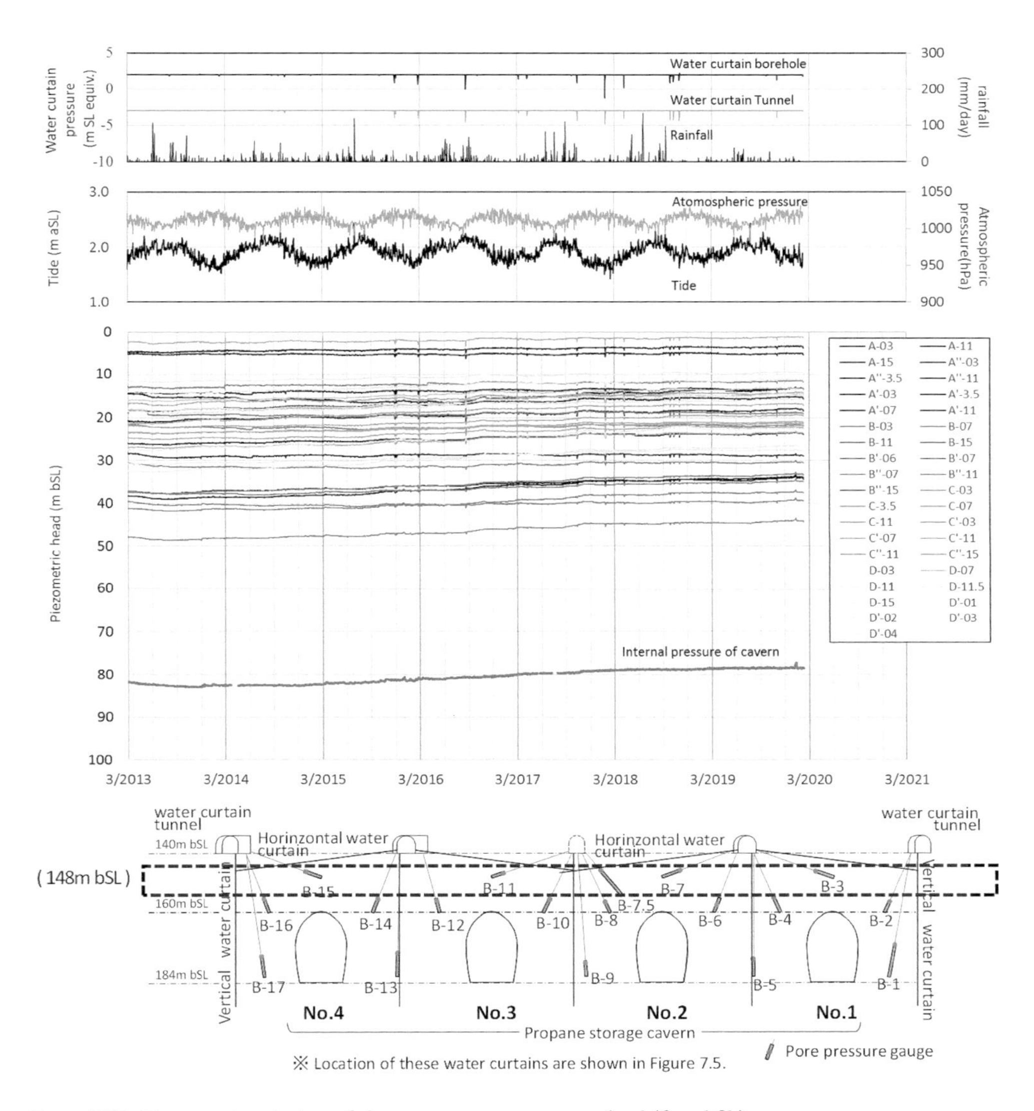

Figure 7.21 Temporal variation of the pore water pressure (at 148 m bSL).

7.4.1.2.2 RATE OF SEEPAGE FLOW AND SEALING WATER SUPPLY

The seepage rate in the cavern slightly declined from 31 to 28 m^3/h, and the rate of sealing water supply was around 26 m^3/h. The proportion of the sealing water in the seepage slightly increased to 90%, and it is stably supplying the sealing water. The water curtains are considered to constantly contribute to the water sealing function (Figure 7.25).

7.4.1.2.3 BEHAVIOUR OF GROUNDWATER IN THE STORAGE AREA

The water levels are measured in the groundwater observation holes in the upper part of the storage cavern. On the shore side to the northeast, where the elevation is low, the water

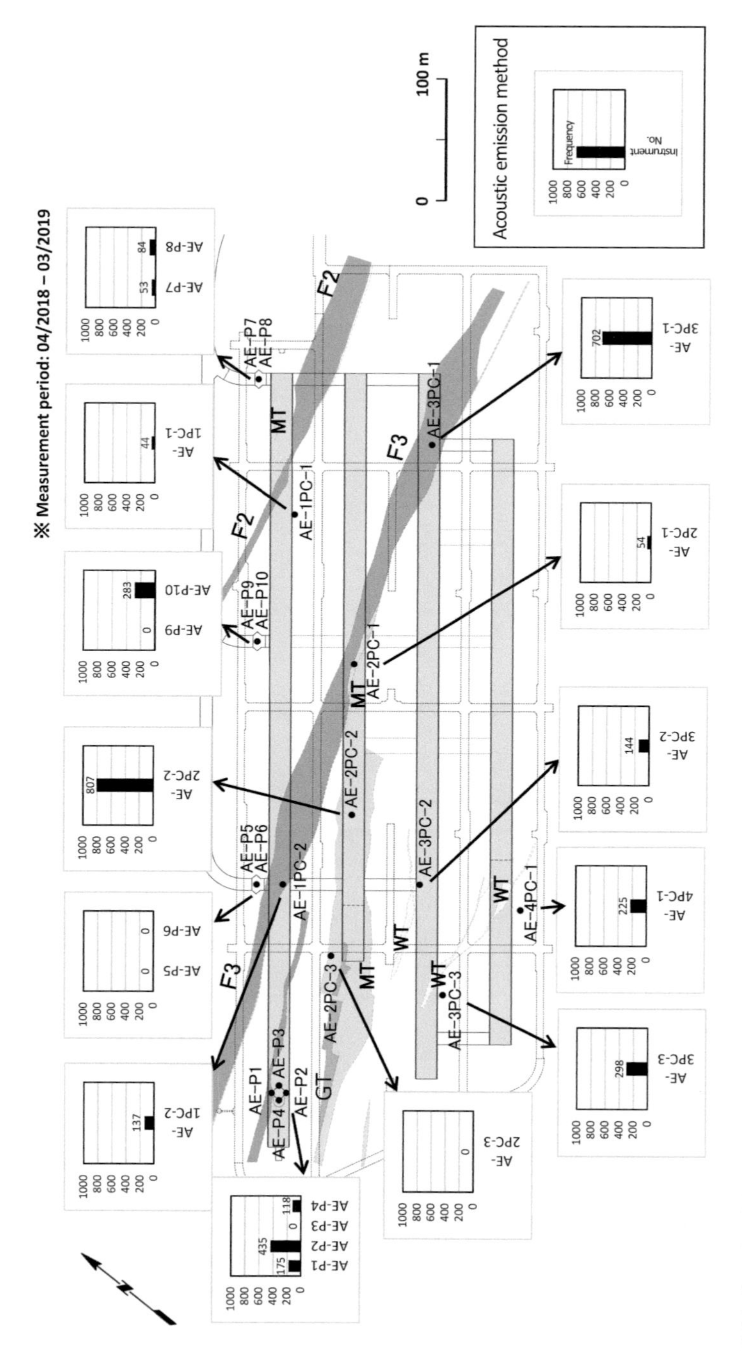

Figure 7.22 Result of AE measurements over entire storage cavern in the Kurashiki facility.

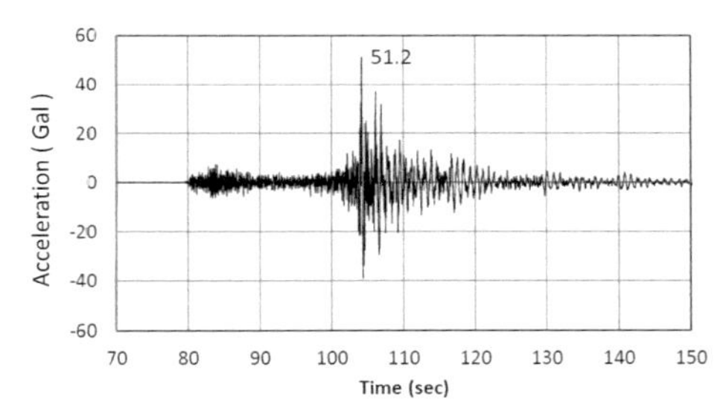

(a) Surface (5.2m bSL) EW component

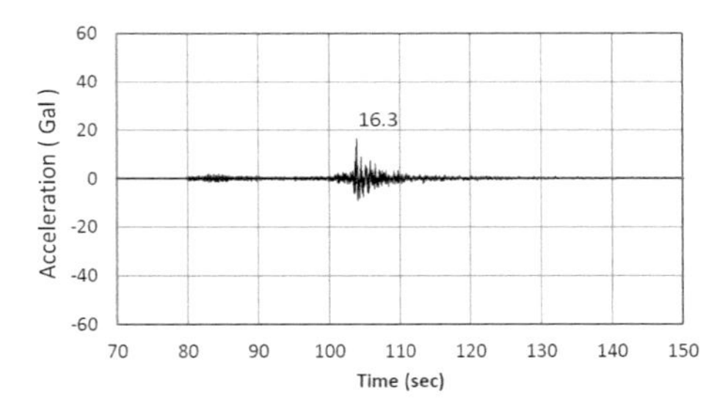

(b) Underground (180m bSL) EW component

Figure 7.23 Temporal variation of the measured acceleration of the E-W component at the Iyo Nada Earthquake on March 14, 2014. Left: at 5.2 m bGL, right: at 180 m bGL.

level was generally 0–5 m above sea level and varied little, while in the hill side to the southwest, where the elevation is high, the water level temporarily rose with rainfall, but was stable in the long term. No tendency of lowering was recognised during the operation (Figure 7.26).

The variation in the pore water pressure near the top of the water curtains (in fresh granite at 105 m below sea level) has remained generally stable since the start of operation except for the influence of rainfall in the elevated hill side (Figure 7.27).

7.4.1.2.4 BEHAVIOUR OF PORE WATER PRESSURE AROUND THE STORAGE CAVERN

Figure 7.28 shows the result of a pore water pressure measurement on the ceiling of the combined butane-propane storage cavern (135 m below sea level).

Since the start of operation, the pore water pressure has generally remained constant, and it has remained high to the piezometric head. Thus, the health of the water sealing function and the air-tightness of the cavern has been maintained.

7.4.1.2.5 MECHANICAL STABILITY OF THE VOID

The result of the AE measurements in the entire storage cavern is shown in Figure 7.29, while Figure 7.30 shows the result of the measurement using geophones installed in the floor of the caverns to monitor the unlikely event of a rockfall. Both the event frequency and the amplitude of the signal were small. Analyses of the waveform of AE events detected no sign to suggest the development of cracks in the rocks. Thus, it is considered that the health of the cavern void has been maintained.

7.4.1.2.6 WATER QUALITY

Figure 7.31 shows the change in the seepage quality in the LPG storage cavern from initial reception. The electric conductivity of the bed drainage tank, which collects seepage in the cavern, was 250 mS/m at the start of the operation and gradually lowered over three years to 50 mS/m in 2015, and has since remained stable. The Na and Cl concentrations also stabilised after lowering with Ec. This was caused by the sea water used to fill the cavern at test

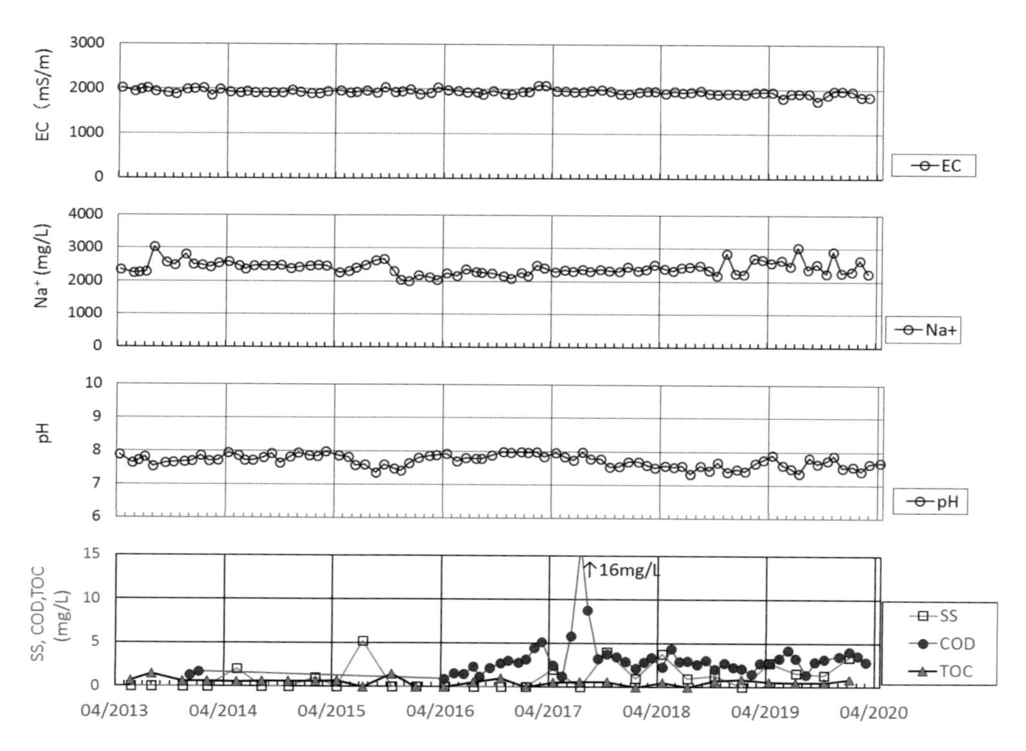

Figure 7.24 Result of water quality measurement and analysis in the Kurashiki facility.

operation and salinity was high at the start of the operation. After the start of the operation, fresh sealing water seeped into the cavern, which diluted the salinity of the water in the cavern.

The acidity has been stable since the start of the operation at about pH 9.5.

7.4.2 Assessment of Water-tightness and Long-Term Prediction with the 3D Heterogeneous Model

7.4.2.1 Example from the Kurashiki Facility

Using the 3-dimensional (3D) analytical model described in Chapter 3, the actual and predicted values of the behaviour of groundwater were compared to assess the health of the water-tightness function.

7.4.2.1.1 COMPARISON BETWEEN ANALYTICAL PREDICTION AND ACTUAL MEASUREMENTS

7.4.2.1.1.1 Flow of Seepage and Sealing Water Supply Figure 7.32 shows a comparison between analytical prediction and actual measurements of the seepage rate and supply rate of sealing water. The level of the sealing water, the internal pressure of the cavern and the level of the liquid-phase LPG changed from the air-tightness test through the operation period. In

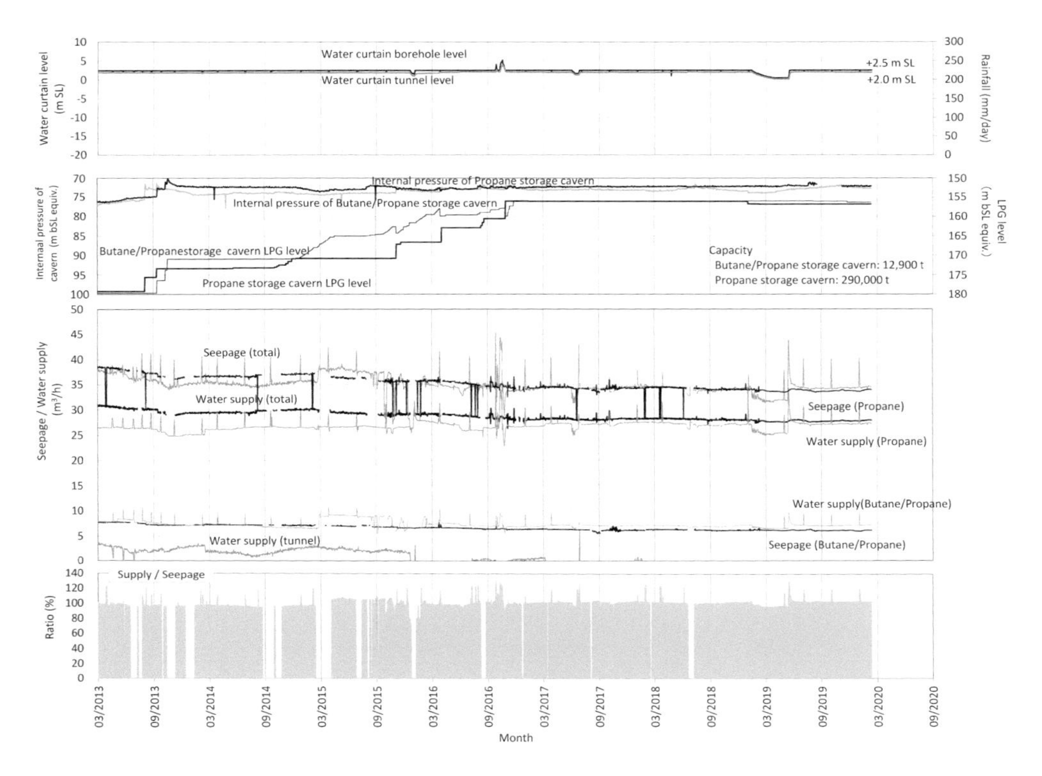

Figure 7.25 Temporal variation of the internal pressure, flow rate of seepage and sealing water supply in the LPG storage cavern in the Namikata facility.

response to these conditions of the rock cavern, the seepage rate and supply rate of the sealing water varied.

The seepage rate was greatest at the start of the air-tightness test at 151.0 m^3/h compared with predicted 154.0 m^3/h. The lowest seepage rate was recorded during the air-tightness test at 70.0 m^3/h (prediction: 61.1 m^3/h). Currently (February 2020), the measured flow rate is 75.0 m^3/h (prediction: 85.6 m^3/h).

Comparing the seepage rate and the supply rate of the sealing water at the start of the operation, the sealing water rate was 95.0 m^3/h while the supply rate of the sealing water was 44.8 m^3/h. Therefore, the natural supply rate is considered 50.2 m^3/h. The ratio (artificial supply)/(natural supply) is 0.89:1, in which the difference is considered negligibly small.

The 3D hydrogeological model was built at the construction stage of the storage cavern. The results of its analyses were repeatedly monitored by prediction and examination for applicability. The actual measurements from construction through to operation largely agreed with predictions, thus 3D analysis is considered applicable to future long-term prediction (Table 7.2).

7.4.2.1.1.2 Pore Water Pressure around the Cavern Some comparisons between the actual measured pore water pressure and the analysis prediction by 3D analysis are shown in Figure 7.33, at three selected locations along cross-sections A and D.

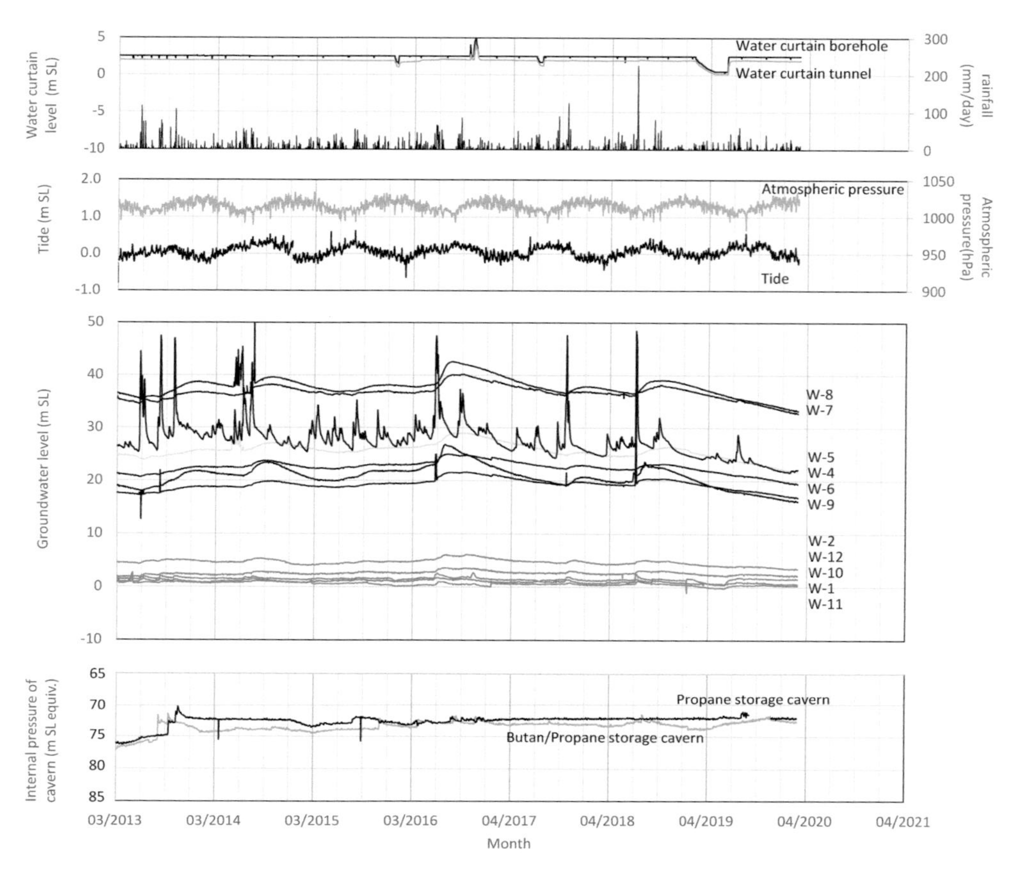

Figure 7.26 Temporal variation of the groundwater level.

The actual pore water pressure, as well as analysis prediction, shows no variation, which evidences stable groundwater.

7.4.2.2 Example from the Namikata Facility

Using the 3D analytical model described in Chapter 3, the actual and predicted values of the behaviour of groundwater were compared to assess the health of the water-tightness function.

7.4.2.2.1 COMPARISON BETWEEN ANALYTICAL PREDICTION AND ACTUAL MEASUREMENTS

7.4.2.2.1.1 Flow of Seepage and Sealing Water Supply Tables 7.3 and 7.4 and Figures 7.34 and 7.35 show comparisons between analytical prediction and actual measurements of the seepage rate and the supply rate of sealing water.

The level of the sealing water, the internal pressure of the cavern and the level of liquid-phase LPG changed from the air-tightness test through the operation period. In response to these conditions of the rock cavern, the seepage rate varied.

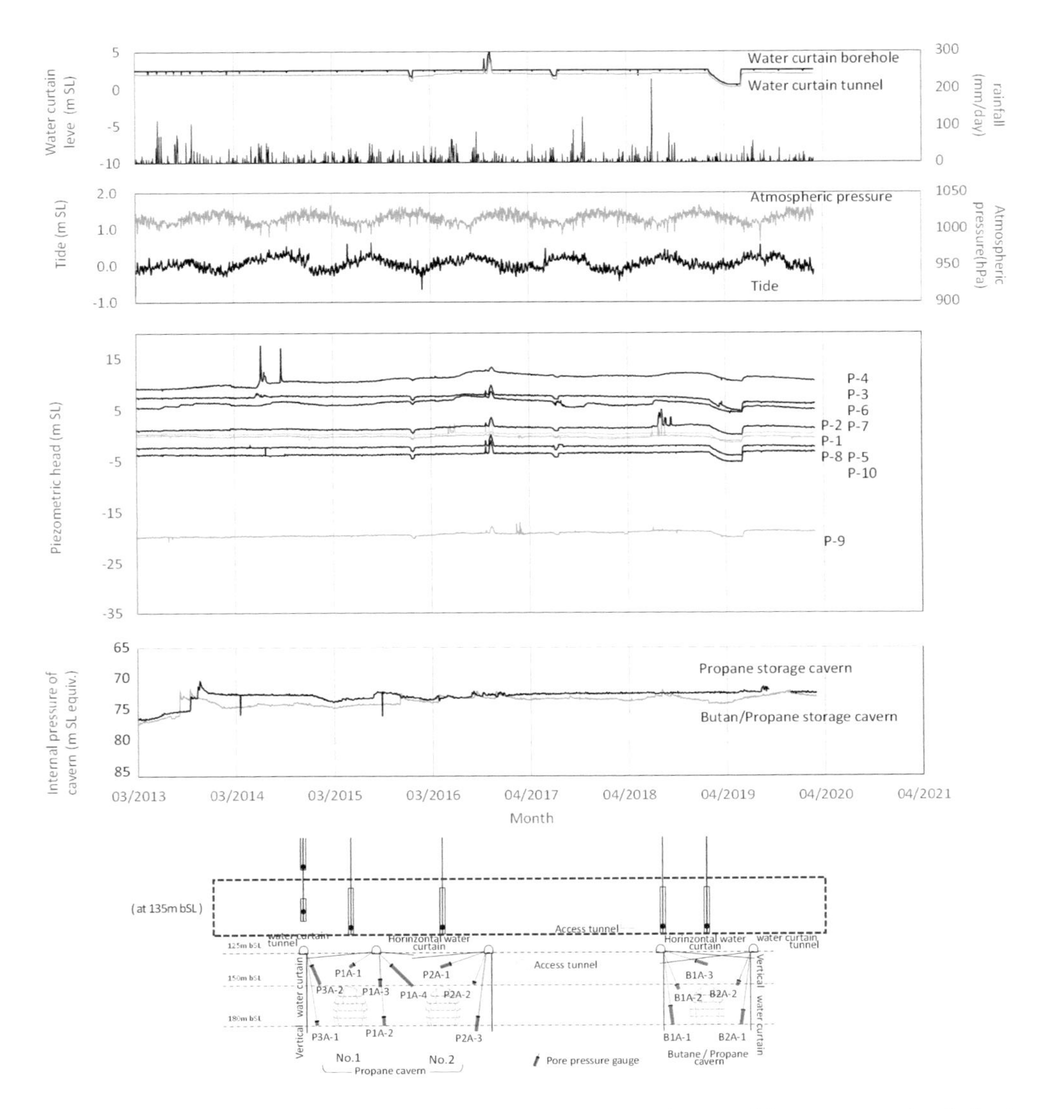

Figure 7.27 Temporal variation of the pore water pressure near the top of the water curtains (in fresh granite at 105 m bSL).

The seepage rate was greatest at the start of the air-tightness test at 11 m^3/h compared with the predicted 12 m^3/h in the combined butane-propane storage and 45 m^3/h compared with the predicted 42 m^3/h in the propane storage. The lowest seepage rate was recorded during the air-tightness test at 4.5 m^3/h (prediction: 3.9 m^3/h) in the combined butane-propane storage and 18.5 m^3/h (prediction: 13.0 m^3/h) in the propane storage. Currently (March 2020), the measured flow rate is 6.1 m^3/h (prediction: 7.6 m^3/h) in the combined butane-propane storage and 24.3 m^3/h (prediction: 27.8 m^3/h) in the propane storage.

Comparing the seepage rate and the supply rate of sealing water at the start of the operation, the rate of sealing water was 95.0 m^3/h while the supply rate of sealing water was

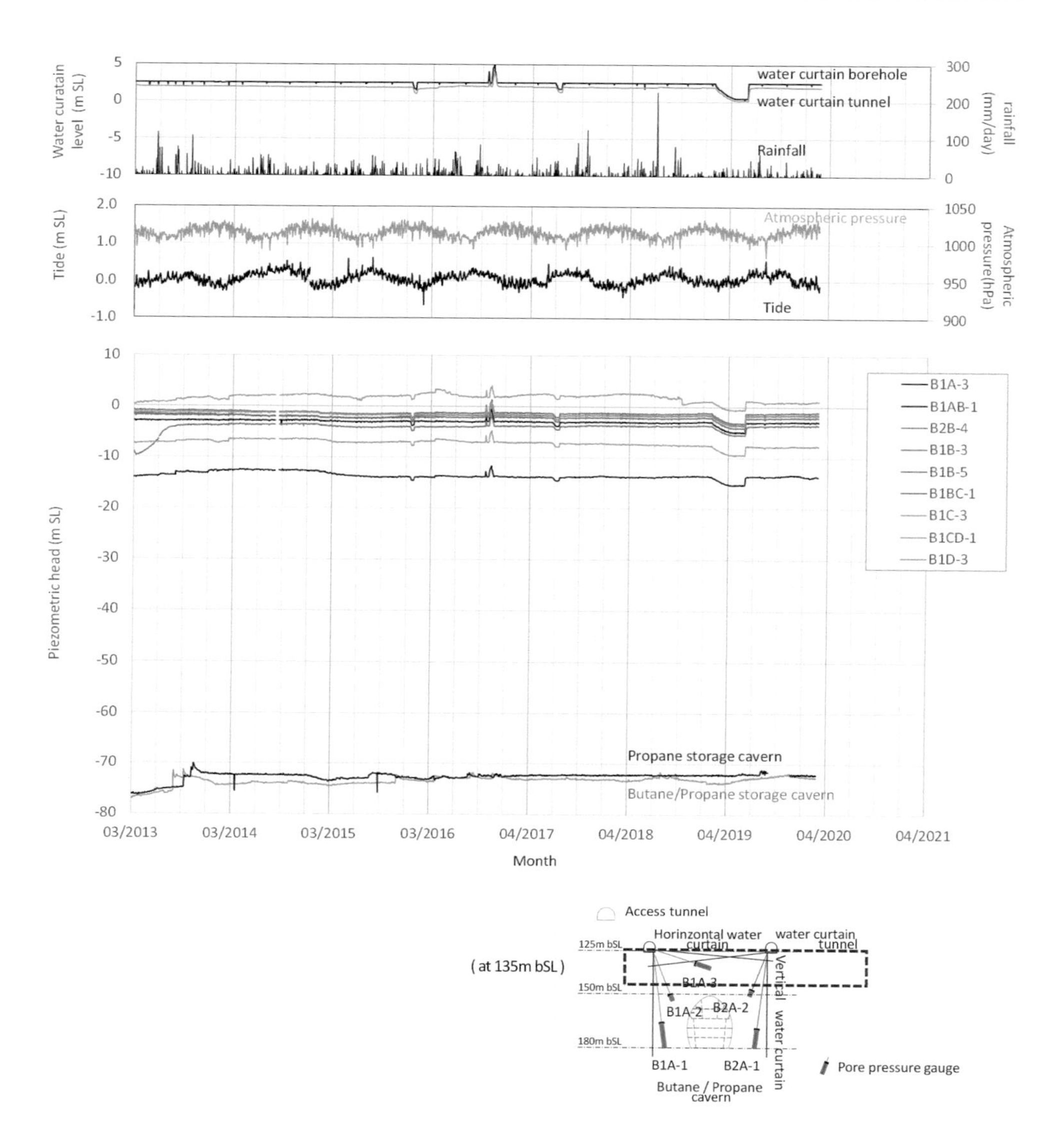

Figure 7.28 Temporal variation of the pore water pressure (butane-propane storage cavern at 135 m bSL).

44.8 m^3/h. Therefore, the natural supply rate is considered to be 50.2 m^3/h. The ratio (artificial supply)/(natural supply) is 0.89:1, in which the difference is considered negligibly small.

As seen, the prediction from the analyses using the hydrogeological model constructed for this site is considered applicable to the assessment of the health of the water sealing function.

7.4.2.2.1.2 Pore Water Pressure around the Cavern Some comparisons between the actual measured pore water pressure and the analysis prediction using the 3D analysis are shown in

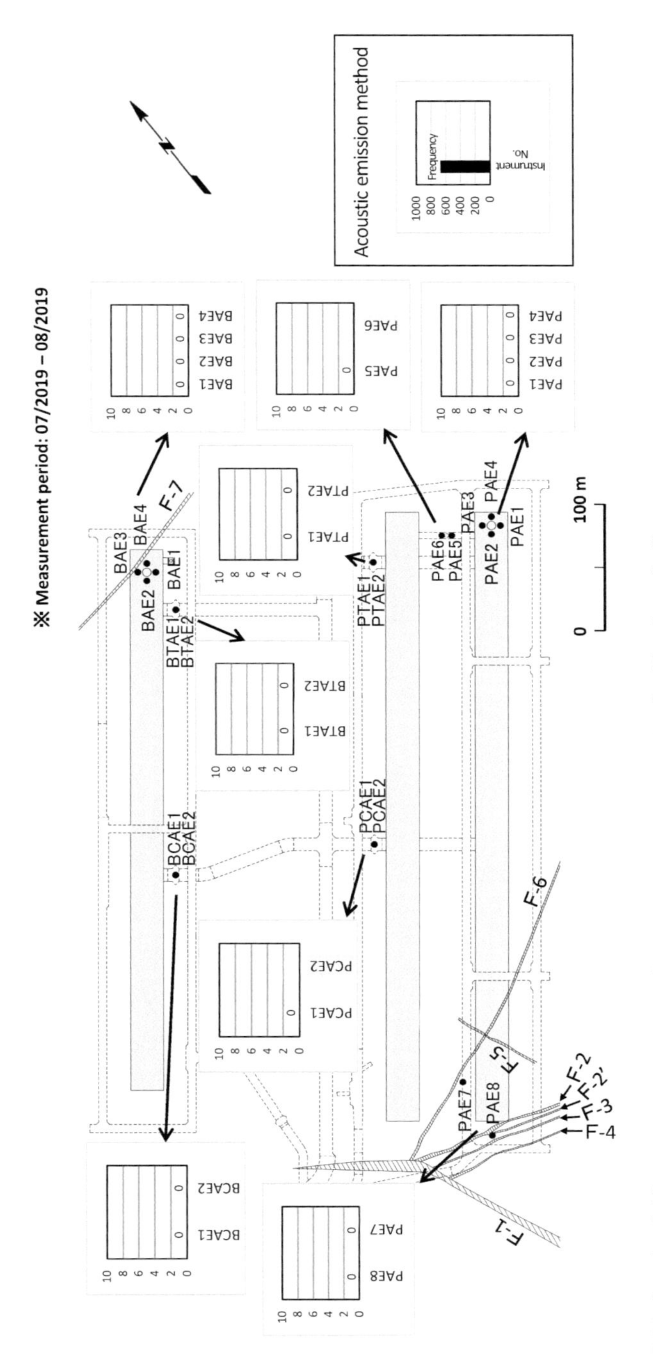

Figure 7.29 Result of AE measurements over the entire storage cavern in the Namikata facility.

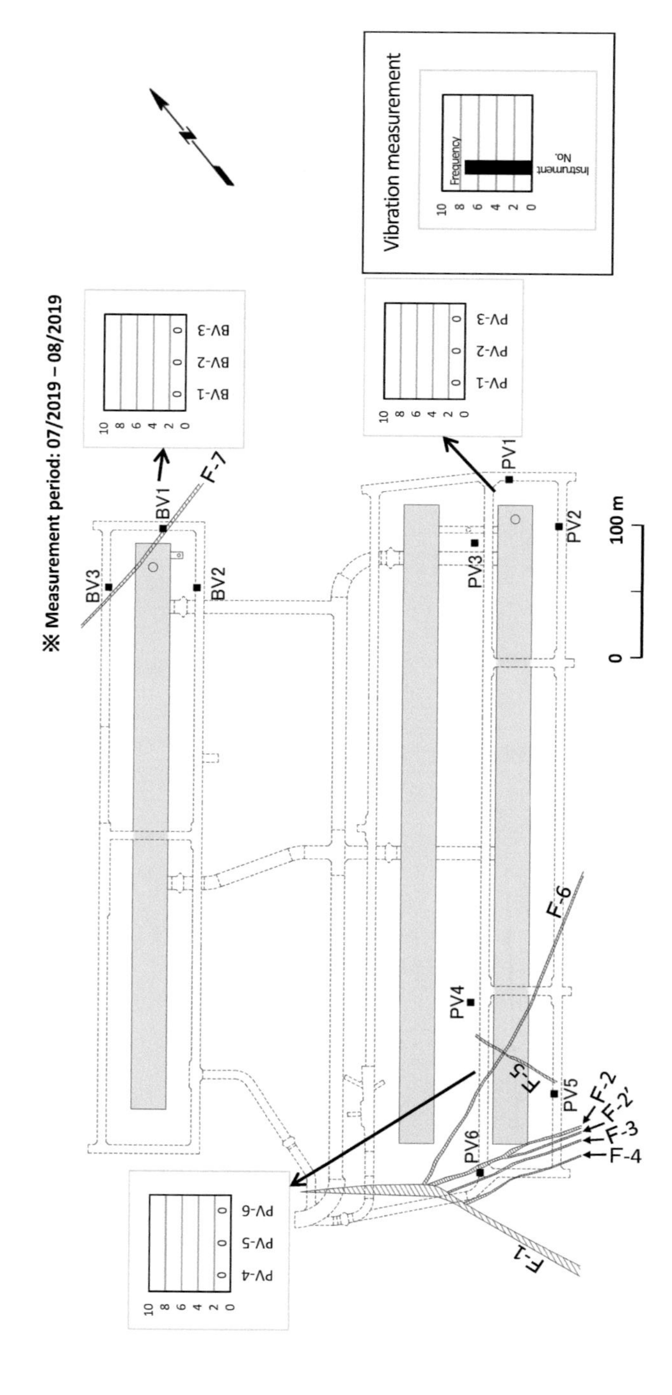

Figure 7.30 Result of vibration measurements over the entire storage cavern in the Namikata facility.

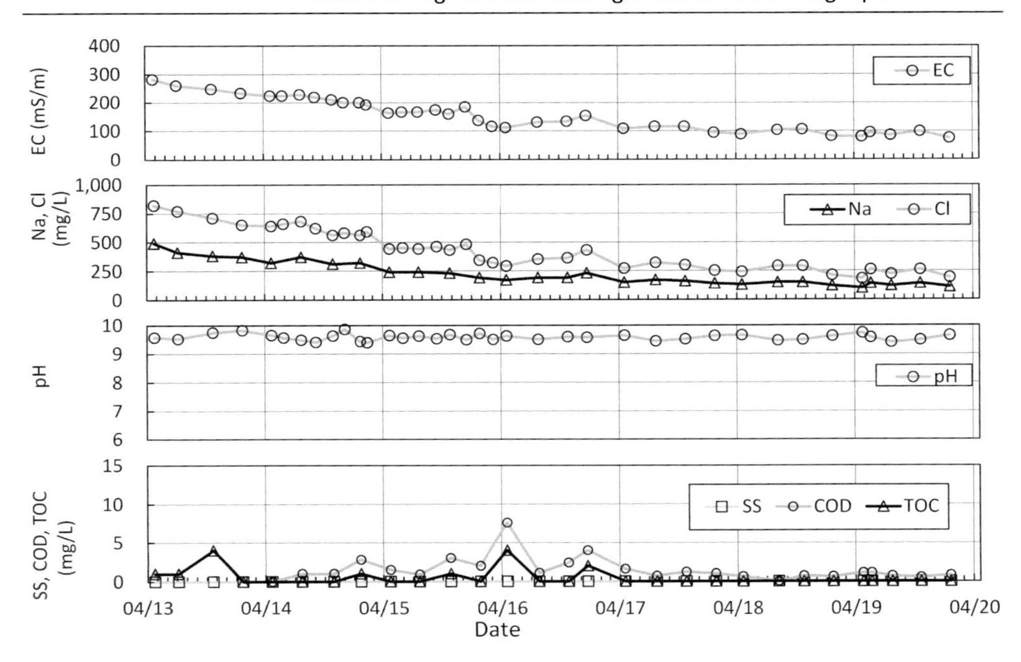

Figure 7.31 Result of water quality measurement and analysis in the Namikata facility.

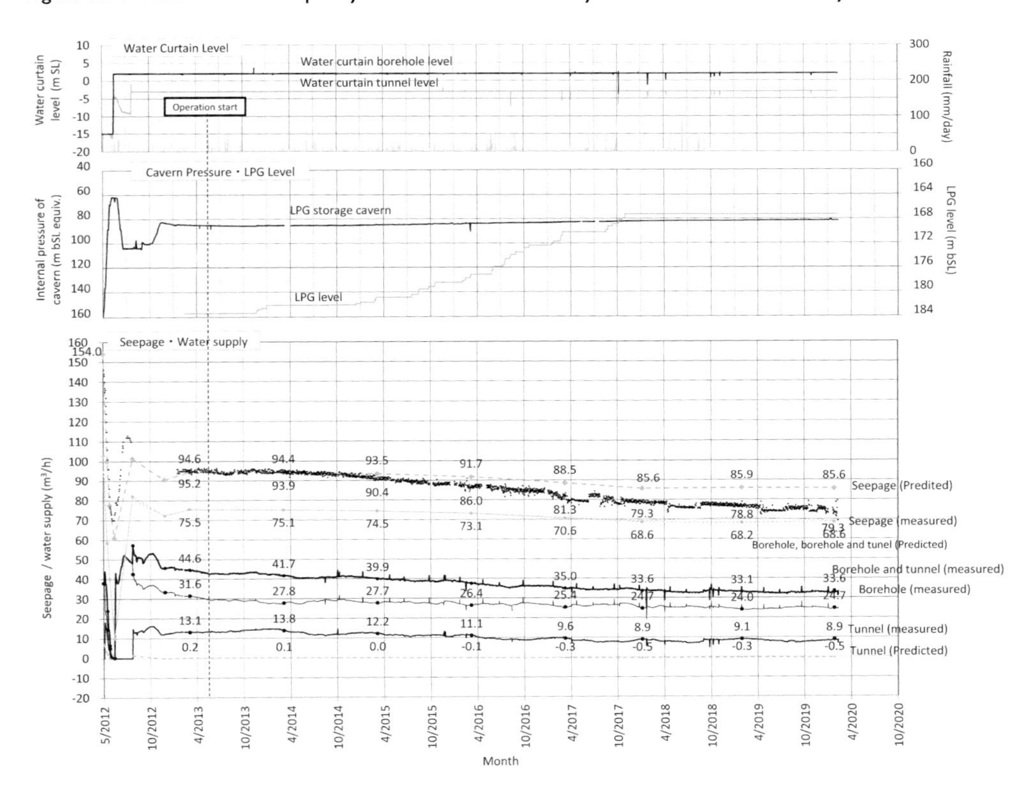

Figure 7.32 Comparisons between analytical prediction and actual measurements of the flow rate of seepage and the supply rate of sealing water.

Table 7.2 Comparisons between analytical prediction and actual measurements of flow rate of seepage and supply rate of sealing water

		Before test	*At air-tightness test*	*At start of operation*	*Current*
Water curtain level (m SL)		−15.0	−15.0	+2.0	+2.0
Internal pressure of cavern (kPa)		Atmospheric pressure	960	738	769
LPG level (m)		–	–	0.5	16.7
Seepage water flow rate (m³/h)	Measurement	151.0	70.0	95.0	75.0
	Predicted	154.0	61.1	94.6	85.6
Water supply (m³/h)	Measurement	38.0	0.0	44.8	33.6
	Predicted	102.8	10.1	75.4	68.6

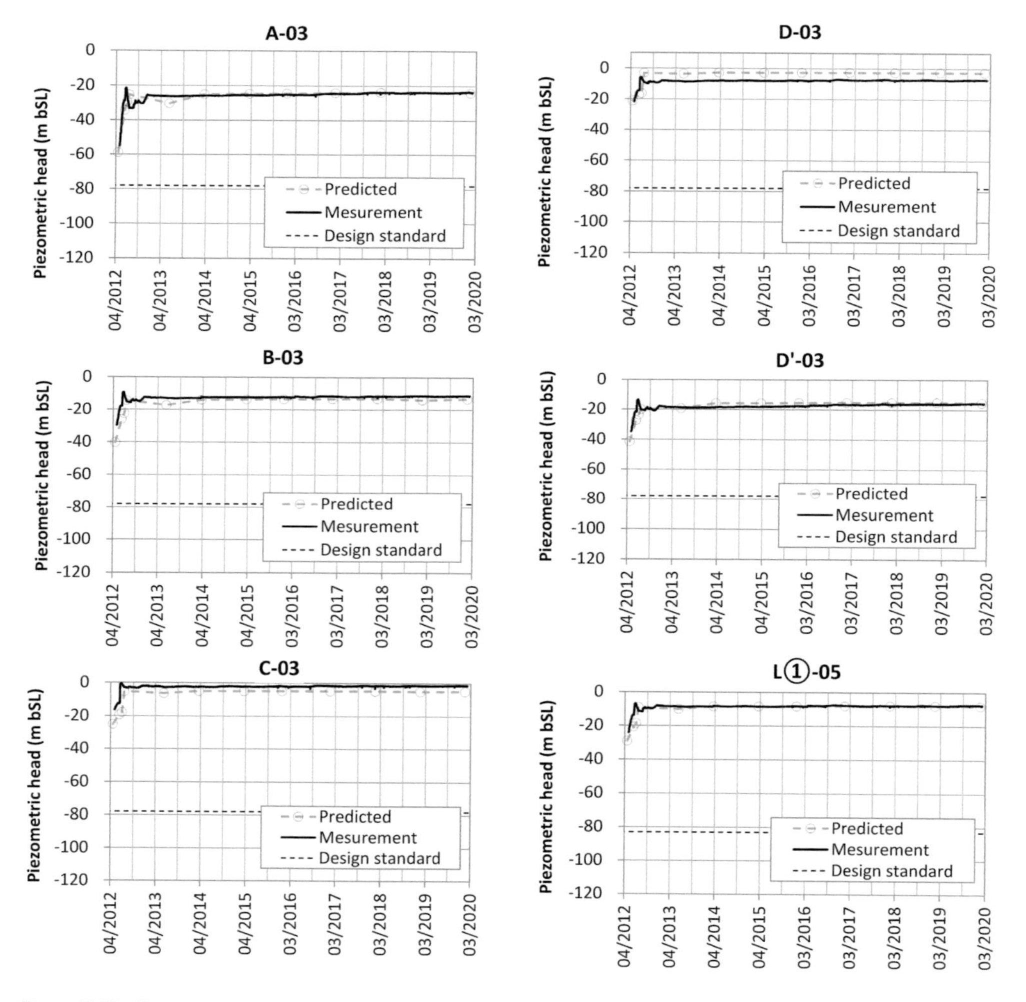

Figure 7.33 Comparisons between actual measured pore water pressure and prediction by the 3D model analysis.

Table 7.3 Comparisons between analytical prediction and actual measurements of flow rate of seepage and supply rate of sealing water in the combined butane-propane storage

		Before test	*At air-tightness test*	*At start of operation*	*Current*
Water curtain level (m SL)		−15.0	−15.0	+2.5	+2.5
Internal pressure of cavern (kPa)		Atmospheric pressure	980	690	731
LPG level (m)		–	–	0.9	25.1
Seepage water flow rate (m^3/h)	Measurement	11.0	4.5	7.7	6.1
	Predicted	12.4	3.9	7.8	7.5
Water supply (m^3/h)	Measurement	12.0	3.9	7.9	7.6
	Predicted	12.4	3.9	7.5	7.1

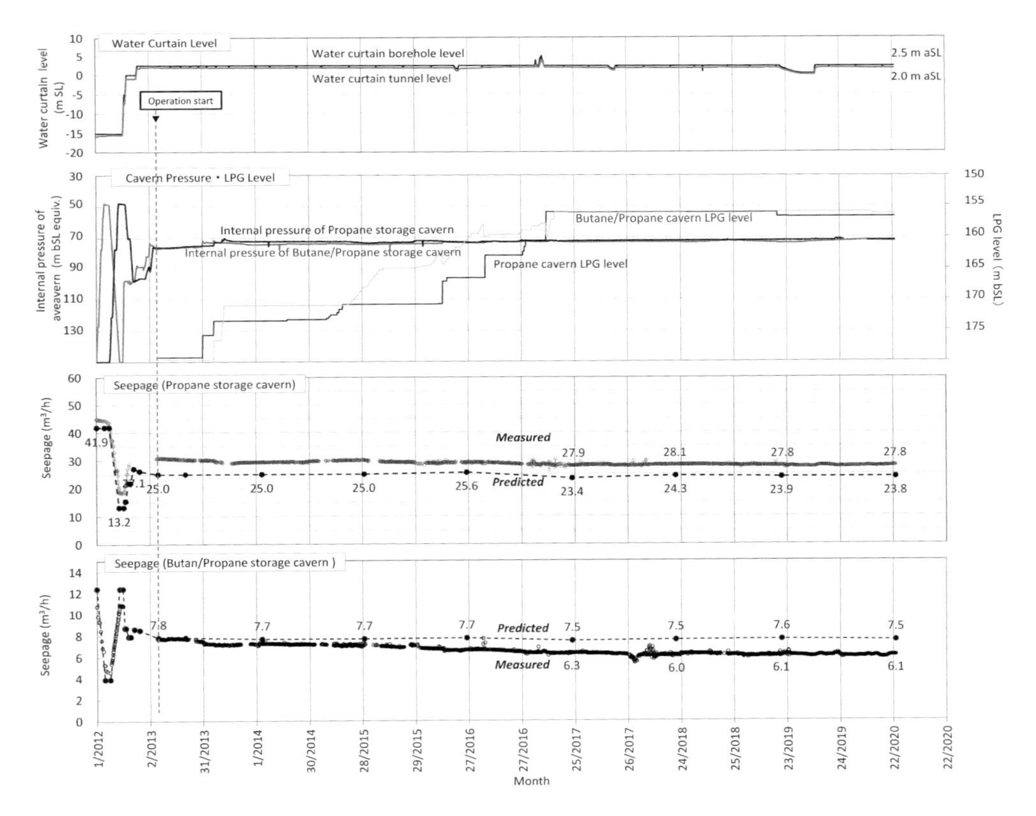

Figure 7.34 Comparisons between analytical prediction and actual measurements of the flow rate of seepage from an air-tightness test to the present.

Table 7.4 Comparisons between analytical prediction and actual measurements of flow rate of seepage and supply rate of sealing water in the propane storage

		Before test	*At air-tightness test*	*At start of operation*	*Current*
Water curtain level (m SL)		−15.0	−15.0	+2.5	+2.5
Internal pressure of cavern (kPa)		Atmospheric pressure	980	699	739
LPG level (m)		–	–	0.5	24.5
Seepage water flow rate (m³/h)	Measurement	45.0	18.5	30.8	27.9
	Predicted	41.9	13.2	25.0	23.8
Water supply (m³/h)	Measurement	42.0	13.0	25.0	24.3
	Predicted	37.1	11.1	24.1	23.4

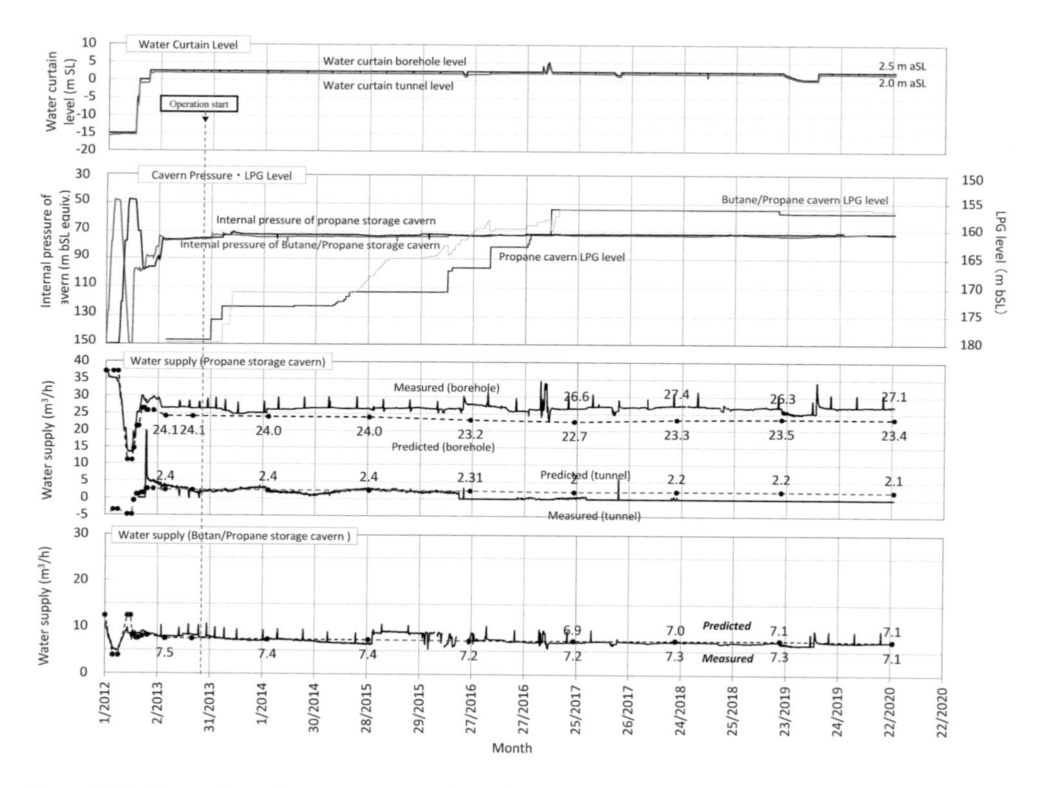

Figure 7.35 Comparisons between analytical prediction and actual measurements of the supply rate of sealing water from an air-tightness test to the present.

Figure 7.36, at three selected locations along the combined butane-propane storage and the propane storage. The actual pore water pressure, as well as the analysis prediction, shows no temporal variation, which evidences the stable state of the groundwater.

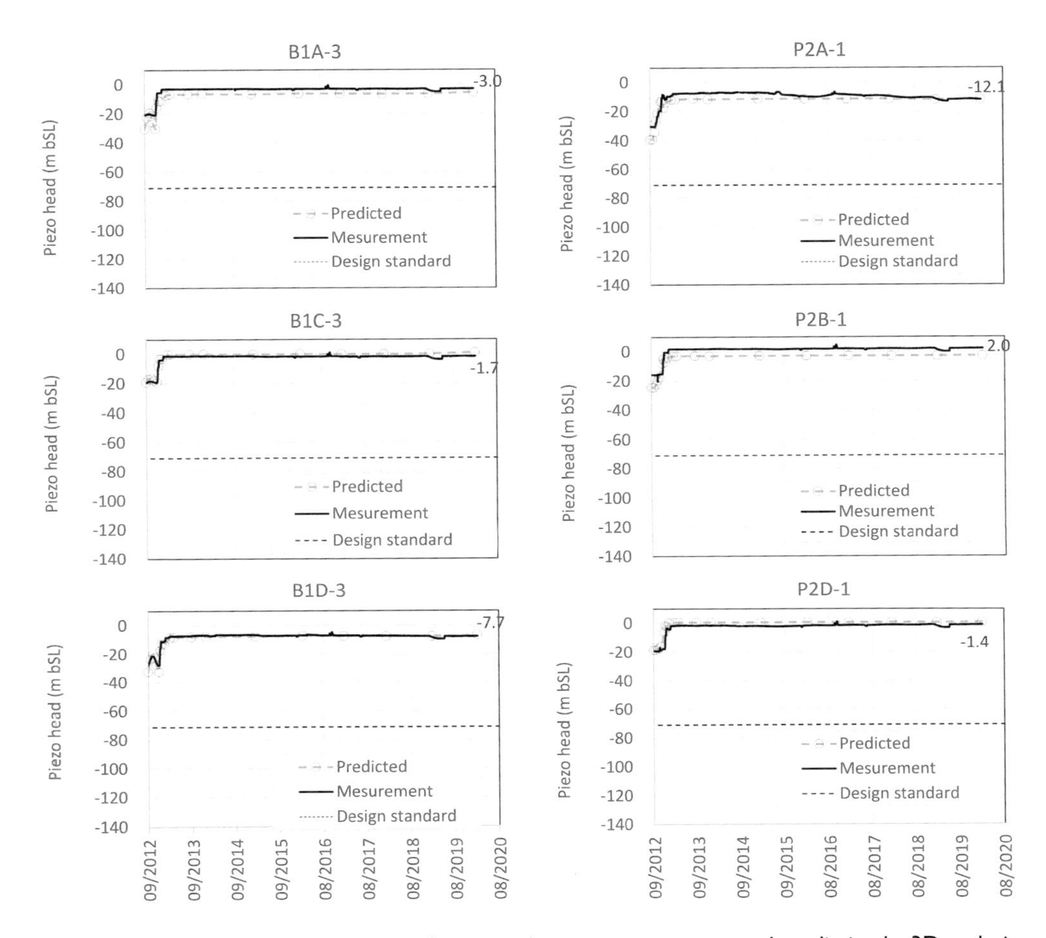

Figure 7.36 Comparisons between actual measured pore water pressure and prediction by 3D analysis.

In the six years since the storage operation began, scrupulous analysis of the behaviour of the surrounding pore water pressure and others has not shown clear change, and the groundwater is considered stable. The behaviour of the groundwater was analysed comparing with the prediction by the 3D heterogeneous model. This analysis ensured the water-tightness functionality. The long-term prediction of the groundwater behaviour assures that the groundwater will maintain its stability. During this operation period, the dynamic behaviour of the surrounding rocks was monitored by AE measurement and accelerometer observation and no abnormal movements were observed. The facility is stably operating. The operation equipment and the safety management system of the facility are well documented.

References and Further Readings

Aoki K., Miyashita K., Hanamura T., Tajima T. (1986) The first test plant of underground crude oil storage in unlined cavern. *Large Rock Caverns*, Vol.1, 3–14.

Berest P. (1989) Accidents of underground oil and gas storages - Case histories and prevention. *Proceedings of the International Conference on Storage of Gases in Rock Caverns*, Trondheim, 289–301.

Bergman M. (1977) Storage in excavated rock caverns. *Proceedings of the First International Symposium*, Stockholm, 832.

Johansen P.M., Madsen O.K. (1989) The Rafnes propane storage cavern −12 years of successful operation. *Proceedings of the International Conference on Storage of Gases in Rock Caverns*, Trondheim, 303–306.

Lindblom U.E. (1989) The performance of a water curtain during 10 years of operation. *Proceedings of the International Conference on Storage of Gases in Rock Caverns*, Trondheim, 347–354.

Murakami K., Shirasagi S., Yamamoto T., Descour J.M., Aoki K. (2004) A study of effective receiver array and adequate wave for 3-dimensional tunnel seismic reflective survey. *Proceedings of the ISRM International Symposium 2004 (3rd ARMS, KYOTO)*, Kyoto, 115–120.

Nilsen B., Olsen J. (1989) Storage of gases in rock caverns. *Proceedings of the International Conference on Storage of Gases in Rock Caverns*, Trondheim, 398.

Palmqvist K., Ullgren S (1989) Design and construction of an LPG rock cavern nearby an existing LPG rock storage. *Proceedings of the International Conference on Storage of Gases in Rock Caverns*, Trondheim, 307–316.

Index

M

N

O